Systems & Control: Foundations & Applications

Fritz Colonius
Uwe Helmke
Dieter Prätzel-Wolters
Fabian Wirth
Editors

Advances in Mathematical Systems Theory

A Volume in Honor of Diederich Hinrichsen

With 16 Figures

Birkhäuser
Boston • Basel • Berlin

Fritz Colonius
Institute for Mathematics
University of Augsburg
86135 Augsburg
Germany

Uwe Helmke
Mathematical Institute
University of Würzburg
Am Hubland
97074 Würzburg
Germany

Dieter Prätzel-Wolters
Department of Mathematics
University of Kaiserslautern
Postfach 3049
67653 Kaiserslautern
Germany

Fabian Wirth
Center for Technomathematics
University of Bremen
D-28334 Bremen
Germany

Library of Congress Cataloging-in-Publication Data
Advances in mathematical systems theory : a volume in honor of D. Hinrichsen / editors, Fritz Colonius . . . [et al.].
p. cm. — (Systems and control)
Includes bibliographical references and index.
ISBN 0-8176-4162-9 (alk. paper)
1. System theory—Congresses. I. Hinrichsen, Diederich, 1939– II. Colonius, Fritz. III. Systems & Control
Q295 .A43 2000
003—dc21 00-044510
CIP

Printed on acid-free paper.

Birkhäuser ®

ISBN 0-8176-4162-9
ISBN 3-7643-4162-9 SPIN 10748587

Production managed by Louise Farkas; manufacturing supervised by Jerome Basma.
Typeset by the editors in LaTeX.
Printed and bound by Sheridan Books, Inc., Ann Arbor, MI.
Printed in the United States of America.

9 8 7 6 5 4 3 2 1

Diederich Hinrichsen

Dedicated to Diederich Hinrichsen
On the Occasion of his 60th Birthday

Contents

Preface

This volume contains the lectures presented at the workshop on "Advances in Mathematical Systems Theory," held on the island of Borkum, Germany (April 20-23, 1999). The book will be of interest to graduate students and researchers interested in control theory and mathematical systems theory, who will find in-depth analysis and presentations from diverse perspectives interacting in this lively area. The editors are proud to dedicate this volume to Diederich Hinrichsen on the occasion of his 60th birthday in acknowledgment of his major contributions to linear systems theory and control theory and his long-term achievements in establishing mathematical systems theory in Germany. We all owe much to him as a teacher, colleague, and friend.

The editors thank the Graduiertenkolleg "Komplexe Dynamische Systeme" at the University of Bremen as well as the European "Nonlinear Control Network" for providing financial support that enabled this workshop.

Augsburg, Germany — Fritz Colonius
Würzburg, Germany — Uwe Helmke
Kaiserslautern, Germany — Dieter Prätzel-Wolters
Bremen, Germany — Fabian Wirth

Introduction

The workshop “Advances in Mathematical Systems Theory” took place in honor of Diederich Hinrichsen on the occasion of his 60th birthday. The following chapters are based on invited lectures and cover a wide range of topics in linear and nonlinear systems theory including parameterization problems, behaviors of linear systems and convolutional codes, as well as complementarity systems and hybrid systems. For controllability and stabilizability of infinite-dimensional systems (allowing also for hysteresis nonlinearities) functional analytic as well as algebraic approaches are presented. A focal point of this volume is stability and robustness of linear and nonlinear systems using the concepts of stability radii and spectral value sets.

The contributions were written by former Ph.D. students as well as close colleagues of Diederich Hinrichsen. As such the chapters form a representative cross–section through Hinrichsen's scientific work in mathematical systems theory.

Diederich Hinrichsen was born in 1939 in Nürnberg. He studied mathematics and German philology at the University of Hamburg. Among his teachers were famous mathematicians such as E. Artin, E. Witt, H. Bauer, and L. Collatz. His scientific interests were widely spread and in addition to mathematics also included literature, politics, and economics. Only one year after his Staatsexamen, he got his Ph.D. at the University of Erlangen in 1966 under the supervision of H. Bauer. His main research area at that time was abstract potential theory with special focus on extensions of the Cauchy–Weil theorem related to the Choquet boundary. An intermediate period followed, including research visits and appointments in Paris, Hamburg, and Bielefeld. Also, his lifelong attachment to Cuba, where his research interests shifted to applied mathematics, dates back to this time.

In 1973 he was appointed professor of mathematics at the University of Bremen. Being strongly influenced by Rosenbrock's work in linear systems theory, he quickly realized the scientific potential of the emerging field of mathematical systems and control theory. Although this field was already flourishing in other countries at that time, it was both Hinrichsen's and Hans W. Knobloch's special merit to establish the field of mathematical systems and control theory in Germany. A very important event in this process was the foundation of the “Forschungsschwerpunkt Dynamische Systeme” in the late 1970s in Bremen. Hinrichsen, together with Ludwig Arnold, Hans–Friedrich Münzner, Heinz–Otto Peitgen, and Peter Richter, succeeded in establishing a research center for dynamical systems at the

University of Bremen concentrating on finite– and infinite–dimensional linear systems, stochastic dynamical systems, and nonlinear dynamics. This center attracted a large number of Ph.D. students to the field. It had a tremendous impact on the further development of mathematical systems theory in Germany and played a central role in establishing the Bremen research group on control theory internationally. Hinrichsen was the leader of the linear systems theory group in that center right from the beginning. Creating a very fruitful scientific climate, he managed to build up an outstanding, successful research group with many international visitors and a lot of young productive Ph.D. students. Since then, seven of them have been appointed lecturers and professors of mathematics in Germany, the United Kingdom, and Switzerland.

His own research interests focused on algebraic systems theory, parameterization problems in control and linear algebra, infinite-dimensional systems, and stability analysis. In his research, he aimed at both a rigorous as well as a profound understanding of the mathematical issues of the field that could serve as a solid basis for future work. Some of his major achievements have been directed towards the development of a comprehensive structure theory of linear systems. In a different direction, together with Antony J. Pritchard, he did pioneering work on the concepts of stability radii and spectral value sets. In a long–lasting collaboration resulting in a large number of joint research papers they built up a fundamental robustness theory covering deterministic and stochastic aspects of dynamical systems.

Diederich Hinrichsen is a man of great vision and talents. His scientific achievements in linear systems theory made him one of the leading figures. He established mathematical systems theory in Germany and has deeply influenced the research and successful careers of his many students.

Ph.D. Students of Diederich Hinrichsen

F. Colonius, *Hereditäre Differenzierbare Systeme mit Funktionenraum-Endbedingungen und punktweisen Steuerbeschränkungen: Notwendige Optimalitätsbedingungen und Erreichbarkeit*, 1979.

D. Prätzel-Wolters, *Feedback Morphisms Between Linear Systems — a Unified Approach to State Space Systems, Transfer Functions and System Matrices*, 1981.

D. Salamon, *On Control and Observation of Neutral Systems*, 1982.

U. Helmke, *Zur Topologie des Raumes linearer Kontrollsysteme*, 1983.

A. Linnemann, *Polzuweisung und Störungsentkopplung in linearen Systemen — Ein numerisch orientierter Ansatz*, 1984.

H. Logemann, *Funktionentheoretische Methoden in der Regelungstheorie unendlich-dimensionaler Systeme,* 1986.

A. Ilchmann, *Contributions to Time-Varying Linear Control Systems,* 1987.

W. Manthey, *Die Bruhat-Zellzerlegung von Räumen rationaler Funktionen und Hankel-Matrizen,* 1991.

H. Glüsing-Lüerßen, *Gruppenaktionen in der Theorie singulärer Systeme,* 1991.

B. Jacob, *Time-Varying Infinite Dimensional State-Space Systems,* 1995.

F. Wirth, *Robust Stability of Discrete Time Systems Under Time-Varying Perturbations,* 1995.

E. Gallestey Alvarez, *Theory and Numerics of Spectral Value Sets,* 1998.

A. Fischer, *Aspects of the Robust Stability of Infinite-Dimensional Positive Systems,* 1999.

A. Ribalta Standford, *Rational Approximation of Transfer Functions of Infinite-Dimensional Systems,* 2000.

T. Damm, *Rational Matrix Equations for Stochastic Control,* in progress.

M. Karow, *Stability Radii and Spectral Value Sets for Real Matrix Perturbations,* in progress.

Publications of D. Hinrichsen July 1999

Books and Proceedings

B1 *Analisis Matematico,* 3 vols., Habana 1972–73.

B2 *Topologia General,* Habana (with J.L. Fernandez), 1977.

B3 *Lecture Notes in Mathematical Systems Theory,* Universities of Bremen and Warwick (with P.E. Crouch, A.J. Pritchard, and D. Salamon), 1981.

B4 Feedback and control of linear and nonlinear systems, *Proc. Joint Workshop on Feedback Systems,* Lecture Notes in Control and Information Sciences 39, Springer–Verlag, Berlin (with A. Isidori, eds.), 1982.

B5 Special issue on parametrization problems, *IMA J. Math. Control Inf.* 3 (with J.C. Willems, eds.), 1986.

B6 Control of uncertain systems, *Proc. International Workshop Bremen 1989,* Birkhäuser, Boston (with B. Mårtensson, eds.), 1990.

B7 *Proc. Annual Meeting GAMM 1998,* 3 vols., ZAMM 79, Supplement 1–3 (with A. Fischer, H. J. Rath, and M. Wanschura, eds.), 1999.

Papers in Journals, Proceedings, and the Like

1. Einige Eigenschaften lokal-kompakter konvexer Mengen und ihrer projektiven Limiten, *Proc. Coll. Convexity, Copenhagen 1965*, 143–153 (with H. Bauer), 1967.

2. Adapted integral representations by measures on Choquet boundaries, *Bulletin AMS* 72:888–891, 1966.

3. Randintegrale und nukleare Funktionenräume, *Annales de l'Institut Fourier de l'Université de Grenoble* 17:225–271, 1967.

4. Représentations intégrales et espaces fonctionnels nucléaires, *Séminaire Brelot-Choquet-Deny* 11:14-01-14, 1966/67.

5. Technischer Fortschritt, Produktivkraft Wissenschaft, Bildungs-Ökonomie, *Materialien zur politischen Ökonomie des Ausbildungssektors,* E. Altvater and F. Huisken, eds., Erlangen, 11–51 (with K. Köhler), 1971.

6. Zum Problem der Reduktion komplizierter auf einfache Arbeit, *Materialien zur politischen Ökonomie des Ausbildungssektors*, E. Altvater and F. Huisken, eds., Erlangen, 303–341, 1971.

7. Una generalizacion del lema de Neyman-Pearson, *Investigaciones Operacionales* 10:24–29, 1973.

8. Some theses concerning the application of mathematical system theory in the social sciences, *System Theory in the Social Sciences,* H. Bossel, S. Klaczko, and N. Müller, eds., Birkhäuser, Basel, 386–400, 1976.

9. Optimal control of hereditary differential systems, *Proc. Conference Recent Theoretical Developments in Control*, Leicester 1976, M.J. Gregson, ed., Academic Press, London, 215–239 (with F. Colonius), 1978.

10. Differentiable state space description of nonlinear time-variant hereditary differential systems, *Proc. IFAC Symposium. Control of Distributed Parameter Systems*, S.P. Banks, and A.J. Pritchard, eds., Pergamon Press, Oxford, (with F. Colonius), 1978.

11. Optimal control of functional differential systems, *SIAM J. Control Optimization* 16:861–879 (with F. Colonius), 1978.

12. Choice of techniques in joint production models, *Operations Research-Verfahren* 34:155–171 (with U. Krause), 1979.

13. Characterizations of strict system equivalence in time domain, *Lectures in Appl. Math.* 18:137–156 (with D. Prätzel-Wolters), 1980.

14. Solution modules and system equivalence, *Int. J. Control* 32:777–802 (with D. Prätzel-Wolters), 1980.

15. A substitution theorem for joint production models with disposal processes, *Operations Research-Verfahren* 41:287–291 (with U. Krause), 1981.

16. Hermite invariants and canonical forms for reachable system matrices, *Proc. 3rd IMA Conference Control Theory*, W.D. Collins et al., eds., Academic Press, London, 31–49. (with D. Prätzel-Wolters), 1981.

17. Parametrization of (C,A)–invariant subspaces, *Syst. Control Letters* 1:192–199 (with H.F. Münzner and D. Prätzel-Wolters), 1981.

18. Generalized Hermite matrices and complete invariants of strict system equivalence, *SIAM J. Control Optimization* 21:289–305 (with D. Prätzel-Wolters), 1983.

19. Unique representation in convex sets by extraction of marked components, *Linear Algebra Appl.* 51:73–96 (with U. Krause), 1983.

20. Normalformen vom Hermite-Typ und die Berechnung dominanter Hermite-Indizes strukturierter Systeme, *Regelungstechnik* 32:124–130 (with A. Linnemann), 1984.

21. Canonical forms and orbit spaces of linear systems, *IMA J. Math. Control Inf.* 3:167–184 (with U. Helmke), 1986.

22. A condensed form approach to disturbance decoupling problems, *IEE Colloquium Digest*:1/1-1/5, 1984/76.

23. A canonical form for static linear output feedback, *Proc. Int. Symposium Math. Theory Networks and Systems 1983*, P. Fuhrmann, ed., Lecture Notes in Control and Information Sciences 58, Springer-Verlag, Berlin, 441–462 (with D. Prätzel-Wolters), 1984.

24. A canonical form for multinomial systems, *Syst. Anal. Model. Simul.* 1:357–371, 1984.

25. The $0(n)$-echelon form and its application to multinomial systems, *Revista Ciencias Matematicas* 6:45-64, 1985.

26. Canonical forms and parametrization problems in linear systems theory, *Proc. 4th IMA Int. Conference Control Theory*, P.A. Cook, ed., Academic Press, London, 45–71, 1985.

27. State and input transformations for reachable systems —- a polynomial approach", *Contemporary Math.* 47:217-239 (with D. Prätzel-Wolters), 1985.

28. A wild quiver in linear systems theory, *Linear Algebra Appl.* 91:143–175 (with D. Prätzel-Wolters), 1987.

29. Stability radii of linear systems, *Syst. Control Letters* 7:1–10 (with A.J. Pritchard), 1986.

30. Stability radius for structured perturbations and the algebraic Riccati equation, *Syst. Control Letters* 8:105–113 (with A.J. Pritchard), 1986.

31. The Bruhat decomposition of finite Hankel matrices, *Syst. Control Letters* 7:173–182 (with W. Manthey and D. Prätzel-Wolters), 1986.

32. Metrical and topological aspects of linear systems theory, *Syst. Anal. Model. Simul.* 4:3–36, 1987.

33. The Bruhat parametrization of infinite real Hankel matrices of rank $\leq n$, *Proc. 25th IEEE Conference Decision and Control*, Athens, 527–529 (with W. Manthey), 1986.

34. Jordan echelon forms for linear state and input transformation, *Proc. 25th Conference Decision and Control*, Athens, 1938–1943 (with D. Prätzel-Wolters), 1986.

35. Robustness of stability of time-varying linear systems, *J. Diff. Equations* 82(2):219–250 (with A. Ilchmann and A.J. Pritchard), 1989.

36. Optimization problems in the robustness analysis of linear state space systems, in *Approximation and Optimization*, A. Gomez et al., eds., Lecture Notes in Mathematics 1354, 54–78, Springer-Verlag, Berlin (with M. Motscha), 1988.

37. Orbit closures of singular systems under restricted system equivalence, in *Linear Circuits, Systems and Signal Processing: Theory and Application*, C.I. Byrnes, G.F. Martin, and R.E. Saeks, eds., North-Holland, Amsterdam, 301–306 (with J. O'Halloran), 1988.

38. A Jordan control canonical form for singular systems, *Int. J. Control* 48:1769–1785 (with H. Glüsing-Lüerßen), 1988.

39. An algorithm for the computation of the structured stability radius with applications, *Automatica* 25:771–775 (with B. Kelb, and A. Linnemann), 1989.

40. A Jordan canonical form for reachable linear systems, *Linear Algebra Appl.* 122/123/124:489–524 (with D. Prätzel-Wolters), 1989.

41. New robustness results for linear systems under real perturbations, *Proc. 27th IEEE Conference Decision and Control*, Austin, 1375–1379 (with A.J. Pritchard), 1988.

42. On the robustness of root locations of polynomials under complex and real perturbations, *Proc. 27th IEEE Conference Decision and Control*, Austin, 1410–1414 (with A.J. Pritchard), 1988.

43. An application of state space methods to obtain explicit formulae for robustness measures of polynomials, *Proc. Int. Workshop Robustness in Identification and Control*, M. Milanese et al., eds., 183–206, Birkhäuser, Boston (with A.J. Pritchard), 1989.

44. Robustness measures for linear systems with applications to stability radii of Hurwitz and Schur polynomials, *Int. J. Control* 55:809–844 (with A.J. Pritchard), 1992.

45. A cell decomposition of the space of real Hankels of rank $\leq n$ and some applications, *Linear Algebra Appl.* 122/123/124:331–355 (with U. Helmke and W. Manthey), 1989.

46. A complete characterization of orbit closures of controllable singular systems under restricted system equivalence, *SIAM J. Control and Optimization* 28:602–623 (with J. O'Halloran), 1990.

47. An improved error estimate for reduced order models of discrete time systems, *IEEE Trans. Auto. Control* 35:317–320 (with A.J. Pritchard), 1990.

48. The complex stability radius of time-varying linear systems, in N. K.Nichols, and D. H. Owens, eds., *Mathematics of Control Theory,* based on the *Proc. 5th Int. IMA Conference on Control Theory,* Glasgow, 1988, Clarendon Press, Oxford, 281–290 (with A. Ilchmann, and A.J. Pritchard), 1992.

49. Topological aspects of the partial realization problem, *Math. Control, Signals, Systems* 5:117–149 (with W. Manthey, and U. Helmke), 1992.

50. Robustness measures for linear state space systems under complex and real parameter perturbations, in *Perspectives in Control Theory, Proc. of the Sielpia Conference,* Sielpia, Poland, 1988, Birkhäuser, Boston, 54–74, (with A. J. Pritchard), 1990.

51. Parametrized Riccati equations and the problem of maximizing the complex stability radius, *Proc. Workshop on The Riccati Equation in Control, Systems, and Signals"*, S. Bittanti, ed., Pitagora Editrice, Bologna, 136–142 (with A. J. Pritchard), 1989.

52. Modellreduktion mit Hilfe balancierter Realisierungen, (Teil 1) *Automatisierungstechnik* 38:416–422, (1990), (Teil 2) *Automatisierungstechnik* 38:460-466 (with H.-W. Philippsen), 1990.

53. A note on the geometry of partial realization, in *Realization and Modelling in System Theory*, Proc. MTNS-89, Birkhäuser, Boston, 157–165 (with W. Manthey and U. Helmke), 1990.

54. The moduli space of controllable singular systems under restricted system equivalence, in *Realization and Modelling in System Theory*, Proc. MTNS-89, Birkhäuser, Boston, 149–155 (with J. O'Halloran), 1990.

55. The complex stability radius of discrete-time systems and symplectic pencils, *Proc. 28th Conference Decision and Control*, Tampa, 2265–2270 (with N.K. Son), 1989.

56. On the orbit closure problem for matrix pencils and controllable singular systems, *Proc. 28th Conference Decision and Control*, Tampa, 1324–1325 (with J. O'Halloran), 1989.

57. Robustness of stability of linear state space systems with respect to time–varying, nonlinear and dynamic perturbations, *Proc. 28th Conference Decision and Control*, Tampa, 52–53 (with A.J. Pritchard), 1989.

58. A note on some differences between real and complex stability radii, *Syst. Control Letters* 14:401–409 (with A.J. Pritchard), 1990.

59. Real and complex stability radii: A survey, in: *Control of Uncertain Systems, Proc. Int. Workshop Bremen 1989*, D. Hinrichsen and B. Mårtensson, eds., Birkhäuser, Boston, 119–162 (with A.J. Pritchard), 1990.

60. Riccati equation approach to maximizing the complex stability radius by state feedback, *Int. J. Control* 52:769–794 (with A.J. Pritchard), 1990.

61. Stability radii of linear discrete–time systems and symplectic pencils, *Int. J. Robust Nonlinear Control* 1:79–97 (with N.K. Son), 1991.

62. On the robustness of stable discrete time linear systems, in *New Trends in Systems Theory, Proc. Conference*, Genova, 1990, Birkhäuser, Boston, 393–400 (with A.J. Pritchard), 1991.

63. Destabilization by output feedback, *Diff. Integral Equations* 5:357–386 (with A.J. Pritchard), 1992.

64. On spectral variations under bounded real matrix perturbations, *Numerische Mathematik* 60:509–524 (with A.J. Pritchard), 1992.

65. The orbit closure problem for matrix pencils: Necessary conditions and an application to high gain feedback, in *New Trends in Systems Theory, Proc. Conference,* Genoa, 1990, Birkhäuser, Boston, 388–392 (with J. O'Halloran), 1991.

66. Orbit closures of matrix pencils and system limits under high gain feedback, *Proc. 29th Conference Decision and Control*, Honolulu, 55–60 (with J. O'Halloran), 1990.

67. New results on orbit closures of singular systems, in *Recent Advances in the Mathematical Theory of Systems, Control, Networks and Signal Processes*, Proc. MTNS-91 Kobe, Osaka, 133–138 (with J. O'Halloran), 1992.

68. Robust stability of linear time–varying systems with respect to multi–perturbations, *Proc. European Control Conference,* Grenoble, 1366–1371 (with A.J. Pritchard), 1991.

69. Orbit closures of singular matrix pencils, *J.Pure Appl. Algebra* 81:117–137 (with J. O'Halloran), 1992.

70. A note on the degeneration of systems under pencil equivalence, *Proc. 30th Conference Decision and Control*, Brighton, 1431–1432 (with J. O'Halloran), 1991.

71. Robust exponential stability of time–varying linear systems under time–varying parameter perturbations, *Int. J. Robust Nonlinear Control* 3:63–83 (with A. J. Pritchard), 1993.

72. Robust stability of linear systems described by higher order dynamic equations, *IEEE Trans. Auto. Control* 38:1430–1435 (with G. Pappas), 1993.

73. Robust stability of linear evolution operators on Banach spaces, *SIAM J. Control and Optimization* 32:1503–1541 (with A.J. Pritchard), 1994.

74. Spectral value sets: A graphical tool for robustness analysis, *Syst. Control Letters* 21:127-136 (with B. Kelb), 1993.

75. On stability radii of infinite dimensional time-varying discrete-time systems, *IMA J. Math. Control Inf.* 11:253–276 (with F. Wirth), 1994.

76. The degeneration of reachable singular systems under feedback-transformations, *Kybernetika* 30:387–391 (with H. Glüsing-Lüerßen), 1994.

77. On a cell decomposition for Hankel matrices and rational functions, *J. Reine Angew. Mathematik* 451:15–50 (with W. Manthey), 1994.

78. Robust stability of time–invariant infinite dimensional systems with respect to time–varying perturbations, *Proc. 2nd European Control Conference,* Groningen, 1089–1093 (with A.J. Pritchard), 1993.

79. A pencil approach to high gain feedback and generalized state space systems, *Kybernetika* 31:109–139 (with J. O'Halloran), 1995.

80. A note on the orbit closure problem for the generalized feedback group, in *Systems and Networks: Mathematical Theory and Applications, Vol. II Invited and Contributed Papers*, Akademie-Verlag, Berlin, 221–224 (with J. O'Halloran), 1994.

81. Stability radii and spectral value sets for real matrix perturbations, in *Systems and Networks: Mathematical Theory and Applications, Vol. II Invited and Contributed Papers*, Akademie-Verlag, Berlin, 217–220 (with B. Kelb), 1994.

82. Stability of uncertain systems, in *Systems and Networks: Mathematical Theory and Applications, Vol. I Key Invited Lectures*, Akademie-Verlag, Berlin, 159–182 (with A.J. Pritchard), 1994.

83. Stability margins for systems with deterministic and stochastic uncertainty, *Proc. 33rd Conference Decision and Control,* Florida, 3825–3826 (with A.J. Pritchard), 1994.

84. Stability radii of systems with stochastic uncertainty and their optimization by output feedback, *SIAM J. Control Optimization* 34:1972–1998 (with A.J. Pritchard), 1996.

85. Stability of polynomials with conic uncertainties, *Math. Control, Signals, Syst.* 8:97–117 (with V.L. Kharitonov), 1995.

86. On convex directions for stable polynomials, *Automatika i Telemechanika* 3:81–92 (with V.L. Kharitonov), 1997.

87. A root locus characterization of convex directions, *Proc. 3rd. European Control Conference*, Rome, 2999–3003 (with V.L. Kharitonov), 1995.

88. Maximization of stability radii via dynamic output feedback for systems with stochastic uncertainty, *Proc. 3rd. European Control Conference*, Rome, 984–988 (with A.J. Pritchard), 1995.

89. Stability radii for infinite dimensional systems with stochastic uncertainty, *Proc. 3rd European Control Conference*, Rome, 3291–3293 (with A.J. Pritchard), 1995.

90. On Fischer–Frobenius transformations and the structure of rectangular block Hankel matrices, *Linear Multilinear Algebra* 41:255–288 (with U. Helmke and W. Manthey), 1996.

91. Stability radii of positive discrete-time systems, *Proc. 3rd Int. Conference on Approximation and Optimization in the Caribbean,* Puebla, *Aportaciones Matematicas, Serie Comunicaciones 24* (with N.K. Son), 1995.

92. Robust stability of positive continuous time systems, *Numer. Funct. Anal. Optimization* 17:649–659 (with N.K. Son), 1996.

93. Robust stability of positive linear systems, *Proc. 34th Conference Decision and Control*, New Orleans, 1423–1424 (with N.K. Son), 1995.

94. Limits under high gain proportional (and derivative) feedback, *Proc. 34th Conference Decision and Control*, New Orleans, 2730–2734 (with J. O'Halloran), 1995.

95. Stability radii of positive dynamical systems, *ZAMM* 76(2):555–556 (with N.K. Son), 1996.

96. On convex stability directions for real quasipolynomials, in *Stability Theory, Proc. Centennial Hurwitz on Stability Theory,* Ascona,1995, R. Jeltsch and M. Mansour, eds., International Series of Numerical Mathematics 121, Birkhäuser, Basel, 43–52 (with L. Atanassova, V.L. Kharitonov), 1996.

97. On structured singular values and robust stability of positive systems under affine perturbations, *Vietnam J. Math.* 24:113–119 (with N.K. Son), 1996.

98. Attenuation of multiperturbed stochastic systems by output feedback, *Proc. 3rd Int. Symposium Methods and Models in Automation and Robotics*, Miedzyzdroje, Poland, vol. 1, 367–372 (with A. El Bouthouri, and A.J. Pritchard), 1996.

99. Limits of generalized state space systems under proportional and derivative feedback, *Math. Control, Signals Syst.* 10:97–124 (with J. O'Halloran), 1997.

100. Convex directions for complex Hurwitz stable polynomials and quasipolynomials, *Proc. 4th European Control Conference*, Brussels, TU-A-D-6 (with L. Atanassova and V.L. Kharitonov), 1997.

101. Stability of positive systems under blockdiagonal perturbations, *Proc. 3rd Int. Symposium Methods and Models in Automation and Robotics*, Miedzyzdroje, Poland, vol. 1, 27-34, 1996. Extended version: μ-Analysis and robust stability of positive linear systems, *Appl. Math. Comp. Sci.* 8:253–268 (with N.K. Son), 1998.

102. Stability radii of positive discrete-time systems under affine parameter perturbations, *Int. J. Robust Nonlinear Control* 8:1169–1188 (with N.K. Son), 1998.

103. Lyapunov exponents and robust stabilization, in *Open Problems in Mathematical Systems and Control Theory,* V. D. Blondel et al., eds., Springer–Verlag, London, 83–88 (with F. Colonius and F. Wirth), 1999.

104. Spectral value sets of infinite-dimensional systems, in *Open Problems in Mathematical Systems and Control Theory,* V.D. Blondel et al., eds., Springer–Verlag, London, 109–113 (with E. Gallestey and A.J. Pritchard), 1999.

105. Robust stability of linear stochastic systems, in *Open Problems in Mathematical Systems and Control Theory,* V. D. Blondel et al., eds., Springer Verlag, London, 125–129 (with A. J. Pritchard), 1999.

106. A Riccati equation for stochastic H_∞, *Proc. 4th European Control Conference*, Brussels, FR-A-L-2 (with A.J. Pritchard), 1997.

107. On the spectral value set problem in infinite dimensions, *Proc. 4th International Symposium on Methods and Models in Automation and Robotics*, Miedzyzdroje, Poland, 109–114 (with E. Gallestey and A.J. Pritchard), 1997.

108. Generalized partial realizations, in *Operators, Systems and Linear Algebra. Three Decades of Algebraic Systems Theory,* U. Helmke, D. Prätzel-Wolters and E. Zerz, eds., Teubner, Stuttgart, 138–156 (with U. Helmke and W. Manthey), 1997.

109. Convex directions for stable polynomials and quasipolynomials: A survey of recent results, in *Stability and Control of Time-Delay Systems,* L. Dugard and E.I. Verriest, eds., Lecture Notes in Control and Information Sciences 228, Springer-Verlag, London 72–88 (with L. Atanassova and V.L. Kharitonov), 1997.

110. Stability radii of discrete-time stochastic systems with respect to blockdiagonal perturbations, *Automatica* 36(7):1033–1040 (with A. El Bouthouri and A.J. Pritchard), 2000.

111. H^∞ type control for discrete-time stochastic systems, *Int. J. Robust Nonlinear Control* 9:923–948 (with A. El Bouthouri and A.J. Pritchard), 1999.

112. Stability radii of Metzler operators, *Vietnam J. Math.* 26:147–163 (with A. Fischer and N.K. Son), 1998.

113. A Riccati approach to disturbance attenuation of discrete time stochastic systems, *Proc. Int. Conference Mathematical Theory of Networks and Systems,* Padua, 641–644 (with A.J. Pritchard, and A. El Bouthouri), 1998.

114. Linear matrix (in)equalities for linear stochastic systems, *Proc. Int. Conference Mathematical Theory of Networks and Systems,* Padua, 593–596 (with T. Damm), 1998.

115. Approximation of spectral value sets, *Proc. Int. Conference Mathematical Theory of Networks and Systems,* Padua, 463–466 (with E. Gallestey and A.J. Pritchard), 1998.

116. On robust stability of positive systems, *Proc. Int. Conference Mathematical Theory of Networks and Systems*, Padua, 161–164 (with A. Fischer), 1998.

117. Stochastic H_∞, *SIAM J. Control Optimization* 36:1504–1538 (with A.J. Pritchard), 1998.

118. On the disturbance attenuation problem for a wide class of time invariant linear stochastic systems, *Stochastics and Stochastics Reports,* 65:255–297 (with A. El Bouthouri and A.J. Pritchard), 1999.

119. Spectral value sets of closed linear operators, *Proc. Royal Soc.,* London, 456:1397–1418 (with E. Gallestey and A.J. Pritchard), 2000.

120. Partial realization by descriptor systems, Report 442, Inst. f. Dynamische Systeme, Universität Bremen (with W. Manthey and U. Helmke), 1999 (submitted).

121. Newton's method for a rational matrix equation occuring in stochastic control, Report 443, Inst. f. Dynamische Systeme, Universität Bremen (with T. Damm), 1999, *Linear Algebra Appl.* (to appear).

Contributors and Speakers

Ludwig Arnold	Fachbereich Mathematik, Universität Bremen, 28334 Bremen, Germany. e-mail: arnold@math.uni-bremen.de
Chris Byrnes	School of Engineering & Applied Science, Washington University, St. Louis, MO 63130-4899, USA. e-mail: ChrisByrnes@seas.wustl.edu
Fritz Colonius	Institut für Mathematik, Universität Augsburg, 86135 Augsburg, Germany. e-mail: colonius@math.uni-augsburg.de
Ruth F. Curtain	Dept. of Mathematics, University of Groningen, P.O. Box 800, 9700AV Groningen, the Netherlands. e-mail: curtain@math.rug.nl
Paul Fuhrmann	Dept. of Mathematics, Ben-Gurion University of the Negev, POB 653, Beer Sheva 84105, Israel. e-mail: paf@math.bgu.ac.il
Heide Gluesing-Luerssen	Dept. of Mathematics, University of Notre Dame, Notre Dame, Indiana 46556-5683, USA. e-mail: gluesing-luerssen.1@nd.edu
Uwe Helmke	Mathematisches Institut, Universität Würzburg, 97074 Würzburg, Germany. e-mail: helmke@mathematik.uni-wuerzburg.de
Michael Heymann	Dept. of Computer Science, Technion, Israel Institute of Technology, Haifa 32000, Israel. e-mail: heymann@cs.technion.ac.il
Achim Ilchmann	Technical University Ilmenau, Weimarer Straße 25, 98693 Ilmenau, Germany. e-mail: ilchmann@mathematik.tu-ilmenau.de
Birgit Jacob	School of Mathematics, University of Leeds, Leeds LS2 9JT, U.K. e-mail: birgit@amsta.leeds.ac.uk

Vladimir L. Kharitonov Departamento de Control Automático, CINVESTAV-IPN, Av. del IPN, 2508, C.P. 07360, México, D.F., Mexico.
e-mail: khar@ctrl.cinvestav.mx

Wolfgang Kliemann Dept. of Mathematics, Iowa State University, Ames, IA 50011, USA.
e-mail: kliemann@iastate.edu

Uli Krause Fachbereich Mathematik, Universität Bremen, 28334 Bremen, Germany.
e-mail: krause@math.uni-bremen.de

Izchak Lewkowicz Dept. of Electrical and Computer Engineering, Ben-Gurion University of the Negev, POB 653, Beer Sheva 84105, Israel.
e-mail: izchak@ee.bgu.ac.il

Feng Lin Dept. of Electrical and Computer Engineering, Wayne State University, Detroit, MI 48202, USA. e-mail: flin@ece.eng.wayne.edu

Hartmut Logemann Dept. of Mathematical Sciences, University of Bath, Bath BA2 7AY, United Kingdom.
e-mail: hl@maths.bath.ac.uk

Guillermo López Lagomasino Departamento de Matemáticas, Universidad Carlos III de Madrid, Spain.
e-mail: lago@math.nc3m.es

Iven M.Y. Mareels Dept. of Electrical and Electronic Eng., The University of Melbourne, Parkville VIC 3052, Australia.
e-mail: i.mareels@unimelb.edu.au

Adam D. Mawby Dept. of Mathematical Sciences, University of Bath, Bath BA2 7AY, United Kingdom.
e-mail: am@maths.bath.ac.uk

George Meyer NASA Ames Research Center, Moffett Field, CA 94035, USA.
e-mail: meyer@tarski.arc.nasa.gov

Henk Nijmeijer Dept. Appl. Mathematics, University of Twente, P.O. Box 217, 7500 AE Enschede, The Netherlands. e-mail: twhenk@math.utwente.nl

Job C. Oostveen Dept of Mathematics, University of Groningen, P.O. Box 800, 9700AV Groningen, the Netherlands.

Andrew D.B. Paice ABB Corporate Research C1, 5405 Baden-Daettwil, Switzerland.
e-mail: andrew.paice@ch.abb.com

Anthony J. Pritchard Dept. of Mathematics, University of Warwick, Coventry CV4 7AL, England.
e-mail: ajp@maths.warwick.ac.uk

Joachim Rosenthal Dept. of Mathematics, University of Notre Dame, Notre Dame, Indiana 46556-5683, USA.
e-mail: rosen@nd.edu

Dietmar Salamon Dept. of Mathematics, ETH Zürich, 8092 Zürich, Switzerland.
e-mail: salamon@math.ethz.ch

J.M. (Hans) Schumacher CWI, Kruislaan 413 1098 SJ Amsterdam, the Netherlands, e-mail: jms@cwi.nl

Paul A. Weiner Dept. of Mathematics, Saint Mary's University of Minnesota, Winona, MN 55987, USA.
e-mail: pweiner@smumn.edu

Jan C. Willems Dept. of Mathematics, University of Groningen, P.O. Box 800, 9700AV Groningen, the Netherlands.
e-mail: j.c.willems@math.rug.nl

Fabian R. Wirth Zentrum für Technomathematik, Universität Bremen, 28334 Bremen, Germany.
e-mail: fabian@math.uni-bremen.de

Hans Zwart Faculty of Mathematical Sciences, University of Twente, P.O.Box 217, 7500 AE Enschede, the Netherlands.
e-mail: H.J.Zwart@math.utwente.nl

Advances in Mathematical Systems Theory
Workshop in Honor of Diederich Hinrichsen
On the occasion of his 60th birthday
April 20–23, 1999, Island of Borkum

Lectures

A.J. Pritchard	*Transitory behavior of uncertain systems*
D. Salamon	*Symplectic topology and J-holomorphic curves*
J.C. Willems	*Dissipative distributed systems*
M. Heymann	*Control of rate-bounded hybrid systems with liveness constraints*
H. Schumacher	*Between mathematical programming and system theory: Linear complementary systems*
H. Nijmeijer	*Cooperative oscillatory behavior of mutually coupled dynamical systems*
U. Krause	*Didi and economics*
P. Fuhrmann	*On conditioned invariant subspaces*
U. Helmke	*The geometry of linear systems*
C.I. Byrnes	*A geometric duality theorem for the spaces of positive real functions*
I. Lewkowicz	*A characterization of convex cones of matrices with constant regular inertia*
J. Rosenthal	*Construction and decoding of convolutional codes*
L. Arnold	*Lyapunov's second method*
W. Kliemann	*Control theory and Lie semigroups*
F. Colonius	*Multistability radii for nonlinear systems*
V.L. Kharitonov	*Robust stability of multivariate polynomials*
F. Wirth	*Robustness analysis for nonlinear systems*
B. Jacob	*Equivalent conditions for exact controllability*
H. Logemann	*Low gain integral control of infinite-dimensional regular linear systems subject to input hysteresis*
A. Ilchmann	*A behavioral approach to time-varying systems*
G. López Lagomasino	*Rational approximation of transfer functions of infinite dimensional unstable control systems*
R. Curtain	*Normalized coprime factorization for strongly stabilizable systems*

Chapter 1

Transitory Behavior of Uncertain Systems

Anthony J. Pritchard

ABSTRACT *The notion of a transitory excursion is introduced as a measure of the distance a stable semigroup (generated by a matrix A) moves away from the origin. Various estimates for the excursion are obtained via the distance of A from the normal matrices, properties of spectral value sets of A, and time-varying Lyapunov equations. It is shown how the excursion can be improved by state feedback. Finally the notion of transitory excursion radius for uncertain systems is introduced and estimates obtained.*

1.1 Introduction

In 1993 Diederich and I became interested in the transitory behavior of systems. Arguing with each other that, although a system may be stable, if the trajectory moves a long way away from an initial state near the origin before eventually approaching the origin, parameter uncertainties in the model or the neglect of dynamics or nonlinear effects may result in catastrophic behaviors and instabilities for the system being modeled. We worked on this for a day or so, feeling, somehow that stability radii, spectral value sets, or Riccati equations might be useful tools in the analysis. Unlike much of our previous work, with the new-found discipline of the computer, on January 12th, 1994, Diederich decided to type up some notes into what he called transitory.tex. And there it has sat in both our machines for the last five years. Although we both thought the problem was an important one and one which would be fruitful and exciting, what with our respective administrative chores, our occasional flurries into research in other areas, and of course the BOOK, we never had time to spend on it.

When the Workshop in Honor of Diederich was organized I agreed, readily, to present a paper. So out came transitory.tex. There was no way in the short time available that I would be able to develop our initial ideas into a rounded complete work. Instead this chapter must be seen as a collection of first ideas on the subject that meander in different directions. Moreover, although the chapter is based on our ideas from five or six years ago, the responsibility for the form in which they are presented here, with all its inconsistencies and possible errors, is mine. I am sure that, with Diederich's mathematical culture, intuition, and rigor, a far better work is in the offing.

In 1993 we were blissfully unaware of the activities of others in this field. In the classical stability analysis of the flow between two stationary parallel plates (*Poiseuille flow*), the Navier–Stokes equations are linearized about a parabolic velocity profile. Squire's theorem states that an unstable eigenmotion in $\mathbb{R}^3$ for some value of the Reynolds number R can only exist if there is an unstable two–dimensional eigenmotion for some lower value of R. This is used to justify the study of two dimensional perturbations that are governed by the Orr–Sommerfeld equation. The analysis of this equation gives rise to a critical value of the Reynolds number, $R_c = 5772$ at which instability should occur. This contradicts experiments in the laboratory where turbulence occurs at values of R as low as 1000. Over one hundred years ago it was recognized that before their eventual decay there was a transient growth of disturbances [14]. But it is only fairly recently [1]; [4]; [9] that the extent of this growth was discovered and the results are startling. If only two dimensional perturbations are considered the amplification is small, but in the three–dimensional case amplification can be as much as many hundreds. This has led to speculation that a mechanism for instability may be that the linear growth of perturbations might be sufficient to activate nonlinear terms. In 1990 Trefethen [11] introduced sets that he called *pseudospectra*, and he later used them to quantify the transient growth. There is an excellent review of the results of Trefethen and his coworkers in [12]; see also [13]. Motivated by different considerations it turns out that these sets are the same as *spectral portraits* introduced by Godunov [3] and almost the same as *unstructured spectral value sets* introduced by Diederich and myself [7], see also [2]. Some of these results are briefly reviewed in Section 1.3.

1.2 Transient Excursions

Throughout this section it is assumed that $A \in \mathbb{K}^{n\times n}$, $\mathbb{K} = \mathbb{R}$ or $\mathbb{C}$, and $\sigma(A) \subset \mathbb{C}_-$.

Definition 1.1. *The* transient bound *of the semigroup generated by A, $(e^{At})_{t\geq 0}$ is defined to be*

$$M_0(A) := \max_{t\geq 0} \|e^{At}\|,$$

where $\|\cdot\|$ is an operator norm.

Suppose the maximum in the above expression is achieved at t_0 and x_0^m is such that $\|e^{At_0}x_0^m\| = \|e^{At_0}\|\|x_0^m\|$; then for every $x_0 \in \mathbb{K}^n$, the solution $x(t)$ of the differential equation $\dot{x} = Ax$, $x(0) = x_0$ satisfies $\|x(t)\| \leq M_0(A)\|x_0\|$, $t \geq 0$ and for $x_0 = x_0^m$, we have $\|x(t_0)\| = M_0(A)\|x_0^m\|$. So

$M_0(A)$ is a measure of the relative (with respect to $\|x_0\|$) maximum excursion of the trajectory $x(\cdot)$ from the origin.

Clearly $M_0(A) \geq 1$ and if $M_0(A) = 1$, then $(e^{At})_{t\geq 0}$ is said to be a *contraction* semigroup. Since contraction semigroups have the best possible transient bounds it is interesting to determine necessary and sufficient conditions for A to generate such semigroups. These can be expressed in terms of the dissipativity of A with respect to any semiscalar inner product on $\mathbb{K}^n$.

Definition 1.2. *Suppose $\|\cdot\|$ is a norm on $\mathbb{K}^n$; then a* semiscalar product *$[\cdot,\cdot]$ on $\mathbb{K}^n$ is such that for each pair of vectors $x, y \in \mathbb{K}^n$, $[x,y] \in \mathbb{K}$ and for all $x, y, z \in \mathbb{K}^n$, $\lambda \in \mathbb{K}$,*

$$[x+y,z] = [x,z] + [y,z], \quad [\lambda x, y] = \lambda[x,y], \quad [x,x] = \|x\|^2, \quad |[x,y]| \leq \|x\|\|y\|.$$

Definition 1.3. *A matrix $A \in \mathbb{K}^{n\times n}$ is* dissipative *with respect to a semiscalar product $[\cdot,\cdot]$ if*

$$\Re[Ax,x] \leq 0 \quad \textit{for all } x \in \mathbb{K}^n.$$

The following theorem can be found in [16, IX.8].

Theorem 1.1. *Let $A \in \mathbb{K}^{n\times n}$; then A generates a contraction semigroup on $\mathbb{K}^n$ if and only if A is dissipative (with respect to any semiscalar inner product on $\mathbb{K}^n$).*

Note that in the above theorem the semiscalar inner product with respect to which A is dissipative is arbitrary (but by definition strongly linked to the norm on $\mathbb{K}^n$). As a corollary we obtain the following.

Corollary 1.1. *Suppose that $\mathbb{K}^n$ is a Hilbert space and $\alpha \in \mathbb{R}_+$, then*

$$\|e^{At}\| \leq e^{-\alpha t}, \qquad t \geq 0 \tag{1.1}$$

if and only if

$$A + A^* \leq -2\alpha I_n, \tag{1.2}$$

where A^* is the adjoint of A with respect to the inner product on $\mathbb{K}^n$.

Proof: A is dissipative with respect to the inner product $\langle\cdot,\cdot\rangle$ on $\mathbb{K}^n$ if and only if

$$\Re\langle Ax,x\rangle = 1/2[\langle Ax,x\rangle + \langle x,Ax\rangle] = 1/2\langle(A+A^*)x,x\rangle \leq 0, \quad x \in \mathbb{K}^n,$$

that is, if and only if $A + A^* \le 0_n$. The corollary now follows from the theorem since (1.1) is satisfied if and only if

$$\|e^{At}e^{\alpha t}\| \le 1 \Leftrightarrow (A + \alpha I_n) + (A + \alpha I_n)^* \le 0_n, \quad t \ge 0.$$

□

Remark 1.1. *Suppose that $\mathbb{K}^n$ is a Hilbert space with inner product $\langle \cdot, P\cdot \rangle_{\mathbb{K}^n}$, where $P \in \mathcal{H}_n$ (Hermitian matrices), $P \succ 0$, and $\langle \cdot, \cdot \rangle$ is the usual Euclidean inner product. Then the adjoint of A, $A^* = P^{-1}\overline{A}^\top P$, where $\overline{A}^\top$ is the complex conjugate and transpose of A. So A generates a contraction semigroup on $\mathbb{K}^n$ if and only if*

$$PA + \overline{A}^\top P \preceq 0.$$

We see therefore that any matrix A with $\sigma(A) \subset \mathbb{C}_-$ generates a contraction semigroup with respect to some norm induced by a particular inner product.

The property of generating a contraction semigroup is robust. To see this let $\mathcal{NC}$ be the set of all $n \times n$ matrices that do not generate a contraction semigroup.

Proposition 1.1. *Suppose that $\mathbb{K}^n$ is a Hilbert space and $A \in \mathbb{K}^{n\times n}$ generates a contraction semigroup. Then the distance $dist(A, \mathcal{NC})$ of A from the set $\mathcal{NC}$ is given by*

$$dist(A, \mathcal{NC}) = -1/2(\lambda_{\max}(A + A^*)).$$

Proof: $A + \Delta$ generates a contraction semigroup if and only if

$$\lambda_{\max}[A + \Delta + (A + \Delta)^*] = \sup_{\|x\|\le 1} \langle [A + \Delta + (A + \Delta)^*]x, x\rangle \le 0.$$

But

$$\begin{aligned} \sup_{\|x\|\le 1} \langle [A + \Delta + (A + \Delta)^*]x, x\rangle &\le \sup_{\|x\|\le 1} \langle [A + A^*]x, x\rangle + \sup_{\|x\|\le 1} \langle [\Delta + \Delta^*]x, x\rangle \\ &= \lambda_{\max}(A + A^*) + \lambda_{\max}(\Delta + \Delta^*). \end{aligned}$$

Now suppose that $\Delta \in \mathbb{K}^{n\times n}$ is such that $\|\Delta\| < -1/2(\lambda_{\max}(A + A^*))$. Then

$$\lambda_{\max}(A + A^*) + \lambda_{\max}(\Delta + \Delta^*) \le \lambda_{\max}(A + A^*) + 2\|\Delta\| < 0.$$

Hence $A + \Delta$ generates a contraction semigroup. Conversely, choose $z \in \mathbb{K}^n$, $\|z\| = 1$ such that $\lambda_{\max}(A + A^*) = \langle (A + A^*)z, z\rangle$; that is, z is an eigenvector of $(A + A^*)$ for the eigenvalue $\lambda = \lambda_{\max}(A + A^*)$ and define $\Delta = -1/2\lambda z\langle \cdot, z\rangle$. Then $\Delta = \Delta^*$, $\|\Delta\| = -1/2\lambda$, and for every $\delta > 1$,

$$\begin{aligned}\langle [A + \delta\Delta + (A + \delta\Delta)^*]z, z\rangle &> \langle [A + \Delta + (A + \Delta)^*]z, z\rangle \\ &= \langle (A + A^*)z, z\rangle + \langle (\Delta + \Delta^*)z, z\rangle \\ &= \lambda - \lambda = 0.\end{aligned}$$

Hence $A + \delta\Delta$ does not generate a contraction semigroup for all $\delta > 1$ □

From now on it is assumed that $\mathbb{K}^n$ is a Hilbert space. An important subset of the matrices that generate contraction semigroups are the normal matrices $\mathcal{N}$.

Lemma 1.1. *Suppose $A \in \mathcal{N}$ satisfies*

$$\alpha(A) = -sup\{\Re\lambda; \lambda \in \sigma(A)\}. \tag{1.3}$$

Then

$$\|e^{At}\| \leq e^{-\alpha(A)t}, \qquad t \geq 0. \tag{1.4}$$

This desirable property (1.4) is captured in the following definition.

Definition 1.4. *A matrix A is said to be* stable with normal transitory behavior *if for some $\alpha > 0$, we have*

$$\|e^{At}\| \leq e^{-\alpha t}, \qquad t \geq 0. \tag{1.5}$$

Using the perturbation result below (an immediate consequence of Gronwall's Lemma) a condition for A to be stable with normal transitory behavior can be obtained.

Lemma 1.2. *Suppose that e^{At} is a semigroup on $\mathbb{K}^n$ satisfying*

$$\|e^{At}\| \leq Me^{-\alpha t}, \qquad t \geq 0 \tag{1.6}$$

for some $M \geq 1$ and $\alpha \in \mathbb{R}$. If $\Delta \in \mathbb{K}^{n\times n}$, then

$$\|e^{(A+\Delta)t}\| \leq Me^{-(\alpha - M\|\Delta\|)t}, \qquad t \geq 0. \tag{1.7}$$

By Lemma 1.2 the following corollary holds.

Corollary 1.2. *Suppose $N \in \mathcal{N}$ has exponential decay rate $\alpha(N)$. If $A \in \mathbb{K}^{n\times n}$ satisfies $\delta = \|A - N\|$, then*

$$\|e^{At}\| \leq e^{-(\alpha(N)-\delta)t}, \qquad t \geq 0. \tag{1.8}$$

Proof: Setting $\Delta = A - N$ we have $A = N + \Delta$ and $\|\Delta\| = \delta$. Thus (1.8) follows from Lemma 1.2 (applied to N instead of A) and (1.4). □

Since the set $\mathcal{N}$ of normal matrices is closed in $\mathbb{K}^{n\times n}$, for a given A, there exists an $N \in \mathcal{N}$ that is the closest to A:

$$\|A - N\| = \operatorname{dist}(A, \mathcal{N}) =: \delta(A).$$

The set of all the normal operators N with this property is compact. By the continuity of the spectrum there exists a closest normal matrix of maximal decay rate (minimal growth rate). Given any A we denote by N_A one such closest normal matrix with maximal decay rate (this matrix may not be uniquely determined). As a consequence of Corollary 1.2 we have the following.

Corollary 1.3. *A matrix is stable with normal transitory behavior if*

$$\alpha(N_A) > \delta(A) \tag{1.9}$$

Another class of matrices for which there are well-known estimates of the transient behavior are the diagonizable ones.

Lemma 1.3. *Suppose that A is diagonizable and $A = TDT^{-1}$ with diagonal D for some $T \in Gl(n, \mathbb{C})$; then*

$$\|e^{At}\| \leq \|T\|\|T^{-1}\|e^{-\alpha(A)t}, \qquad t \geq 0. \tag{1.10}$$

Reddy et al. [10] found that for critical Poiseuille flow $\|T\|\|T^{-1}\|$ is of order 10^8 and rises exponentially with the Reynolds number R. So the operator is far from normal and this is another reason for being suspicious of the traditional stability analysis of Poiseuille flow. Lemma 1.3 does not imply that the transient bound of a diagonizable matrix is necessarily $\|T\|\|T^{-1}\|$, since it may be possible to obtain an estimate of the form $Me^{-\alpha t}$ with $\alpha < \alpha(A)$ and $M < \|T\|\|T^{-1}\|$.

For a general A and every $\alpha \in [0, \alpha(A)]$ let M_α be the smallest M satisfying

$$\|e^{At}\| \leq Me^{-\alpha t}, \qquad t \geq 0.$$

In particular, $M_0(A)$ corresponds to the transient bound. Now for every $\alpha \in [0, \alpha(A))$ let t_α be such that

$$\|e^{At_\alpha}\| = M_\alpha e^{-\alpha t_\alpha}$$

(such a $t_\alpha \geq 0$ exists). Then

$$M_\alpha e^{-\alpha t_\alpha} \leq M_0(A) = \|e^{At_0}\| \leq M_\alpha e^{-\alpha t_0}$$

and hence

$$M_0(A)e^{\alpha t_0} \leq M_\alpha \leq M_0(A)e^{\alpha t_\alpha}. \tag{1.11}$$

As a final result of this section we relate the transient bound of a matrix A to estimates for the solution of time-varying Lyapunov inequalities.

Proposition 1.2. *Given an Hermitian matrix $P_0 \in \mathcal{H}_n$, $P_0 \succ 0$ such that*

$$\dot{P} - A^*P - PA \succeq 0, \quad P(0) = P_0 \tag{1.12}$$

has a solution satisfying $P(t) \preceq M^2\,\underline{\sigma}(P_0)I_n$, $t \geq 0$, then $M \geq M_0(A)$, the transient excursion of the matrix A. For the special case $P_0 = I_n$, the solution of the equation in (1.12) satisfies $P(t) \preceq M_0(A)^2 I_n$, $t \geq 0$.

Proof: If $P(\cdot)$ satisfies (1.12), then $P(t) \succeq e^{A^*t}P_0e^{At}$. But

$$M^2\,\underline{\sigma}(P_0) \geq \|P(t)\| \geq \|e^{A^*t}P_0e^{At}\| \geq \underline{\sigma}(P_0)\|e^{At}\|^2.$$

For $P_0 = I_n$, the solution of the equation in (1.12) is $P(t) = e^{A^*t}e^{At}$ and hence $P(t) \preceq M_0(A)^2 I_n$, $t \geq 0$. □

If for some $Q \in \mathcal{H}_n$, $Q \succeq 0$,

$$\dot{P} - A^*P - PA - Q = 0, \quad P(0) = P_0$$

then

$$P(t) = e^{A^*t}P_0e^{At} + \int_0^t e^{A^*(t-s)}Qe^{A(t-s)}ds.$$

So the smallest bound for $P(t)$ is obtained for $Q = 0$. This suggests that we should have restricted our considerations to the equality in (1.12). We have chosen not to do this because this precludes the possibility of constant solutions. Indeed since $\sigma(A) \subset \mathbb{C}_-$, there exists a $P_0 \succeq 0$, such that (1.12) has the constant solution $P(t) = P_0$. Then if $P_0 \succ 0$ we may choose $M^2\,\underline{\sigma}(P_0) = \|P_0\| = \overline{\sigma}(P_0)$ and hence we obtain the bound

$$\overline{\sigma}(P_0)^{1/2}\,\underline{\sigma}(P_0)^{-1/2} = (\|P_0\|\|P_0^{-1}\|)^{1/2} \geq M_0(A). \tag{1.13}$$

So this gives rise to the following.

Question : *How should one choose $C \in \mathbb{K}^{p\times n}$ with (A, C) observable, such that for the unique solution P_0 of the Lyapunov equation*

$$A^*P + PA + C^*C = 0$$

the expression $\|P_0\|\|P_0^{-1}\|$ is minimized.

A slight generalization of this result is to seek solutions of the form $P(t) = P_0 e^{-2\alpha t}$, where $\alpha(A) > \alpha > 0$. Then we obtain the same bound for $M_0(A)$ as above but now P_0 is the solution of the Lyapunov equation

$$(A + \alpha I_n)^* P + P(A + \alpha I_n) + C^* C = 0$$

for some $C \in \mathbb{K}^{p \times n}$ with (A, C) observable. In particular if (1.2) holds then we may choose $P_0 = I_n$ and hence obtain $M_0(A) = 1$ as anticipated in Corollary 1.1.

1.3 Pseudospectra and Spectral Value Sets

Consider the system $\dot{x} = Ax$ driven by a signal $u(t) = e^{st}u$, so that

$$\dot{x}(t) = Ax(t) + e^{st}u, \ t > 0, \ s \in \mathbb{C}_-, \ u \in \mathbb{K}^n.$$

For a suitably chosen initial state the response is

$$x(t) = e^{st}(sI_n - A)^{-1}u.$$

So the degree of amplification

$$\sup_{u \neq 0} \frac{\|x(t)\|}{\|u(t)\|} = \sup_{u \neq 0} \frac{\|e^{st}(sI_n - A)^{-1}u\|}{\|e^{st}u\|} = \|(sI_n - A)^{-1}\|.$$

If $s \in \sigma(A)$ is an eigenvalue we set the amplification $\|(sI_n - A)^{-1}\| = \infty$. Trefethen coined the expression *pseudospectrum* for the following set.

Definition 1.5. *The* ε-pseudospectrum *is*

$$\Lambda_\varepsilon(A) := \{s \in \mathbb{C};\ \|(sI_n - A)^{-1}\| \geq \varepsilon^{-1}\}.$$

Unstructured spectral value sets were introduced by Diederich and myself in 1990 [7].

Definition 1.6. *The* unstructured spectral value set *at level* δ *is*

$$\sigma_{\mathbb{K}}(A, \delta) = \bigcup_{\Delta \in \mathbb{K}^{n \times n},\ \|\Delta\| < \delta} \sigma(A + \Delta).$$

A characterization of real spectral value sets is given in terms of the real μ-function [6], but their computation is difficult. In contrast for the complex case we have (see [5]) the following.

Proposition 1.3. *The complex spectral value set at level* δ *is*

$$\sigma_{\mathbb{C}}(A, \delta) = \{s \in \mathbb{C};\ \|(sI_n - A)^{-1}\| > \delta^{-1}\}.$$

The boundary of the set $\sigma_{\mathbb{C}}(A,\delta)$ *is given by*

$$\partial\sigma_{\mathbb{C}}(A,\delta) = \{s \in \mathbb{C};\ \|(sI_n - A)^{-1}\| = \delta^{-1}\} =: C_\delta.$$

And $\sigma_{\mathbb{C}}(A,\delta)$ *is the union of those connected components of* $\mathbb{C}\backslash cl(C_\delta)$ *that contain an eigenvalue of* A.

This proposition connects pseudospectra and spectral value sets.

Trefethen has made use of pseudospectra in order to obtain bounds on the transient behavior. In fact he has used the equality

$$e^{At} = \frac{1}{2\pi\imath}\int_\Gamma e^{st}(sI - A)^{-1}ds,$$

where Γ is a contour enclosing $\sigma(A)$ to obtain the upper bound.

Proposition 1.4.

$$\|e^{At}\| \le \frac{1}{2\pi\varepsilon}\int_{\partial\Lambda_\varepsilon(A)} e^{t\Re s}|ds|.$$

Also, using

$$(sI - A)^{-1} = \int_0^\infty e^{-st}e^{At}dt, \quad \Re s > 0,$$

he obtains the following lower bound for the transient excursion.

Proposition 1.5.

$$M_0(A) \ge \sup_{\varepsilon \ge d_{\mathbb{C}}(A)} \varepsilon^{-1}\alpha_\varepsilon(A), \quad \alpha_\varepsilon(A) = \sup_{s\in\Lambda_\varepsilon(A)} \Re s,$$

where $d_{\mathbb{C}}(A)$ *is the complex distance of* A *from the set of unstable matrices.*

Remark 1.2. *(i) In the discrete time case when* $\sigma(A) \subset \mathbb{D}_1$*, the following estimates are sometimes known as Kreiss's theorem [15].*

$$\sup_{|s|>1}(|s|-1)\|(sI-A)^{-1}\| \le \sup_{t\in\mathbb{N}}\|A^t\| \le en \sup_{|s|>1}(|s|-1)\|(sI-A)^{-1}\|.$$

(ii) Although the problem of transient excursions has been posed on finite-dimensional spaces many of the ideas and results can be extended to infinite-dimensional ones. If

$$\alpha(A) = sup_{s\in\sigma(A)}\Re s,$$

then for infinite-dimensional spaces it is not necessarily the case that for any $\omega > \alpha(A)$*, there exists* M_ω *such that* $\|e^{At}\| \leq M_\omega e^{\omega t}$, $t \geq 0$*. However, it can be shown [12] that if*

$$\alpha_\varepsilon(A) = \sup_{s \in \Lambda_\varepsilon(A)} \Re s$$

for any $\omega > \lim_{\varepsilon \to 0} \alpha_\varepsilon(A)$*, there exists* M_ω *such that* $\|e^{At}\| \leq M_\omega e^{\omega t}$, $t \geq 0$*.*

In the previous section, in Corollary 1.3, the distance of a matrix from normality was used to get a sufficient condition for a matrix to be stable with normal transitory growth. A similar condition in terms of spectral value sets can also be found.

Proposition 1.6. *If* $\delta(A)$ *is the distance of* A *from normality, then a sufficient condition for* A *to be stable with normal transitory behavior is*

$$\overline{\sigma_{\mathbb{C}}(A, \delta(A))} \subset \mathbb{C}_{-\delta(A)}, \tag{1.14}$$

where $\mathbb{C}_\gamma$ *denotes, for every* $\gamma \in \mathbb{R}$*, the open left half plane* $\{s \in \mathbb{C}; \Re s < \gamma\}$*.*

Proof: There exists Δ, with $\|\Delta\| \leq \delta(A)$ such that $A + \Delta = N$ is normal. But then by (1.14), $\alpha(N) > \delta(A)$ and so the the result follows from Corollary 1.2. □

If $\sigma(A) \subset \mathbb{C}_{-\delta(A)}$ (otherwise (1.14) is clearly not valid) condition (1.14) is equivalent to

$$\sigma_{\mathbb{C}}(A, \delta(A)) \cap -\delta(A) + \imath\mathbb{R} = \emptyset. \tag{1.15}$$

It follows from Proposition 1.3 that (1.14) holds if and only if

$$\max_{\omega \in \mathbb{R}} \|((-\delta(A) + \imath\omega)I_n - A)^{-1}\| < \delta(A)^{-1}. \tag{1.16}$$

Now it is clear that

$$\overline{(\sigma_{\mathbb{C}}(A, \delta))} \subset \mathbb{C}_{-\delta}$$

for small $\delta > 0$. If this is not the case for some δ' (e.g., $\delta' = \alpha(A)$) then it will not be satisfied for any $\delta \geq \delta'$. Hence there exists exactly one $\delta_0 > 0$ such that

$$\overline{(\sigma_{\mathbb{C}}(A, \delta))} \subset \mathbb{C}_{-\delta},\ \delta < \delta_0 \quad \text{and} \quad \overline{(\sigma_{\mathbb{C}}(A, \delta_0))} \not\subset \mathbb{C}_{-\delta_0}, \delta \geq \delta_0. \tag{1.17}$$

This δ_0 is uniquely determined by the equation

$$\begin{aligned} [d_{\mathbb{C}}(A + \delta_0 I)]^{-1} &= \max_{\omega \in \mathbb{R}} \|((-\delta_0 + \imath\omega)I_n - A)^{-1}\| \\ &= \max_{s \in -\delta_0 + \imath\mathbb{R}} \|(sI_n - A)^{-1}\| = \delta_0. \end{aligned} \tag{1.18}$$

So the condition for A to be stable with normal transitory behavior given in Proposition 1.6 is $\delta(A) < \delta_0$. This seemingly nice result suffers because there is not a good computational scheme for determining $\delta(A)$. Since $N \in \mathcal{N}$ implies $NN^* = N^*N$, $\delta(A)$ is the norm of the smallest Δ such that

$$AA^* - A^*A + \Delta A^* + A\Delta^* - \Delta^* A - A^*\Delta + \Delta\Delta^* - \Delta^*\Delta = 0.$$

But I am unaware of a study of such quadratic equations.

Remark 1.3. *For large δ one observes that the spectral value sets $\sigma_{\mathbb{C}}(A,\delta)$ of any matrix A become more and more like the spectral value sets of normal operators. This can be explained as follows. For every $\Delta \in \mathbb{C}^{n\times n}$ and $N \in \mathcal{N}$,*

$$A + \Delta = N + \Delta + (A - N).$$

Hence

$$\sigma_{\mathbb{C}}(N, \delta - \delta(A)) \subset \sigma_{\mathbb{C}}(A,\delta) \subset \sigma_{\mathbb{C}}(N, \delta + \delta(A)). \tag{1.19}$$

For large δ, $\sigma_{\mathbb{C}}(N, \delta - \delta(A))$ is a union of disks of radius $\delta - \delta(A)$ around the eigenvalues of N. Similarly, $\sigma_{\mathbb{C}}(N, \delta + \delta(A))$ is a union of disks of radius $\delta + \delta(A)$ around the eigenvalues of N. The relative *difference of the radii tends to zero:*

$$\frac{\delta + \delta(A) - [\delta - \delta(A)]}{\delta} \to 0 \quad as \quad \delta \to \infty\,.$$

Question : *Can the distance from normality be recovered from the asymptotic behavior of the spectral value sets as $\delta \to \infty$? More precisely, for every $\delta > 0$ let $R(\delta)$ be the smallest number $R > 0$ such that $\sigma_{\mathbb{C}}(A,\delta)$ is contained in the union of the disks with radius R around the eigenvalues of A. Do we have the following?*

$$R(\delta) - \delta \to \delta(A) \quad as \quad \delta \to \infty.$$

1.4 State Feedback

In this section the possibility of improving the transient bound by stabilizing state feedback is considered. The output feedback problem is not discussed, but it is expected that results can be obtained in a fashion similar to that of H_∞– theory, where the state feedback problem is often regarded as a first step. Suppose $B \in \mathbb{K}^{n\times m}$, $F : [0,\infty) \to \mathbb{K}^{m\times n}$ is piecewise continuous and the controlled dynamics are

$$\dot{x} = (A + BF(t))x, \qquad x(0) = x_0 \in \mathbb{K}^n.$$

If $\varphi_F(\cdot,\cdot)$ is the evolution operator generated by $A + BF(t)$ its transient bound is defined to be

$$M_0(A + BF(\cdot)) = \max_{t \geq 0} \|\varphi_F(t,0)\|.$$

We have the following lemma.

Lemma 1.4. *Suppose $R \in \mathcal{H}_n$, $R \succ 0$ and consider the differential Lyapunov equation*

$$\dot{Z} - AZ - ZA^* + BRB^* = 0,\ t \geq 0. \tag{1.20}$$

The unique solution of (1.20) on R_+ with initial value $Z(0) = I_n$ is given by

$$Z(t) = e^{At}\left[I_n - \int_0^t e^{-As}BRB^*e^{-A^*s}ds\right]e^{A^*t}. \tag{1.21}$$

Let $I^+ = [\,0, t_+) = \{t \geq 0;\ Z(t) \succ 0\}$. Then $0 < t_+ \leq \infty$ and the initial value problem

$$\dot{X} + A^*X + XA - XBRB^*X = 0,\ X(0) = I_n \tag{1.22}$$

has the solution $P(t) = Z(t)^{-1}$ on I^+. Moreover $\lim_{t\to t_+} \|P(t)\| = \infty$.
If $F(t) = -RB^*P(t)$, $t \in I^+$, then

$$\varphi_F(t,0)\varphi_F(t,0)^* \preceq Z(t),\ \ t \in I^+. \tag{1.23}$$

Proof: It is easy to see that

$$Z(t) = \varphi_F(t,0)\varphi_F(t,0)^* + \int_0^t \varphi_F(t,s)BRB^*\varphi_F(t,s)^*ds,\ t \in I^+.$$

□

We see from (1.21) and (1.23) that $\|\varphi_F(t,0)\| \leq \|e^{At}\|$, $t \in I^+$ and if the pair (A, B) is controllable, the inequality is strict for $t \in (0, t_+)$. This suggests a possible way of reducing the transient bound $M_0(A)$ to a more acceptable level $\mathbf{M}$.

Proposition 1.7. *Suppose that $\mathbf{M} \geq 1$ is an acceptable transient bound and there exists a $T \in (0, t_+)$ such that*

$$\|Z(T)\| \leq \mathbf{M}^2/M_0(A)^2,\ \ \|Z(t)\| \leq \mathbf{M}^2,\ t \in [0,T] \tag{1.24}$$

where $Z(\cdot)$ is as in Lemma 1.4. Then for

$$F(t) = \begin{cases} -RB^*P(t) & t \in [0,T] \\ 0 & t > T \end{cases}$$

where $P(\cdot)$ is as in Lemma 1.4 one has $M_0(A+F(\cdot)) \le \mathbf{M}$.

Proof: By Lemma 1.4

$$\|\varphi_F(t,0)\|^2 \le \|Z(t)\| \le \mathbf{M}^2, \quad t \in [0,T].$$

But $\varphi_F(t,0) = e^{A(t-T)}\varphi_F(T,0),\ t > T$. Hence by (1.24)

$$\|\varphi_F(t,0)\| \le M_0(A)\mathbf{M}/M_0(A) = \mathbf{M}, \qquad t > T.$$

□

In order to check whether or not the conditions in the Proposition 1.7 hold, one needs to compute the solution of the differential Lyapunov equation (1.20) with initial value $Z(0) = I_n$ and monitor whether or not it is possible to find T such that $\|Z(T)\| \le \mathbf{M}^2/M_0(A)^2$ and on the interval $[0,T]$, $Z(t)$ does not transcend the value $\mathbf{M}^2$ and $\underline{\sigma}(Z(t)) > 0$. In order to get more insight into this let $\mathcal{C}_R(t)$ be the controllability Gramian of $(-A, BR^{1/2})$ on the interval $[0,t]$. Then from (1.21) we see that $Z(t) \succ 0$ provided

$$\mathcal{C}_R(t) = \int_0^t e^{-As}BRB^*e^{-A^*s}ds \prec I_n. \tag{1.25}$$

In particular we have

$$t_+ = \sup\{t \in \mathbb{R}_+; \det(I_n - \mathcal{C}_R(s)) \ne 0,\ s \in [0,t]\}.$$

Now assume that (A,B) is controllable, then

$$\mathcal{C}_R(T) \succeq \underline{\sigma}(R)\int_0^T e^{-As}BB^*e^{-A^*s}ds \succ 0.$$

For any $R \succ 0$ such that (1.25) holds with $t = T$, let

$$\beta\,(R) = \underline{\sigma}(\mathcal{C}_R(T))/\overline{\sigma}(\mathcal{C}_R(T))\,.$$

Then $(1-\beta\,(R))^{1/2}$ is a measure of the eccentricity of the ellipsoid described by $\{x \in \mathbb{K}^n;\ \langle x, \mathcal{C}_R(T)x\rangle = 1\}$. Now $\beta\,(R) = \beta\,(\alpha R), \alpha > 0$ and provided $\alpha < \overline{\sigma}(\mathcal{C}_R(T))^{-1}$; then (1.25) will continue to hold at $t = T$ when R is replaced by αR. In which case from (1.21) one has

$$\|Z(T)\| \le (1 - \alpha\underline{\sigma}(\mathcal{C}_R(T)))\,\|e^{AT}\|^2.$$

The choice $\alpha = \overline{\sigma}(\mathcal{C}_R(T))^{-1}$ yields

$$\|Z(T)\| \le (1 - \beta\,(R))\,\|e^{AT}\|^2\,.$$

So, as in (1.13), in order to obtain a good bound, one would try to choose R to make the eccentricity small, that is, $\beta(R)$ close to one.

We will now give an interpretation of the solution $P(t)$ of the initial value problem (1.22) in terms of the following finite time optimal control problem:

Minimize
$$J(x^0,u) = \langle x(T), P(T)x(T)\rangle + \int_0^T \langle u(t), R^{-1}u(t)\rangle dt$$
subject to
$$\dot{x} = Ax + Bu, \quad x(0) = x^0.$$

Let $F(t) = -RB^*P(t), t \in [0,T]$. Then we have for every control $u(\cdot) \in L^2(0,T;\mathbb{K}^m)$,

$$\begin{aligned}
&\frac{d}{dt}\langle x(t), P(t)x(t)\rangle \\
= &\langle x(t), (\frac{d}{dt}P(t) + A^*P(t) + P(t)A)x(t)\rangle + 2\Re\langle x(t), P(t)Bu(t)\rangle \\
= &\langle u(t) - F(t)x(t), R^{-1}(u(t) - F(t)x(t))\rangle - \langle u(t), R^{-1}u(t)\rangle, \quad t \in [0,T].
\end{aligned}$$

Hence integrating from 0 to T yields

$$J(x^0,u) = \langle x^0, P(0)x^0\rangle + \int_0^T \langle u(t) + F(t)x(t), R^{-1}(u(t) + F(t)x(t))\rangle dt.$$

So the control $u(t) = -RB^*P(t)x(t)$ minimizes the cost functional $J(x^0,u)$ and the optimal cost is $\|x^0\|^2$. Note that this is a characterization *a posteriori* since the cost functional $J(x^0,u)$ depends on the final value $P(T)$.

1.5 Transient Excursions of Uncertain Systems

Suppose

$$(A,D,E) \in \mathbb{K}^{n\times n} \times \mathbb{K}^{n\times \ell} \times \mathbb{K}^{q\times n}, \quad \text{with} \quad \sigma(A) \subset \mathbb{C}_-$$

are given and $\Delta \in \mathbb{K}^{\ell\times q}$ is unknown. Consider the uncertain system

$$\dot{x} = (A + D\Delta E)x := A_\Delta x,$$

where $\|\Delta\| < r_{\mathbb{K}}(A;D,E;\mathbb{C}_+)$, the stability radius of A with respect to the perturbation structure (D,E).

Definition 1.7. *Given an acceptable bound* $\mathbf{M} \geq M_0(A)$, *the* radius of acceptable transient behavior *of A at level $\mathbf{M}$ with respect to the perturbation structure D,E is*

$$r_{\mathbb{K}}(A;D,E;\mathbf{M}) = \inf\{\|\Delta\|\,;\ \exists\Delta\,\exists t \geq 0,\ \|e^{(A+D\Delta E)\,t}\| > \mathbf{M}\}.$$

Theorem 1.2. *Suppose there exists $P_0 \in \mathcal{H}_n$, $Q \in \mathcal{H}_q$, $R \in \mathcal{H}_\ell$, $P_0 \succeq 0$, $Q \succeq 0$, $R \succeq 0$ such that*

$$\dot{P} - A^*P - PA - E^*QE - PDRD^*P = 0, \quad P(0) = P_0, \tag{1.26}$$

has a solution satisfying $P(t) \preceq \mathbf{M}^2\underline{\sigma}(P_0)I_n$, $t \geq 0$; then $r_{\mathbb{K}}(A; D, E; \mathbf{M}) \geq (\underline{\sigma}(Q)\underline{\sigma}(R))^{1/2}$.

Proof: If P satisfies (1.26), then

$$\begin{aligned}\dot{P} - (A + D\Delta E)^*P - P(A + D\Delta E) &= E^*QE + PDRD^*P - (D\Delta E)^*P - PD\Delta E \\ &= (D^*P - R^{-1}\Delta E)^*R(D^*P - R^{-1}\Delta E) + E^*(Q - \Delta^*R^{-1}\Delta)E.\end{aligned}$$

Assume that $\underline{\sigma}(Q)\underline{\sigma}(R) \geq \|\Delta\|^2$; then

$$V(t) := (D^*P - R^{-1}\Delta E)^*R(D^*P - R^{-1}\Delta E) + E^*(Q - \Delta^*R^{-1}\Delta)E \succeq 0, \; t \geq 0$$

and

$$P(t) = e^{A_\Delta^* t}P_0 e^{A_\Delta t} + \int_0^t e^{A_\Delta^*(t-s)}V(s)e^{A_\Delta(t-s)}ds.$$

So

$$\mathbf{M}^2\underline{\sigma}(P_0)I_n \succeq P(t), \, t \geq 0 \Rightarrow \mathbf{M}^2 I_n \succeq e^{A_\Delta^* t}e^{A_\Delta t}.$$

□

Remark 1.4.

(i) *In the case $Q = 0, R = 0, P_0 = I$, $P(t) = e^{A^*t}e^{At}$.*

(ii) *The choice $P_0 = I$, $Q = \alpha^2 I_q$, $R = \beta^2 I_\ell$ yields $r_{\mathbb{K}}(A; D, E; \mathbf{M}) \geq \alpha\beta$ and may be optimal.*

(iii) *Let $R = I_\ell$, $Q = r_{\mathbb{K}}(A; D, E; \mathbb{C}_+)^2 I_q$; then there exists a steady-state solution. So for $\mathbf{M}$ sufficiently large $r_{\mathbb{K}}(A; D, E; \mathbf{M}) = r_{\mathbb{K}}(A; D, E; \mathbb{C}_+)$.*

Example 1.1. Suppose A is normal, $A = T^*\text{diag}\,(\lambda_1, \lambda_2, \ldots, \lambda_n)T$, $P_0 = I$, $Q = \alpha^2 I_q$, $R = \beta^2 I_\ell$ and $D = E = I$, then if $\hat{P} = TPT^*$, (1.26) becomes

$$\dot{\hat{P}} - \text{diag}\,(\lambda_1, \lambda_2, \ldots, \lambda_n)^*\hat{P} - \hat{P}\text{diag}\,(\lambda_1, \lambda_2, \ldots, \lambda_n) - \alpha^2 I - \beta^2\hat{P}^2 = 0, \; \hat{P}(0) = I.$$

Suppose $\hat{P}(t) = \text{diag}\,(p_1(t), p_2(t), \ldots, p_2(t))$; then

$$\dot{p_i} - (\lambda_i + \lambda_i^*)p_i - \alpha^2 - \beta^2 p_i^2 = 0, \quad p_i(0) = 1.$$

Let $-(\lambda_i + \lambda_i^*) = 2\gamma_i$, $\gamma_1 \leq \gamma_2 \leq \ldots \leq \gamma_n$ and choose $\alpha^2 = \beta^2 = \gamma_1$; then $p_1(t) \equiv 1$. And

$$\dot{p_i} = 0, \text{ when } p_i = \gamma_1^{-1}[\gamma_i \pm \sqrt{\gamma_i^2 - \gamma_1^2}], \text{ and } \dot{p_i} = 2(\gamma_1 - \gamma_i) \leq 0, \text{ when } p_i = 1.$$

Hence $p_i(t) \leq 1, t \geq 0$ and so $1 \geq \|\hat{P}(t)\| = \|P(t)\|$, $t \geq 0$. Thus $r_{\mathbb{K}}(A; I, I; 1) \geq \gamma_1 = -1/2(\lambda_{\max}(A + A^*))$, a result already anticipated in Proposition 1.1.

Finally Theorem 1.2 is used to prove a nonlinear result. Consider the nonlinear equation

$$\dot{x} = Ax + DN(Ex), \quad x(0) = x^0, \tag{1.27}$$

where N is locally Lipschitz $N : \mathbb{K}^q \mapsto \mathbb{K}^\ell$ satisfies $N(0) = 0$ and

$$\|N(z)\| < \rho\|z\| \,, \quad z \in \mathbb{K}^q \,, \quad z \neq 0 \,. \tag{1.28}$$

Then there exists a unique solution $x(\cdot)$ of (1.27) and if $P(\cdot)$ satisfies the Riccati equation (1.26), (in the following calculation we abbreviate $y(t - s) := Ex(t - s)$),

$$\begin{aligned}
&\frac{d}{ds}\langle x(t-s), P(s)x(t-s)\rangle \\
&= \langle x(t-s), \dot{P}(s)x(t-s)\rangle - 2\Re\langle Ax(t-s) + DN(y(t-s)), P(s)x(t-s)\rangle \\
&= \langle y(t-s), Qy(t-s)\rangle + \langle D^*P(s)x(t-s), RD^*P(s)x(t-s)\rangle \\
&- 2\Re\langle D^*P(s)x(t-s), N(y(t-s))\rangle \\
&= \langle [D^*P(s)x(t-s) – R^{-1}N(y(t-s))], R[D^*P(s)x(t-s) – R^{-1}N(y(t-s))]\rangle \\
&+ \langle y(t-s), Qy(t-s)\rangle - \langle N(y(t-s)), R^{-1}N(y(t-s))\rangle.
\end{aligned}$$

Now suppose that $\underline{\sigma}(Q)\underline{\sigma}(R) \geq \|N\|^2$; then the right-hand side of the above expression is positive and integrating from 0 to t, yields

$$\langle x^0, P(t)x^0\rangle \geq \langle x(t), P_0 x(t)\rangle.$$

Thus under the conditions of Theorem 1.2, $\|x(t)\| \leq \mathbf{M}\|x^0\|$, $t \geq 0$.

References

[1] K. M. Butler and B. F. Farrell, Three dimensional optimal perturbations in viscous shear flow, *Phys. Fluids A* 4:1637–1650, 1992.

[2] E. Gallestey, D. Hinrichsen, and A. J. Pritchard, Spectral value sets of closed linear operators, *Proc. Royal Soc. London A* 456:1397–1418, 2000.

[3] S. K. Godunov, Spectral portraits of matrices and criteria of spectrum dichotomy, *Proc. Int. Conference on Computer Arithmetic, Scientific Computation and Mathematical Modelling. SCAN-91*, Oldenburg, 1991.

[4] L. H. Gustavsson, Energy growth of three dimensional disturbances in plane Poiseuille flow, *J. Fluid Mech.* 224:241–260, 1991.

[5] D. Hinrichsen and B. Kelb, Spectral value sets: a graphical tool for robustness analysis, *Syst. Control Letters* 21:127–136, 1993.

[6] D. Hinrichsen and B. Kelb, Stability radii and spectral value sets for real matrix perturbations, *Proc. Conference MTNS*, Regensburg, U. Helmke and R. Mennicken, eds., 1993.

[7] D. Hinrichsen and A. J. Pritchard, On the robustness of stable discrete time linear systems, *New Trends in Systems Theory*, Proc. Conference. Genova, 393–400, 1990.

[8] T. Kato, *Perturbation Theory for Linear Operators*, Springer-Verlag, Berlin, 1976.

[9] S. C. Reddy and D. S. Henningson, Energy growth in viscous channel flows, *J. Fluid Mech.* 252:209–238, 1993.

[10] S. C. Reddy, P. Schmidt, and D. S. Henningson, Pseudospectra of the Orr-Sommerfeld operator, *SIAM J. Appl. Math.* 53:15–47, 1993.

[11] L. N. Trefethen, Approximation theory and linear algebra, *Algorithms for Approximation II*, Chapman and Hall, London, 1990.

[12] L. N. Trefethen, Pseudospectra of linear operators, *SIAM Rev.* 39:383–406, 1997.

[13] L. N. Trefethen, A. E. Trefethen, S. C. Reddy, and T. A. Driscoll, Hydrodynamic stability without eigenvalues, *Science* 261:578-584, 1993.

[14] W. Thomson, Stability of fluid motion-rectilinear motion of viscous fluid between two parallel plates, *Philos. Mag.* 24:188, 1887.

[15] E. Wegert and L. N. Trefethen, From Buffon needle problem to the Kreiss matrix theorem, *Amer. Math. Monthly* 101:132–139, 1994.

[16] K. Yosida, *Functional Analysis*, Springer-Verlag, Berlin, 1974.

Chapter 2

Robust Stability of Multivariate Polynomials

Vladimir L. Kharitonov

ABSTRACT This chapter deals with stability analysis of multivariate polynomials. The maximal class of polynomials is introduced such that polynomials of this class preserve stability under small variations of the coefficients. Basic properties of such stable polynomials are discussed. Some robust stability conditions are given.

2.1 Introduction

The robustness of the stability property of multivariate polynomials has been under investigation for more than 10 years, see [1] and [3].

In [3] an extension of the Hurwitz stability result for univariate interval polynomials was proved for the case of real bivariate scattering Hurwitz stable polynomials. Shortly after this publication a natural extension of the result was obtained for the case of m-variate polynomials that are scattering Hurwitz stable [1]. In [2] some basic results on robust stability analysis of univariate polynomials were extended to the case of strict sense stable m-variate polynomials. In particular for the case of m-variate strict sense stable polynomials such fundamental principle tools as the edge theorem and the zero exclusion principle were obtained.

In [9] some explicit expressions for the stability radius of strict sense stable bivariate polynomials were presented.

A good example of a systematic application of the frequency domain technique for robustness analysis of strict sense stable bivariate polynomials can be found in [10].

It should be mentioned that for the case of polynomials with several variables there is a large variety of classes of stable polynomials that may be treated in one sense or another as extensions of the class of Hurwitz stable univariate polynomials. From this variety only two classes are involved in the robustness study; the class of scattering Hurwitz stable polynomials (SHS) and the class of strict sense stable polynomials (SSS). The latter class is a proper subclass of the former one. The (SHS) class is the biggest one which coincides with the class of Hurwitz polynomials in the case of one variable. So, it may seem quite natural to try to extend basic results from robustness analysis from Hurwitz polynomials to this class of multivariate polynomials. On the other hand there is a principal difference between (SHS) polynomials and Hurwitz ones. For the latter, small variations of the coefficient do not destroy stability, but for the former, arbitrary small coefficient variations may do so. The same phenomenon takes place for strict sense stable multivariate polynomials, too; see [7]. Of course such extraordinary fragility of the stability properties is unacceptable for practical problems especially when one deals with robustness. This observation was one of the principal motivations for the introduction of a new class of stable multivariate polynomials that do not suffer from this deficiency [7].

In [8] extensions of some known results on robust stability obtained previously for scattering Hurwitz stable polynomials in [3] and for strict sense stable polynomials in [2] have been proved for the new class.

One of the principal tools for robustness analysis of univariate polynomial families is the zero exclusion principle which is based on analysis of the value set of a polynomial family under investigation. This principle assures that under some modest assumptions the family is robustly stable if the value set of the family does not contain the origin of the complex plane for all frequencies. It happens to be that this principle remains true for the case of scattering Hurwitz stable polynomials and for the case of strict sense stable polynomials.

Unfortunately, simple examples illustrate that for the case of stable polynomials continuous variations of the coefficients may result in the loss of stability without any appearance of zeros on the essential boundary. This means that for this class of polynomials one can not directly apply the principle. In other words, it is necessary to look for a new version of the principle that is valid for the new class of stable polynomials in order to have in hands this powerful tool of robustness analysis. The main aim of this chapter is to develop some necessary modifications of the frequency domain technique for stability study of multivariate polynomials.

In the next section some basic notions and notations are introduced and several standard definitions are recalled. In Section 2.2.1 some useful properties of stable polynomials are discussed, and in Section 2.4 a modernization of the zero exclusion principle is proposed.

2.2 Basic Notions and Definitions

In this section we first recall some basic notions from the theory of multivariate polynomials and introduce some basic classes of stable polynomials.

A multivariate polynomial is a finite sum of the form

$$p(s_1, \dots, s_m) = \sum_{k_1=0}^{n_1}, \dots, \sum_{k_m=0}^{n_m} a_{k_1,\dots,k_m} s_1^{k_1}, \dots, s_m^{k_m}. \tag{2.1}$$

Here $(s_1, \ldots, s_m)$ is the vector of independent variables and $a_{k_1,\ldots,k_m}$ are real coefficients. In the following the vector $(s_1, \ldots, s_m)$ is frequently denoted by $\mathbf{s}$, and

$$\mathbf{a} = (a_{00,\ldots,0}, a_{10,\ldots,0}, \ldots, a_{n_1 n_2,\ldots,n_m})$$

is the coefficient vector of $p(\mathbf{s})$.

The partial degree, n_k, of $p(s_1, s_2, \ldots, s_m)$ with respect to the variable s_k is the maximal degree of the variable that appears in terms of (2.1) with nonzero coefficients. By $\deg(p)$ we denote the vector $\mathbf{n} = (n_1, n_2, \ldots, n_m)$ of the partial degrees.

A nonconstant polynomial $q(s_1, s_2, \ldots, s_m)$ is a divisor of (2.1), if there exists a polynomial $r(s_1, s_2, \ldots, s_m)$ such that

$$p(s_1, s_2, \ldots, s_m) = q(s_1, s_2, \ldots, s_m) r(s_1, s_2, \ldots, s_m).$$

A polynomial is said to be prime if it has no divisors different from itself multiplied by a constant.

A vector $\mathbf{s}^{(0)} = (s_1^{(0)}, s_2^{(0)}, \ldots, s_m^{(0)})$ with complex components is called a root of (2.1) if

$$p(s_1^{(0)}, s_2^{(0)}, \ldots, s_m^{(0)}) = 0.$$

Two polynomials are called coprime if they have no common divisors. For one-variable polynomials this means that they have no common roots, but this is not the case for polynomials of several variables: Two multivariate polynomials may be relatively prime but at the same time have common roots.

The polynomial (2.1) may be rewritten in the following form.

$$p(s_1, s_2, \ldots, s_m) = \sum_{j=0}^{n_k} a_j^{(k)} (s_1, \ldots, s_{k-1}, s_{k+1}, \ldots, s_m) \, s_k^j, \tag{2.2}$$

where the coefficients $a_j^{(k)} (s_1, \ldots, s_{k-1}, s_{k+1}, \ldots, s_m)$, $j = 0, 1, \ldots, n_k$ are polynomials of $(m-1)$ variables. In this decomposition, the main coefficient $a_{n_k}^{(k)} (s_1, \ldots, s_{k-1}, s_{k+1}, \ldots, s_m)$ is an $(m-1)$-variate polynomial with coefficient vector

$$(a_{00,\ldots,n_k,\ldots,0}, a_{10,\ldots,n_k,\ldots,0}, \ldots, a_{n_1,\ldots,n_k,\ldots,n_m}).$$

It is clear that there are m different power decompositions of this form, each one corresponding to one of the variables $s_1, s_2, \ldots, s_m$.

Given a set of nonnegative integer numbers $(n_1, n_2, \ldots, n_m)$ let us define the set

$$\mathbf{P}_{n_1,\ldots,n_m} = \{ p(s_1, \ldots, s_m) | \deg(p) = (n_1, \ldots, n_m) \}, \tag{2.3}$$

of constant degree m-variate real polynomials.

2.2.1 Classes of Stable Polynomials

The polynomial

$$p(-\mathbf{s}) = p(-s_1, -s_2, \dots, -s_m)$$

is called conjugate to $p(\mathbf{s})$. If $p(-\mathbf{s}) = cp(\mathbf{s})$ for $c = 1$ or $c = -1$, then $p(\mathbf{s})$ is called paraconjugate.

Lemma 2.1. *The greatest common divisor of $p(\mathbf{s})$ and $p(-\mathbf{s})$ is a paraconjugate polynomial.*

Define two polydomains

$$\Gamma_m^{(+)} = \{\, (s_1, s_2, \dots, s_m) \mid Re\{s_j\} > 0, \qquad j = 1, 2, \dots, m \,\}, \tag{2.4}$$

$$\Gamma_m^{(0)} = \{\, (s_1, s_2, \dots, s_m) \mid Re\{s_j\} \geq 0, \qquad j = 1, 2, \dots, m \,\}, \tag{2.5}$$

and its essential boundary

$$\Omega^{(m)} = \{(i\omega_1, i\omega_2, \dots, i\omega_m) \,|\omega_k \in \mathcal{R},\ k = 1, 2, \dots, m\}\,. \tag{2.6}$$

Definition 2.1. *A polynomial $p(\mathbf{s}) \in \mathbf{P}_{n_1,n_2,\dots,n_m}$ is called* wide sense stable (WSS) *if*

$$p(\mathbf{s}) \neq 0, \qquad \forall \mathbf{s} \in \Gamma_m^{(+)}.$$

Definition 2.2. *A (WSS) $p(\mathbf{s}) \in \mathbf{P}_{n_1,n_2,\dots,n_m}$ is called* scattering Hurwitz stable (SHS) *if it is relatively prime with its conjugate polynomial.*

Definition 2.3. *A polynomial $p(\mathbf{s}) \in \mathbf{P}_{n_1,n_2,\dots,n_m}$ is called* strict sense stable (SSS) *if*

$$p(\mathbf{s}) \neq 0, \qquad \forall \mathbf{s} \in \Gamma_m^{(0)}.$$

The following corollary is immediate from the definitions.

Corollary 2.1. *(SSS) $\subset$ (SHS) $\subset$ (WSS).*

Now we define by induction a new class of m-variate stable polynomials; see [7].

Definition 2.4.

(1) For $m = 1$: given $n_1 \geq 0$, a polynomial $p(s_1)$ of degree n_1 is called stable *if it is Hurwitz stable;*

(2) for $m > 1$: given a vector $\mathbf{n} = (n_1, n_2, \ldots, n_m)$ with nonnegative entire components, a polynomial $p(\mathbf{s}) \in \mathbf{P}_{n_1,\ldots,n_m}$ is called stable *if it is strict sense stable and satisfies the conditions:*

- *the main coefficients,*

$$a_{n_k}^{(k)}(s_1, s_2, \ldots, s_{k-1}, s_{k+1}, \ldots, s_m), \qquad k = 1, 2, \ldots, m,$$

  are $(m-1)$-variate stable polynomials; and

- $$\deg\left(a_{n_k}^{(k)}\right) = (n_1, \ldots, n_{k-1}, n_{k+1}, \ldots, n_m), \qquad k = 1, 2, \ldots, m.$$

As shown in [7], this is the biggest class of polynomials preserving stability under small coefficient perturbations. So it is reasonable to study robustness of the stability property for this class of polynomials. To this end we first recall some basic properties of stable polynomials; see [7].

2.3 Properties of Stable Polynomials

Lemma 2.2. *Let $p(s_1, s_2, \ldots, s_m) \in \mathbf{P}_{n_1,n_2,\ldots,n_m}$ be stable. If all the variables but one are fixed such that they have zero real parts, then the resulting univariate polynomial is Hurwitz stable and has degree equal to the corresponding partial degree of $p(\mathbf{s})$.*

Lemma 2.3. *Let $p(s_1, s_2, \ldots, s_m) \in \mathbf{P}_{n_1,n_2,\ldots,n_m}$ be stable. If one of the variables is fixed such that it has nonnegative real part, then the resulting $(m-1)$-variate polynomial is stable, and has partial degrees equal to the corresponding partial degrees of $p(\mathbf{s})$.*

Lemma 2.4. *Let $p(\mathbf{s}) \in \mathbf{P}_{n_1,n_2,\ldots,n_m}$ be stable, and $n_m > 0$. The polynomial*

$$\widehat{p}(\mathbf{s}) = (s_m)^{n_m} p(s_1, \ldots, s_{m-1}, \frac{1}{s_m})$$

is stable and it holds that $\deg(\widehat{p}) = \deg(p)$.

For (SSS) polynomials the statement of the previous lemma is not true, as may be seen by the following example.

Example 2.1. The polynomial $p(s_1, s_2) = s_1 s_2 + s_1 + 1 \in \mathbf{P}_{1,1}$ is (SSS) but $\widehat{p}(s_1, s_2) = s_1 + s_1 s_2 + s_2$ has a root at $(s_1, s_2) = (0, 0)$ and therefore is not (SSS).

Corollary 2.2. *By applying Lemma 2.4 successively to several variables one may obtain a set of stable polynomials of the same degree.*

Theorem 2.1. *Let $p(s_1, s_2, \dots, s_m) \in \mathbf{P}_{n_1,n_2,\dots,n_m}$ be a stable polynomial. There exists $\varepsilon > 0$ such that every polynomial with a coefficient vector from the ε-neighborhood of the coefficient vector of $p(s_1, s_2, \dots, s_m)$ is stable.*

Theorem 2.2. *Let $p(s_1, s_2, \dots, s_m) \in \mathbf{P}_{n_1,n_2,\dots,n_m}$ be a stable polynomial. If $n_m > 0$, then*

$$\frac{\partial p(s_1, s_2, \dots, s_m)}{\partial s_m}$$

is a stable polynomial from $\mathbf{P}_{n_1,n_2,\dots,n_m-1}$.

Corollary 2.3. *Let $p(s_1, s_2, \dots, s_m) \in \mathbf{P}_{n_1,n_2,\dots,n_m}$ be a stable polynomial. Then all coefficients in the decomposition (2.2) are stable $(m-1)$-variate polynomials from $\mathbf{P}_{n_1,n_2,\dots,n_{m-1}}$.*

For (SSS) polynomials the statement is not true.

Example 2.2. The polynomial $p(s_1, s_2) = s_1 s_2 + s_1 + 1$ belongs to $\mathbf{P}_{1,1}$, and is (SSS). However, the derivative

$$\frac{\partial p(s_1, s_2)}{\partial s_2} = s_1$$

is not a (SSS) polynomial of $\mathbf{P}_{1,0}$.

Theorem 2.3. *All coefficients of a polynomial $p(s_1, s_2, \dots, s_m) \in \mathbf{P}_{n_1,n_2,\dots,n_m}$ that is stable have the same sign.*

Theorem 2.4. *For every stable $p(\mathbf{s}) \in \mathbf{P}_{n_1,n_2,\dots,n_m}$ there exists $\varepsilon > 0$, such that $p(\mathbf{s})$ has no roots in the ε-neighborhood of the essential boundary Ω.*

2.4 Zero Exclusion Principle

2.4.1 Families of Polynomials

From now on $p(\mathbf{s}, \mathbf{a})$ denotes a polynomial with the coefficient vector $\mathbf{a}$.

Definition 2.5. A family *of m-variate polynomials is defined as*

$$\mathcal{F} = \{p(\mathbf{s}, \mathbf{a}) : \mathbf{a} \in Q\}, \tag{2.7}$$

where Q is a given set in the coefficient space.

Definition 2.6. *A family $\mathcal{F}$ is called* polytopic *if Q is a convex polytope in the coefficient space.*

Assumption 2.1. *In the following we always assume that all polynomials from $\mathcal{F}$ have the same degree; that is, $\mathcal{F} \subset \mathbf{P}_{n_1,\ldots,n_m}$.*

It has been already mentioned in the previous section that all coefficients of a stable polynomial have the same sign. So, without any loss of generality one can make the following assumption.

Assumption 2.2. *All polynomials in $\mathcal{F}$ have positive coefficients.*

Definition 2.7. *A family $\mathcal{F} \subset \mathbf{P}_{n_1,\ldots,n_m}$ of multivariate polynomials is said to be* robustly stable *if every member of the family is stable.*

With every family of multivariate polynomials $\mathcal{F}$ one may associate m polytopic subfamilies $\mathcal{F}_1, \mathcal{F}_2, \ldots, \mathcal{F}_m$ of $(m-1)$-variate polynomials formed by the main coefficients in the decomposition (2.2), corresponding to elements from $\mathcal{F}$.

2.4.2 Zero Exclusion Principle

Given a family $\mathcal{F}$, for every fixed complex vector $\mathbf{s}^{(0)} = \left(s_1^{(0)}, \ldots, s_m^{(0)}\right)$ we denote by $V_F\left(\mathbf{s}^{(0)}\right)$ the value set of the family calculated at $\mathbf{s}^{(0)}$:

$$V_F\left(\mathbf{s}^{(0)}\right) = \left\{ p\left(\mathbf{s}^{(0)}\right) \middle| \, p \in \mathcal{F} \right\}. \tag{2.8}$$

For the case of one-variable polynomials the following statement is known as the zero exclusion principle. Assume that the set Q in the coefficient space is pathwise connected and at least one polynomial from $\mathcal{F}$ is Hurwitz stable. Then $\mathcal{F}$ is robustly stable if and only if the origin $s = 0$ is excluded from the value set for all frequencies $\omega \in R$. This principle is one of the basic tools for the robust stability analysis of univariate polynomials so it looks quite natural to check how it works for multivariate polynomials. Let us start with the following example.

Example 2.3. Consider the family of bivariate polynomials

$$\mathcal{F} = \{ s_1 s_2 + \lambda s_1 + s_2 + 1 | \, \lambda \in [0,1]\}.$$

In the coefficient space this family is represented by a segment that is a pathwise connected set. For $\lambda = 1$, the polynomial $s_1 s_2 + s_1 + s_2 + 1 \in \mathcal{F}$ is stable. Direct calculations show that the value set of the family

$$V_F(i\omega_1, i\omega_2) = \{ (1 - \omega_1\omega_2) + i(\lambda\omega_1 + \omega_2) | \, \lambda \in [0,1]\}$$

does not contain zero for all real $\omega_1$and ω_2. But for $\lambda = 0$, polynomial $s_1 s_2 + s_2 + 1 \in \mathcal{F}$ is not stable. This example shows that the zero exclusion principle can not be applied directly to the class of multivariate polynomials which are stable in the sense of Definition 2.4.

It should be mentioned here that the principle remains valid for multivariate polynomials if one considers strict sense stability or scattering Hurwitz stability; see [1]. The following modification of the principle allows its application for the case of stability in the sense of Definition 2.4.

Theorem 2.5. *[Zero Exclusion Principle] Let $\mathcal{F}$ be a family of polynomials of constant degree containing at least one stable polynomial $p_0(\mathbf{s})$. Then $\mathcal{F}$ is robustly stable if and only if*

for $m = 1$: $0 \notin V_{\mathcal{F}}(i\omega_1)$;

for $m > 1$:

1. $0 \notin V_{\mathcal{F}}(i\omega_1, i\omega_2, \ldots, i\omega_m)$, *and*
2. *the subfamilies $\mathcal{F}_1, \mathcal{F}_2, \ldots, \mathcal{F}_m$ of the main coefficients satisfy the conditions of the theorem.*

Proof: Necessity. The proof is performed by induction on the number of variables. The statement is certainly true for $m = 1$. Assume now that it is valid for polynomial families with $(m-1)$ variables and consider the case of polynomial families with m variables. By Definition 2.4, the robust stability of $\mathcal{F}$ implies that $0 \notin V_{\mathcal{F}}(i\omega)$ for all $\omega \in \Omega$ and that $\mathcal{F}_k$ is robustly stable for all $k = 1, 2, \ldots, m$. But now each subfamily $\mathcal{F}_k$ is a family of $(m-1)$-variate polynomials. Applying the induction hypothesis to $\mathcal{F}_k$ one may conclude that it satisfies the conditions of the theorem.

Now let us address sufficiency. Again we resort to induction. The statement is true for polynomial families of one variable. Assume that it holds for the case of $(m-1)$-variate polynomial families and consider the case of m-variate polynomial families. First of all we observe that all polynomials in $\mathcal{F}$ are strict sense stable. This is a consequence of the zero exclusion condition $0 \notin V_{\mathcal{F}}(i\omega)$ for all $i\omega \in \Omega$; see [2] for a proof. Now stability of $p_0(\mathbf{s})$ implies stability of all main coefficients of the polynomial. This means that each subfamily $\mathcal{F}_k$ contains at least one stable polynomial. Having in mind that $\mathcal{F}_k$ is a family of $(m-1)$-variate polynomials one may conclude, applying the induction hypothesis, that $\mathcal{F}_k$ is robustly stable. By assumption all polynomials from $\mathcal{F}_k$ have the same degree $\mathbf{n}_k = (n_1, \ldots, n_{k-1}, n_{k+1}, \ldots, n_m)$. □

2.4.3 Edge Theorem

There is a specific class of polynomial families for which stability analysis may be performed without resorting to the study of the corresponding value set. This is the class of polytopic families of polynomials.

Theorem 2.6. *[Edge Theorem] A polytopic family $\mathcal{F} \subset \mathbf{P}_{n_1,\dots,n_m}$ of constant degree polynomials is stable if and only if the set of polynomials belonging to the edges of the family $Ed(\mathcal{F})$ are all stable.*

Proof: Necessity is trivial since every edge belongs the family $\mathcal{F}$.

The sufficiency part of the proof is given in an inductive way.

First, observe that for $m = 1$ the statement follows immediately from the edge theorem [5].

Assume now that the statement holds for the case of $(m-1)$-variate polynomials and consider the case of polynomials with m variables.

From Theorem 12 in [2] we know that the strict sense stability of the edges implies that of the whole family $\mathcal{F}$.

Consider now the family $\mathcal{F}_k$ of main coefficients $a_{n_k}^{(k)}(s_1, s_2, \dots, s_{k-1}, s_{k+1}, \dots, s_m)$ in (2.2), corresponding to polynomials in $\mathcal{F}$. It is clear that this family has the same polytopic structure as the original family $\mathcal{F}$, and that the edges of the new family are generated by edges of the original one. It follows from the assumptions that all edges of $\mathcal{F}$ are stable. In its turn by Definition 2.4 this implies that all the edges of $\mathcal{F}_k$ are stable. So in accordance with the induction assumption this implies the stability of $\mathcal{F}_k$. From the coefficient positivity condition we know that all coefficients of polynomials from $\mathcal{F}$ are positive numbers and henceforth all polynomials in $\mathcal{F}_k$ have the same degree $\mathbf{n}_k = (n_1, \dots, n_{k-1}, n_{k+1}, \dots, n_m)$. In other words every polynomial in $\mathcal{F}$ satisfies all conditions of Definition 2.4. □

The stability analysis of an edge polynomial family may be sometimes reduced to stability of the extreme polynomials of the family. To this end one may employ the notion of convex direction for multivariate polynomials in the same fashion as has been done for univariate Hurwitz polynomials in [11].

Definition 2.8. *An m-variate polynomials $g(\mathbf{s})$ is said to be a* convex direction *if the following condition is satisfied. Given any stable polynomial $f(\mathbf{s}) \in \mathbf{P}_{n_1,\dots,n_m}$ such that $f(\mathbf{s}) + g(\mathbf{s})$ is also stable and $f(\mathbf{s}) + \lambda g(\mathbf{s}) \in \mathbf{P}_{n_1,\dots,n_m}$ for all $\lambda \in [0,1]$, then $f(\mathbf{s}) + \lambda g(\mathbf{s})$ is stable for all $\lambda \in [0,1]$.*

The following extension of the Rantzer condition has been recently obtained in [4].

Theorem 2.7. *A polynomial $g(\mathbf{s})$ is an m-variate convex direction if the inequality*

$$\sum_{k=1}^{m} |\omega_k| \frac{\partial \arg[g(i\omega)]}{\partial \omega_k} \leq \left| \frac{\sin\{2\arg[g(i\omega)]\}}{2} \right| \tag{2.9}$$

holds for all values $\omega_k \in \mathcal{R}$, $k = 1, 2, \dots, m$, where $g(i\omega) \neq 0$.

Example 2.4. Consider polynomials

$$g(\mathbf{s}) = (1 \pm s_k), \qquad k = 1, 2, \ldots, m.$$

Direct calculations show that all these polynomials are convex directions. Moreover they remain convex directions after multiplication by an arbitrary monom $s_1^\alpha s_2^\beta \cdot, \ldots, \cdot s_m^\gamma$. It is worth mentioning that every monom $s_1^\alpha s_2^\beta \cdot, \ldots, \cdot s_m^\gamma$ is also convex direction.

It has been shown in [8], using the idea of convex directions, that the robust stability of an interval family of multivariate polynomials and that of a diamond family follows from the stability of a finite number of vertex polynomials of these families.

2.4.4 Stability Radius

Given a stable polynomial $p_0(\mathbf{s}) \in \mathbf{P}_{n_1,\ldots,n_m}$ with the coefficient vector $\mathbf{a}_0$, define the family

$$B = \left\{ p(\mathbf{s}) \mid \|\mathbf{a} - \mathbf{a}_0\|_2 \leq r \right\}.$$

The following definition is a direct extention of the stability radius concept introduced in [6].

Definition 2.9. *The minimal value of r for which the family B is not robustly stable is called the* stability radius *of* $p_0(\mathbf{s})$.

Having in mind the inductive nature of the definition of stable multivariate polynomials there is a possibility that one of the subfamilies of main coefficients of B may have a stability radius less than that of the original family. It can be easily shown that it is not the case.

Theorem 2.8. *The stability radius of every subfamily* B_k, $k = 1, 2, \ldots, m$, *of main coefficients is less than or equal to that of the original family B.*

Corollary 2.4. *The stability radius of a stable polynomial* $p_0(\mathbf{s})$ *is equal to*

$$\inf_{\omega_1,\ldots,\omega_m} \left\{ \frac{(\mathrm{Re}\, p_0(i\omega_1,\ldots,i\omega_m))^2}{\Phi(\omega_1,\ldots,\omega_m)} + \frac{(\mathrm{Im}\, p_0(i\omega_1,\ldots,i\omega_m))^2}{\Psi(\omega_1,\ldots,\omega_m)} \right\}^{1/2},$$

where

$$\Phi(\omega_1,\ldots,\omega_m) = \sum_{k=0}^{\left[\frac{n_1+,\cdots,+n_m}{2}\right]} \sum_{i_1+,\cdots,+i_m=2k} \left(\omega_1^{i_1},\ldots,\omega_m^{i_m}\right)^2,$$

and

$$\Psi(\omega_1, \ldots, \omega_m) = \sum_{k=0}^{\left[\frac{n_1+,\cdots,+n_m-1}{2}\right]} \sum_{i_1+,\cdots,+i_m=2k+1} \left(\omega_1^{i_1}, \ldots, \omega_m^{i_m}\right)^2.$$

References

[1] S. Basu, On the multidimensional generalization of robustness of scattering Hurwitz property of complex polynomials *IEEE Trans. Circuits Syst.*, 36:1159–1167, 1989.

[2] S. Basu, On boundary implications of stability and positivity properties of multidimensional systems, *Proc. of IEEE* 78:614–626, 1990.

[3] N.K. Bose, Robust multivariable scattering Hurwitz interval polynomial, *Linear Algebra Appl.* 98:123–136, 1988.

[4] N.K. Bose, Edge property from end-points for scattering Hurwitz polynomials, *Automatica* 32:4, 655–657, 1996.

[5] A.C. Bartlett, C.V. Hollot, and L. Huang, Root locations of an entire polytope of polynomials: it suffices to check the edges, *Math. Control, Signals Syst.* 1:61–71, 1988.

[6] D. Hinrichsen and A.J. Pritchard, Real and complex stability radii: A survey; in *Control of Uncertain Systems*, Birkhäuser, Boston, 119–162, 1990.

[7] V.L. Kharitonov and J.A. Torres Muñoz, Robust stability of multivariate polynomials part 1: Small coefficient perturbations, *Multidimensional Syst. Signal Processing* 10:7–20, 1998.

[8] V.L. Kharitonov, J.A. Torres Muñoz, and M.-I. Ramirez Sosa, Robust stability of multivariate polynomials Part 2: Polytopic coefficient variations, *Multidimensional Syst. Signal Processing* 10:1–12, 1999.

[9] J. Kogan, Computation of stability radius for families of bivariate polynomials, *Multidimensional Syst. Signal Processing* 4:151–165, 1993.

[10] B.T. Polyak and S.B. Shmulyian, Frequency domain criteria for robust stability of bivariate polynomials, *IEEE Trans. Circuits and Sys. 1: Fundamental Theory Appl.* 41:161–167, 1994.

[11] A. Rantzer, Stability conditions for polytopes of polynomials, *IEEE Trans. Auto. Control* 37:79–89, 1992.

Chapter 3

Robustness of Nonlinear Systems and Their Domains of Attraction

Andrew D.B. Paice
Fabian R. Wirth[1]

ABSTRACT In this chapter we consider the problem of analyzing the robustness of stability of nonlinear systems with respect to time-varying perturbations. We show that generically the stability radii of a singular fixed point of the nonlinear system and that of the corresponding linearization coincide. A brief introduction to a method for the calculation of the linear stability radius is presented. Furthermore, we consider the problem of determining a robust domain of attraction for the fixed point of a perturbed system under the assumption that the perturbations do not destroy exponential stability. We discuss some topological properties of the robust domain of attraction and present an approximation scheme for its determination.

3.1 Introduction

The robustness analysis of linear systems via a state space approach has been one of the main interests in the scientific work of Didi Hinrichsen. Both authors of this chapter have been introduced to the topic by him and their perspective in the area has been greatly influenced by his work. A significant step in this area was the introduction of the *stability radius* as a measure of robustness by Hinrichsen and Pritchard [16]. This methodology has subsequently been extended to several classes of linear systems and perturbations; see the survey [17] and also [18, 22].

In this chapter we consider the problem of analyzing the robustness of stability of nonlinear systems with time-varying perturbations. The key idea is to define a stability radius for the perturbed nonlinear system, and then to examine the related stability radii for the linearized system. Following the approach outlined in [4, 6, 24] we assume there exists a fixed point x^* of the nonlinear system, and that it is singular with respect to the perturbations, that is, not perturbed under the perturbation class consid-

[1]F. Wirth was supported by the Nonlinear Control Network, TMR-Project.

ered. For this fixed point we define the exponential stability radius. It was shown in [6] that lower and upper bounds of this stability radius can be obtained by studying the linearization in x^*. We show that generically the stability radii of the nonlinear system and its linearization coincide. A brief introduction to a method for the calculation of the linear stability radius is presented.

Having thus obtained some understanding of the local problem, we go on to consider the problem of determining a robust domain of attraction for the fixed point. For nonlinear systems one basic question is that of the determination of domains of attraction of asymptotically stable fixed points. This question has received considerable attention over the last decade; see for examples [26, 3, 20]. We discuss some topological properties of the robust domain of attraction and present an approximation scheme for its determination.

In both cases, the study of the local problem and the study of the robust domain of attraction, it is seen that the tools of optimal control theory may be applied to yield methods of calculating both the local stability radius and the robust domain of attraction.

We proceed as follows. In Section 3.2 we introduce a stability radius for nonlinear systems with time–varying perturbations. The concept of a *robust domain of attraction* is introduced, and some remarks on the ways in which the robust domain of attraction may shrink as the admissible size of perturbations increases are made. In Section 3.3 we develop a local robustness theory based on the linearization of the system. It is shown that generically the linear stability radii coincide, demonstrating that generically one need only consider the linearization in order to determine the nonlinear robustness properties of the system. In Section 3.4 a method of calculating the nonlinear stability radius based on the methods of discounted optimal control is presented.

In Section 3.5 we introduce the concept of the robust domain of attraction and a few properties are discussed. This object has also been studied in [2], where a generalization of Zubov's method to the perturbed case is presented. In the following section we then analyze the linearization of the nonlinear systems, finding a ball of initial conditions yielding trajectories that robustly converge to the origin. The determination of the domain of attraction, however, is clearly a nonlinear problem; thus in Section 3.6 we characterize the robust domain of attraction in terms of an optimal control problem, and present approximations to this problem whose value functions are computable as viscosity solutions of Hamilton–Jacobi–Bellman equations. In order to improve these approximations, we suggest how to use the information provided by the linearization. From this we obtain an algorithm for which we prove convergence.

In Section 3.7 we summarize the results and give a short outlook on remaining problems.

3.2 Preliminaries

In this paper we study nonlinear systems of the form

$$\begin{aligned} \dot{x}(t) &= f_0(x(t)), \quad t \in \mathbb{R} \\ x(0) &= x_0 \in \mathbb{R}^n, \end{aligned} \tag{3.1}$$

which are exponentially stable at a fixed point that we take to be 0. By this we mean that there exists a neighborhood U of 0 and constants $c > 1, \beta < 0$ such that the solutions $\varphi(t; x, 0)$ of (3.1) satisfy $\|\varphi(t; x, 0)\| \leq ce^{\beta t}$ for all $x \in U$. Under the assumptions of local exponential stability it is of interest to know the domain of attraction of 0, defined by

$$\mathcal{A}(0) := \{x \in \mathbb{R}^n \mid \varphi(t; x, 0) \to 0, t \to \infty\} .$$

Assume that (3.1) is subject to perturbations of the form

$$\dot{x}(t) = f_0(x(t)) + \sum_{i=1}^{m} d_i(t) f_i(x(t)) =: F(x(t), d(t)), \quad t \in \mathbb{R}, \tag{3.2}$$

where the perturbation functions f_i leave the fixed point invariant; that is, $f_i(0) = 0, i = 0, 1, \ldots, m$. We assume that the vector fields f_i are locally Lipschitz continuous and continuously differentiable in 0. The *unknown* perturbation function d is assumed to take values in $D \subset \mathbb{R}^m$, where D is compact, convex, with nonempty interior, and $0 \in \operatorname{int} D$. Denote $\mathcal{D} := \{d \in \ell^\infty(\mathbb{R}, \mathbb{R}^m) \mid d(t) \in D \text{ a. a.}\}$. Solutions to the initial value problem (3.2) with $x(0) = x_0$ for a particular d are denoted $\varphi(t; x_0, d)$.

It is our aim to analyze two robustness problems related to this setup. The first one is local in nature, as we study the corresponding time-varying stability radius at 0, defined by

$$\begin{aligned} r_{tv}(f_0, (f_i)) := \inf\{\alpha > 0 \mid \ & \exists\, d \in \alpha\mathcal{D} \text{ such that (3.2)} \\ & \text{is not exponentially stable at } 0\} . \end{aligned}$$

For time-invariant perturbations this problem has been studied in [24].

The second robustness problem considered is that of the domain of attraction of the unperturbed system. Given $d \in \mathcal{D}$, the domain of attraction of 0 at time $t_0 = 0$ for (3.2) is

$$\mathcal{A}_d(0) := \{x \in \mathbb{R}^n \mid \varphi(t; x, d) \to 0, t \to \infty\} .$$

A robust domain of attraction may now be defined. In the definition of the robust domain of attraction we do assume that the perturbed system is locally exponentially stable for all perturbations; that is, $r_{tv}(f_0, (f_i)) > 1$.

Definition 3.1. *[D-robust Domain of Attraction] Let $D \subset \mathbb{R}^m$ as before and assume that $r_{tv}(f_0, (f_i)) > 1$. The D-*robust domain of attraction *of the* *equilibrium* 0 *of* (3.2), *is*

$$\begin{aligned} \mathcal{A}_D(0) &:= \{x \in \mathbb{R}^n \mid \forall d \in \mathcal{D}, \varphi(t;x,d) \to 0, t \to \infty\} \\ &= \bigcap_{d \in \mathcal{D}} A_d(0) \,. \end{aligned}$$

When studying the robustness properties of the domain of attraction the main problems of interest are:

1. Given $D \subset \mathbb{R}^m$, determine $\mathcal{A}_D(0)$.
2. Given $A \subset \mathcal{A}(0)$ and a perturbation set $D \subset \mathbb{R}^m$, determine the largest α such that $A \subset \mathcal{A}_{\alpha D}(0)$.

In the first case we are most interested in determining the robust domain of attraction, while in the second case we consider a variant of a stability radius problem. Here the focus is on the mechanism by which stability is lost. Note that if the allowable perturbations are increased there are three different scenarios with which the property $A \subset \mathcal{A}_{\alpha D}(0)$ is lost at some minimal α_0.

1. Loss of stability at 0. That is, $A \subset \mathcal{A}_{\alpha D}(0)$ for $\alpha < \alpha_0$ and on the other hand $\mathrm{dist}(A, \partial\mathcal{A}_{\alpha D}(0)) > \delta > 0$ for all $0 < \alpha < \alpha_0$. This is the case if linear systems are considered.
2. Contraction of the domain of attraction. As $\alpha \to \alpha_0$ it holds that $\mathrm{dist}(A, \partial\mathcal{A}_{\alpha D}(0)) \to 0$.
3. Birth of an attractor in $\mathrm{int}\, A$. While $\mathrm{dist}(A, \partial\mathcal{A}_{\alpha D}(0)) > \delta > 0$ for all $0 < \alpha < \alpha_0$, it holds that $A \cap \partial\mathcal{A}_{\alpha_0 D}(0) \neq \emptyset$.

An example for the last scenario is given in the following example.

Example 3.1. Consider the following perturbed system on $\mathbb{R}$:

$$\dot{x} = -x + d(t)x\sin(x)$$

with $D = [-1, 1]$. Then, $\mathcal{A}_D(0) = (-\pi/2, \pi/2)$, while for $0 < \alpha < 1$ we have $\mathcal{A}_{\alpha D}(0) = \mathbb{R}$.

In this chapter we concentrate on the first question, determining the robust domain of attraction, and obtaining estimates of the robust domain of attraction, that is, determining sets that are guaranteed to lie within $\mathcal{A}_{\alpha D}(0)$. We begin with the analysis of the local problem, for which it is to be expected that linearization will play a vital role. This is developed in the following section.

3.3 Linearization Theory

As the functions f_i are continuously differentiable we may study the linearization in 0 associated with the nonlinear system (3.2) given by

$$\dot{x}(t) = A_0 x(t) + \sum_{i=1}^{m} d_i(t) A_i x(t)\,, \quad t \in \mathbb{R}\,, \tag{3.3}$$

where A_i denotes the Jacobian of f_i in 0, $i = 0, \ldots, m$. We abbreviate $A(d) = A_0 + \sum d_i A_i$. Solutions of (3.3) are denoted $\psi(t; x, d)$.

In the analysis of the problem where constant perturbations were considered, it was seen that an examination of the movement of the eigenvalues as the system was perturbed was key in understanding the problem. It was then possible to show that generically the stability radius for the nonlinear system could be determined by considering the linearization. In the time-varying case we are considering here it is necessary to consider the *Bohl exponent* (or equivalently the maximal Lyapunov exponent); see [15, 4, 5]. For $d \in \mathcal{D}$, let $\Phi_d(t,s)$ denote the evolution operator of the time-varying linear system $\dot{x} = A(d(t))x(t)$. The Bohl exponent of system (3.3) given $d \in \mathcal{D}$ is defined by

$$\beta(d) := \inf\{\beta \in \mathbb{R} \mid \exists M : \|\Phi_d(t,s)\| \leq M e^{\beta(t-s)}\, \forall t \geq s \geq 0\}\,;$$

see [9, 15] for an introduction to the properties of the Bohl exponent, in particular for the fact that exponential stability of a linear time-varying system given by $d \in \mathcal{D}$ is equivalent to $\beta(d) < 0$. We denote the maximal Bohl exponent of system (3.3) by

$$\beta(A_0, \ldots, A_m, D) := \max_{d \in \mathcal{D}} \beta(d)\,,$$

where we suppress the dependence on the matrices $(A_0, \ldots, A_m)$ and on the perturbation set D when these are clear from the context. In [1, 4] it is shown that the maximum is indeed attained. Furthermore if $\beta = \beta(A_0, \ldots, A_m, D) < 0$ then there exists for every $\varepsilon > 0$ an $M_\varepsilon \geq 1$ such that

$$\|\Phi_d(t,s)\| \leq M_\varepsilon e^{(\beta+\varepsilon)(t-s)}\,, \quad \forall t \geq s \geq 0, d \in \mathcal{D}\,. \tag{3.4}$$

Define the linear stability radii

$$r_{Ly}(A_0, (A_i)) = \inf\{\alpha > 0 \mid \beta(A_0, \ldots, A_m, \alpha D) \geq 0\}\,, \tag{3.5}$$

$$\bar{r}_{Ly}(A_0, (A_i)) = \inf\{\alpha > 0 \mid \beta(A_0, \ldots, A_m, \alpha D) > 0\}\,. \tag{3.6}$$

The following lemma from [6] states a basic relationship between the nonlinear and the linear stability radii.

Lemma 3.1.

$$r_{Ly}(A_0,(A_i)) \le r_{tv}(f_0,(f_i)) \le \bar{r}_{Ly}(A_0,(A_i)) .$$

The previous lemma gives rise to the question whether $r_{Ly}(A_0,(A_i))$ and $\bar{r}_{Ly}(A_0,(A_i))$ commonly coincide. A simple example shows that this need not always be the case.

Example 3.2. Let $n = 2, m = 1$, and

$$A_0 = \begin{bmatrix} -1 & 0 \\ 0 & 0 \end{bmatrix} \qquad A_1 = \begin{bmatrix} 1 & 0 \\ 0 & 0 \end{bmatrix} .$$

Clearly $r_{Ly}(A_0, A_1) = 0$ and $\bar{r}_{Ly}(A_0, A_1) = 1$.

This example is trivial in the sense that it shows a difference for the two linear stability radii only in the case when the unperturbed linear system is not asymptotically stable. In the discrete-time case it can be shown that only such trivial examples exist. In the continuous time case it is an open question whether there are nontrivial examples. Here we show a genericity result.

We now further develop the properties of r_{Ly} and $\bar{r}_{Ly}$, when considered as functions of $(A_0,(A_i))$. Recall the following definitions of semicontinuity. A function $f : \mathbb{R}^n \to \mathbb{R}$ is upper (lower) semicontinuous when, given $f(x_0) > c$ (resp., $f(x_0) < c$) for some $x_0 \in \mathbb{R}^n$ there exists a neighborhood $U \subset \mathbb{R}^n$ of x_0 such that $\forall x \in U,\ f(x) > c$ (resp., $f(x) < c$).

It is shown in [1] that the maximal Lyapunov exponent is a continuous function from the set of compact subsets of $\mathbb{R}^{n\times n}$ endowed with the Hausdorff metric to $\mathbb{R}$. Here the Lyapunov exponent corresponding to a compact set $M \subset \mathbb{R}^{n\times n}$ is defined as the one given by the differential inclusion

$$\dot{x} \in \{Ax \mid A \in M\} .$$

We may now prove the following semicontinuity properties for r_{Ly} and $\bar{r}_{Ly}$, generalizing the results for time-invariant perturbations.

Lemma 3.2.

(i) $r_{Ly}(A_0;(A_i))$ *is an upper semicontinuous function of* $(A_0;(A_i))$.

(ii) $\bar{r}_{Ly}(A_0;(A_i))$ *is a lower semicontinuous function of* $(A_0;(A_i))$.

Proof: Suppose that $\alpha_0 := r_{Ly}(A_0;(A_i)) > c$. Then $\forall d \in c\mathcal{D},\ \beta(d) \le \varepsilon < 0$. By upper semicontinuity of $\beta(\cdot)$ (see [9]), there exists a neighborhood U of $(A_0;(A_i))$ such that for all $(B_0;(B_i)) \in U$, $\beta(B_0,\ldots,B_m,cD) \le \varepsilon/2 < 0$, so that $r_{Ly}(B_0;(B_i)) > c$. Thus r_{Ly} is upper semicontinuous.

A similar argument establishes lower semicontinuity of $\bar{r}_{Ly}$. □

In some situations it may be interesting to consider an extended version of the stability radius for the linearized system.

$$r^c_{Ly}(A_0, (A_i)) := \inf\{\alpha > 0 \mid \beta(A_0, \ldots, A_m, \alpha D) \geq c\}; \tag{3.7}$$
$$\overline{r}^c_{Ly}(A_0, (A_i)) := \inf\{\alpha > 0 \mid \beta(A_0, \ldots, A_m, \alpha D) > c\}. \tag{3.8}$$

This allows measurement of the robustness of the system with respect to a guaranteed level of exponential convergence or divergence. It is straightforward to show that these new stability radii may be linked to those of (3.5) as follows.

Lemma 3.3. *Let $(A_0, \ldots, A_m) \in \mathbb{R}^{(n\times n)\times(m+1)}$; then*

$$r^c_{Ly}(A_0, (A_i)) = r_{Ly}(A_0 - cI; (A_i)); \tag{3.9}$$
$$\overline{r}^c_{Ly}(A_0, (A_i)) = \overline{r}_{Ly}(A_0 - cI; (A_i)). \tag{3.10}$$

The following lemma is needed in the proof of the main result of this section.

Lemma 3.4. *Let $m, n \in \mathbb{N}$ and $(A_0, \ldots, A_m) \in \mathbb{R}^{(n\times n)\times m}$ be fixed. For the maps*

$$g : c \mapsto r_{Ly}(A_0 - cI, (A_i)), \quad \overline{g} : c \mapsto \overline{r}_{Ly}(A_0 - cI, (A_i)),$$

the following statements hold.

(i) g is upper semicontinuous; $\overline{g}$ is lower semicontinuous.

(ii) g is discontinuous at c_0 iff $\overline{g}$ is discontinuous at c_0 iff $g(c_0) < \overline{g}(c_0)$.

(iii) $g, \overline{g}$ have at most countably many discontinuities.

(iv) $g(c) = \overline{g}(c)$ for all $c \in \mathbb{R}$ with the exception of at most countably many points.

Proof:

(i) Semicontinuity is an immediate consequence of Lemma 3.2.

(ii) Let $g(c_0) < \overline{g}(c_0)$; then by definition and continuity of the maximal Lyapunov exponent $g(c_0 + \varepsilon) \geq \overline{g}(c_0)$ for any $\varepsilon > 0$. Thus the assumption implies discontinuity of g at c_0. Discontinuity of $\overline{g}$ at c_0 follows from $g(c_0) \geq \overline{g}(c_0 - \varepsilon)$. Conversely, let g be discontinuous at c_0. By semicontinuity this implies that for any $\varepsilon > 0$ and a suitable constant C we have $g(c_0) < C < g(c_0 + \varepsilon) \leq \overline{g}(c_0 + \varepsilon)$. Now the right–hand term tends to $\overline{g}(c_0)$ as $\varepsilon \to 0$, proving $g(c_0) < C \leq \overline{g}(c_0)$. A similar argument works for $\overline{g}$.

(iii) This follows from the monotonicity of $g, \overline{g}$.

(iv) This follows from *(ii)* and *(iii)*.

□

With the help of the previous results, it is possible to prove the following genericity result, which is the main result of this section.

Theorem 3.1.

(i) For fixed $m \geq 1$ the set $\mathfrak{L}$ given by

$$\{(A_0, \ldots, A_m) \,|\, r_{Ly}(A_0, (A_i)) = \bar{r}_{Ly}(A_0, (A_i))\}$$

is a countable intersection of open and dense sets. Furthermore, the Lebesgue measure of the complement $\mathfrak{L}^c$ is 0.

(ii) For fixed $m \geq 1$ the set $\mathcal{N}$ of maps $(f_0, \ldots, f_m)$ satisfying

$$r_{Ly}(A_0, (A_i)) = \bar{r}_{Ly}(A_0, (A_i)) = r_{tv}(f_0, (f_i))$$

contains a countable intersection of open and dense sets with respect to the C^1-topology on the space of C^1-maps $(f_0, \ldots, f_m)$ satisfying $f_0(x^*) = x^*, f_i(x^*) = 0$, $i = 1, \ldots, m$.

Proof: *(i):* We introduce the set

$$T_0 := \{(A_0, \ldots, A_m, \alpha) \in \mathbb{R}^{(n \times n) \times (m+1)} \times \mathbb{R} \,|\, \beta(A_0, \ldots, A_m, \alpha D) = 0\},$$

which is clearly closed by the continuity of the maximal Lyapunov exponent.

Note that $(A_0, \ldots, A_m, r_{Ly}) \in T_0$, and $(A_0, \ldots, A_m, \bar{r}_{Ly}) \in T_0$ again by continuity of the maximal Lyapunov exponent. Thus $(A_0, \ldots, A_m) \in \mathfrak{L}^c$ iff $\exists\ a, b \geq 0$, $a \neq b$ such that $(A_0, \ldots, A_m, a), (A_0, \ldots, A_m, b) \in T_0$. Under this condition it follows for all $a < c < b$ that $(A_0, \ldots, A_m, c) \in T_0$. For $k \geq 1$ we denote

$$T_{0,k} := \left\{(A_0, \ldots, A_m, \alpha) \in \mathbb{R}^{(n \times n) \times (m+1)} \times \mathbb{R} \;\middle|\; \left(A_0, \ldots, A_m, \alpha + \frac{1}{k}\right) \in T_0\right\}.$$

Thus $\mathfrak{L}^c = \bigcup_{k=1}^{\infty} Q_k$, where

$$Q_k := \{(A_0, \ldots, A_m) \,| \exists\, a \geq 0 \text{ such that } (A_0, \ldots, A_m, a) \in T_0 \cap T_{0,k}\}.$$

The Q_k are projections of $T_0 \cap T_{0,k}$ onto $\mathbb{R}^{(n \times n) \times (m+1)}$. As $T_0 \cap T_{0,k}$ is closed Q_k is also closed for every $k \geq 1$. Therefore we now need to show that all of the sets Q_k that compose $\mathfrak{L}^c$ are nowhere dense in $\mathbb{R}^{n \times n \times (m+1)}$. For this it is sufficient that in every neighborhood of any point in $\mathfrak{L}^c$ there exists a point that does not belong to $\mathfrak{L}^c$, as any closed set either has interior points or is nowhere dense. As any affine subspace of the form $\{(A_0 - cI, \ldots, A_m) \mid c \in \mathbb{R}\}$ intersects $\mathfrak{L}^c$ in at most countably many points by Corollary 3.4*(iv)* the assertion is proved.

In particular, this shows that $\mathcal{L}^c$ is Lebesgue measurable, and an easy application of Fubini's theorem in conjunction with Corollary 3.4*(iv)* shows that the Lebesgue measure of $\mathcal{L}^c$ is zero.

(ii): Note that for $(f_0, \ldots, f_m) \in \mathcal{N}$ it is sufficient that for the linearized system $(A_0, \ldots, A_m) \in \mathcal{L}$. It is thus sufficient to show that the preimage of an open and dense set under the continuous linear map

$$\{f_0, f_1, \ldots, f_m\} \mapsto \{\frac{\partial f_0}{\partial x}(x^*), \ldots, \frac{\partial f_m}{\partial x}(x^*)\}$$

is open and dense. This, however, is clear by definition of the C^1-topology. □

A consequence of the previous theorem is that other stability radii that might be defined for the nonlinear system, for instance with respect to Lyapunov or asymptotic stability, generically coincide with the exponential stability radius.

3.4 Calculating the Local Stability Radius

In this section we briefly present a method for the calculation of the linear stability radii defined in (3.5) and (3.6). It is based on an idea for the calculation of Lyapunov exponents presented in [13] and in the discrete-time case in [28]. We define the Lyapunov exponent corresponding to an initial condition $x_0 \neq 0$ and a disturbance $d \in \mathcal{D}$ to be

$$\lambda(x_0, d) = \limsup_{t\to\infty} \frac{1}{t} \log \|\psi(t; x_0, d)\| . \tag{3.11}$$

By [4],

$$\beta(A_0, \ldots, A_m, D) = \max\{\lambda(x_0, d) \mid x_0 \neq 0, d \in \mathcal{D}\} ;$$

that is, $\beta(A_0, \ldots, A_m, D)$ is equal to the maximal Lyapunov exponent of the family of time-varying systems given by (3.3) and the condition $d \in \mathcal{D}$. This is the quantity studied in [1] and [4]. We now briefly present a way for the calculation of approximations of Lyapunov exponents. Via projection onto the projective space $\mathbb{P}^{n-1}$ we obtain (in local coordinates, which we take to be vectors of unit length) from system (3.3) the system

$$\dot{s}(t) = \left(A(d(t)) - s(t)^T A(d(t))s(t) \cdot \mathrm{Id}\right) s(t) , \tag{3.12}$$

$$s(0) = s_0 = x_0/\|x_0\| \in \mathbb{P}^{n-1} .$$

Defining the function $q : \mathbb{P}^{n-1} \times D \to \mathbb{R}$, $q(s,d) = s^T A(d)s$ it is an easy calculation (see [4]) that for $s_0 \in \mathbb{P}^{n-1}, d \in \mathcal{D}$ we have

$$||\Phi_d(t,0)s_0|| = \exp\left(\int_0^t q(\eta(s; x_0, d), d(s))ds\right) . \tag{3.13}$$

Thus the Lyapunov exponent is of the form on an average yield along a trajectory of system (3.12); that is,

$$\lambda(x_0, d) = \limsup_{t\to\infty} \frac{1}{t}\int_0^t q(\eta(s; x_0, d), d(s))ds\,,$$

where $\eta(s; x_0, d)$ denotes the trajectory of (3.12). Interpreting this expression as an average yield optimal control problem on $\mathbb{P}^{n-1}$, we introduce the following approximating functional for $\delta > 0$.

$$J_\delta(x_0, d) := \int_0^\infty \delta e^{-\delta s} q(\eta(s; x_0, d), d(s))ds\,, \qquad (3.14)$$

with associated value function $V_{\delta,D}(x) := \sup_{d\in\mathcal{D}} J_\delta(x_0, d)$. For these optimal control problems it is known [12] that

$$\kappa_\delta(D) := \max_{x\in\mathbb{P}^{n-1}} V_{\delta,D}(x) \geq \beta(A_0, \ldots, A_m, D)\,,$$

and $\kappa_\delta(D) \to \beta(A_0, \ldots, A_m, D)$ as $\delta \to 0$, but here we need more details about the corresponding rate of convergence. Recall that a set of matrices $M \subset \mathbb{R}^{n\times n}$ is called irreducible, if only the trivial subspaces $\{0\}$ and $\mathbb{R}^n$ are invariant under all $A \in M$. Recall further that an invariant control set C of system (3.12) is a set with the properties

(i) for all $x \in C$ it holds that $\mathrm{cl}\, O^+(x) := \{y \in \mathbb{P}^{n-1} \mid \exists d \in D, t \geq 0 : y = \eta(t, x, d)\} = \mathrm{cl}\, C$;

(ii) C is a maximal set with property (i).

We have the following relation between these concepts.

Lemma 3.5. *The following statements are equivalent.*

(i) The set $A(D)$ is irreducible.

(ii) Every invariant control set C of (3.12) contains a basis of $\mathbb{R}^n$.

Proof: Assume that C is an invariant control set and let $x_1, \ldots, x_l \in C$ be a basis of $V := \mathrm{span}\, C$. Let $x \in V$ and $d \in \mathcal{D}$ be arbitrary. If $x = \sum_{k=1}^{l} \gamma_k x_k$ then it follows

$$\Phi_d(t, 0)x = \sum_{k=1}^{l} \gamma_k \Phi_d(t, 0)x_k\,.$$

Now by invariance of C the projection of the trajectories $\Phi_d(t, 0)x_k$ onto the projective space remain in C. This shows that any trajectory of (3.3) starting in V remains there for all time. Hence for all $x \in V$, $d \in D$ we have $A(d)x \in V$, so that C spans a subspace invariant under all $A(d), d \in D$. Now the assertion follows. $\square$

If $A(D)$ is not irreducible, then there exists a coordinate transform T such that for any $A \in A(D)$ we have

$$TAT^{-1} = \begin{bmatrix} A_{11} & A_{12} & \dots & \dots & A_{1k} \\ 0 & A_{22} & A_{23} & \dots & A_{2k} \\ 0 & 0 & A_{33} & & \vdots \\ \vdots & & \ddots & \ddots & \vdots \\ 0 & & \dots & 0 & A_{kk} \end{bmatrix},$$

where each of the sets $\mathcal{A}(\mathcal{D})_j := \{A(d)_{jj}; d \in D\}, i = 1, \dots, d$ is irreducible. It is then easy to see that

$$r_{Ly}(A_0, (A_i)) = \min_{j=1,\dots,k} r_{Ly}(A_{0,j}, (A_{i,j}))\,;$$

thus we can constrict all considerations to the irreducible case.

The following result is shown in [13] for the case that the projected system is locally accessible, but the proof easily carries over to the following slightly more general assumption.

(A) The set $A(\alpha D)$ is irreducible and there exists a closed maximal integral manifold of (3.12) that contains exactly one invariant control set.

We present it here because we need a further detail of parameter dependence that is not immediate from [13] (although it is not hard to obtain the result from the proofs presented in that reference).

Remark 3.1. *Assumption (A) is satisfied in particular if (3.12) is locally accessible. However, it is sufficient that in addition to irreducibility there exists a single matrix $A \in A(D)$ with a simple eigenvalue $\lambda_{\max}$ satisfying $\operatorname{Re}\lambda_{\max} > \operatorname{Re}\lambda, \lambda \in \sigma(A) \setminus \{\lambda_{\max}\}$. This may be seen as follows. The existence of a closed maximal integral manifold N is clear as the orbits of system (3.12) are orbits of a Lie group. On the other hand if $\hat{x} \in \mathbb{P}^{n-1}$ is an eigenvector corresponding to the eigenvalue $\lambda_{\max}$ then it follows from standard arguments [8] that $\hat{x} \in \operatorname{cl} C$ for any invariant control set in $\mathbb{P}^{n-1}$. As the system is forward accessible on N and N is closed this implies $\hat{x} \in C$ for any invariant control set contained in N. By maximality of control sets this implies the existence of a unique invariant control set in N. The existence of at least one invariant control set follows by compactness of N. Furthermore the relative interior $\operatorname{int}_N C$ is nonempty, as the system restricted to maximal integral manifolds is locally accessible; see [8].*

Lemma 3.6. *Consider system (3.3) and two constants $0 < \underline{\alpha} \leq \overline{\alpha} < \infty$. Assume that (A) holds for system (3.12) with control range $\underline{\alpha}D$; then there exists a constant $M > 0$ such that for all $\alpha \in [\underline{\alpha}, \overline{\alpha}]$ we have*

$$\kappa_\delta(\alpha D) \in [\beta(\alpha D), \beta(\alpha D) + \delta M] .$$

Proof: Let N be the closed maximal integral manifold that contains exactly one invariant control set of (3.12) with control range $\underline{\alpha}D$. We denote this invariant control set by $C_{\underline{\alpha}}$. Then for every $\alpha \in [\underline{\alpha}, \overline{\alpha}]$ there exists an invariant control set $C_\alpha \supset C_{\underline{\alpha}}$ of (3.12) with control range αD [8]. By Lemma 3.5 we may fix a basis $x_1, \dots, x_n \in \mathrm{int}_N\, C_{\underline{\alpha}}$. Define

$$A := \max_{x \in \mathbb{P}^{d-1}, d \in \overline{\alpha} D} |q(x,d)| .$$

As $C_{\underline{\alpha}}$ is the only invariant control set in the compact manifold N it follows that for all $x \in N$ we have $O^+(x) \cap C \neq \emptyset$ and T defined as follows is finite.

$$T := \max_{k=1,\dots,n} \sup_{x \in N} \inf\{t \geq 0 \mid \exists d \in \underline{\alpha}\mathfrak{D} \text{ such that } \eta(t,x,d) = x_k\} .$$

We claim that for any $k = 1, \dots, n$, $\alpha \in [\underline{\alpha}, \overline{\alpha}]$, and $d \in \alpha\mathfrak{D}$ we have for all $t > 0$,

$$\frac{1}{t}\int_0^t q(\eta(s;x_k,d), d(s))ds \leq \beta(\alpha D) + \frac{2AT}{t} .$$

Assume to the contrary the existence of $d \in \alpha D$ and a $t > 0$ such that

$$\frac{1}{t}\int_0^t q(\eta(s;x_k,d), d(s))ds > \beta(\alpha D) + \frac{2AT}{t} ,$$

and denote $y = \eta(t, x_k, d)$; then there exists a control $d_1 \in \alpha\mathfrak{D}$ such that $x_k = \eta(t_1, y, d_1)$ for some $t_1 \leq T$. Denoting the concatenation of the controls $d(\cdot)|_{[0,t)}$ and $d_1(\cdot)|_{[0,t_2)}$ by d_2 and extending d_2 periodically to $\mathbb{R}$ we obtain with $t_2 = t + t_1$,

$$\lambda(x_k, d_2) = \frac{1}{t_2}\int_0^{t_2} q(\eta(s;x_k,d_2), d_2(s))ds > \frac{t}{t_2}\beta(\alpha D) + \frac{2AT}{t_2} - \frac{At_1}{t_2} \geq \beta(\alpha D) .$$

This contradicts the definition of β. Now the assertion follows as in Section 5 of [13]. □

As $\beta(\alpha D)$ is monotonically increasing in α and $\beta(r_{Ly}(A_0,(A_i))D) = 0$ it follows that

$$c(D) := \liminf_{h \downarrow 0} -\frac{\beta((r_{Ly}(A_0,(A_i)) - h)D)}{h} \geq 0 .$$

The number $c(D)$ may be interpreted as the supremum of the gradients of those linear functions that have their zero in $r_{tv}(A_0,(A_i))$ and are larger than β on some interval of the form $[a, r_{Ly}(A_0,(A_i))]$, where $a < r_{Ly}(A_0,(A_i))$.

Theorem 3.2. *Consider system (3.3) and assume that (A) holds for some $\alpha < r_{Ly}(A_0,(A_i))$; then the following properties hold.*

(i) For all $\delta > 0$ it holds that

$$r_{Ly}(A_0, (A_i)) \geq r_\delta(A_0, \mathcal{D}) := \inf\{\alpha > 0 \mid \kappa_\delta(\alpha D) \geq 0\}\,. \tag{3.15}$$

(ii)
$$r_{Ly}(A_0, (A_i)) = \lim_{\delta \to 0} r_\delta(A_0, \mathcal{D}).$$

(iii) If $c(D) > 0$ then there exist $\bar{\delta} > 0$ and a constant $M > 0$ such that for all $0 < \delta < \bar{\delta}$,

$$r_{Ly}(A_0, (A_i)) - r_\delta(A_0, \mathcal{D}) \leq \delta M\,.$$

Proof:

(i) If $\alpha > r_{Ly}(A_0, (A_i))$, then $0 \leq \beta(\alpha D) \leq \kappa_\delta(\alpha D)$ by Lemma 3.6. Thus $\alpha \geq r_\delta(A_0, \mathcal{D})$.

(ii) If $\alpha < r_{Ly}(A_0, (A_i))$, then $\beta(\alpha D) < 0$ and by Lemma 3.6 there exists a δ_α such that for all $0 < \delta < \delta_\alpha$ it holds that $\kappa_\delta(\alpha D) < 0$, and therefore for $0 < \delta < \delta_\alpha$ it follows that $\alpha \leq r_\delta(A_0, \mathcal{D}) \leq r_{Ly}(A_0, (A_i))$. Letting α tend to $r_{Ly}(A_0, (A_i))$ from below shows the assertion.

(iii) Choose $\varepsilon > 0$ small enough such that $c := c(D) - \varepsilon > 0$. Then there exists an $\eta > 0$ such that for all $\alpha \in [r_{Ly}(A_0, (A_i)) - \eta, r_{Ly}(A_0, (A_i))]$ we have

$$\beta(\alpha D) < c(\alpha - r_{Ly}(A_0, (A_i))).$$

By Lemma 3.6 for every $\alpha \in [r_{Ly}(A_0, (A_i)) - \eta, r_{Ly}(A_0, (A_i))]$ there exists an $M_\alpha > 0$ such that

$$\kappa_\delta(\alpha D) \leq c(\alpha - r_{Ly}(A_0, (A_i))) + M_\alpha \delta\,.$$

Again by Lemma 3.6 the value $M := \sup\{M_\alpha \mid \alpha \in [r_{Ly}(A_0, (A_i)) - \eta, r_{Ly}(A_0, (A_i))]\}$ is finite. Denote the zero of the right-hand side in the above equation by

$$\tilde{r}_\delta := r_{Ly}(A_0, (A_i)) - \frac{M}{c(D) - \varepsilon}\delta \leq r_\delta(A_0, \mathcal{D})\,.$$

Then for all $0 < \delta < \delta'$ small enough so that $M\delta c^{-1} < \eta$ we obtain

$$r_{Ly}(A_0, (A_i)) - r_\delta(A_0, \mathcal{D}) \leq r_{Ly}(A_0, (A_i)) - \tilde{r}_\delta = \frac{M}{c}\delta\,.$$

□

The time-varying stability radius may thus be calculated by applying Theorem 3.2. A description of the actual mathematical background for the calculation of the objects defined in this section can be found in [10, 11, 12] and references therein.

3.5 Robust Domains of Attraction

We now turn our attention to a study of the robust domain of attraction. After determining some properties of such sets, we study the problem of determining a ball about the origin that is guaranteed to be in the robust domain of attraction.

We collect some properties of the robust domain of attraction $\mathcal{A}_D(0)$. It may be surprising that these resemble closely the properties of domains of attraction of fixed points of unperturbed systems [14, §33]. For our proofs, we need the following slight extensions of the results in [14] and [27].

Lemma 3.7. *Let 0 be an asymptotically stable fixed point of the unperturbed system* (3.1). *Let W be an open subset of $\mathcal{A}(0)$ with $0 \in W$ and with the property that it is invariant under* (3.1). *Then W is connected and contractible to zero. If f_0 is of class C^1, then there exists a C^1 diffeomorphism from W to $\mathbb{R}^n$.*

For the proof of the preceding lemma we recall the following basic lemma from [21, Lemma 3].

Lemma 3.8. *Let M be a paracompact manifold such that every compact subset is contained in an open set which is diffeomorphic to $\mathbb{R}^n$. Then M is diffeomorphic to $\mathbb{R}^n$.*

Proof: **(of Lemma 3.7)** Let $V \subset W$ be a connected neighborhood of 0. By definition for each $x \in W$ we have $\varphi(t; x, 0) \in V$ for all t large enough. This shows connectedness of W. In the following we denote the image of a set V under the flow of (3.1) by $\varphi(\cdot; V)$. To complete the proof let K be an arbitrary compact subset of W. We choose a relatively compact neighborhood of K that is invariant under (3.1) as follows. Choose a relatively compact neighborhood V of 0 contained in $\mathcal{A}(0)$ that is invariant under (3.1). This exists as a sublevel set of a Lyapunov function. Then for some $T > 0$ we have $K \subset \varphi(-T; V)$, and $\varphi(-T; V)$ is relatively compact and invariant. Hence $V_2 := W \cap \varphi(-T; V)$ is a relatively compact open neighborhood of K. For $\varepsilon > 0$ small enough the set $\varphi(\varepsilon; V_2)$ is still an open neighborhood of K with the property that $\operatorname{cl} \varphi(\varepsilon; V_2) \subset W$. Now choose a C^r function $\alpha : \mathbb{R}^n \to \mathbb{R}_+$ with the property $\alpha_{|\operatorname{cl} \varphi(\varepsilon; V_2)} \equiv 1$ and $\alpha_{|\mathbb{R}^n \setminus W} \equiv 0$, consider the system

$$\dot{x} = \alpha(x) f_0(x)\,, \tag{3.16}$$

and denote the corresponding flow by $\tilde{\varphi}$. Choose $\varepsilon > 0$ such that $B(0, \varepsilon) \subset W$. Then for some $T > 0$ we have $K \subset \tilde{\varphi}(-T; B(0, \varepsilon))$. On the other hand by construction of α it follows that $\tilde{\varphi}(-T; B(0, \varepsilon)) \subset W$ and $\tilde{\varphi}(-T; B(0, \varepsilon))$ is diffeomorphic to $\mathbb{R}^n$ via the diffeomorphism induced by the flow of (3.16). This completes the proof. □

Let us briefly recall that system (3.2) with a perturbation set D is called locally uniformly asymptotically stable, if for every $\varepsilon > 0$ there exists a $\delta > 0$ such that $\|x\| < \delta$ implies $\|\varphi(t;x,d)\| < \varepsilon$ for all $t > 0, d \in \mathcal{D}$ and if there exists a neighborhood U of 0 such that for all $x \in U$ we have $\varphi(t;x,d) \to 0$ uniformly in d as $t \to \infty$.

Proposition 3.1. *Consider system (3.2) and assume that 0 is locally uniformly asymptotically stable for the perturbation set D; then*

(i) $x_0 \in \mathcal{A}_D(0) \Leftrightarrow \lim_{t\to\infty} \sup_{d\in\mathcal{D}} \|\varphi(t;x_0,d)\| = 0$;

(ii) $\mathcal{A}_D(0)$ *is an open,connected, invariant set;*

(iii) $\mathrm{cl}\,\mathcal{A}_D(0)$ *is an invariant set;*

(iv) $\mathcal{A}_D(0)$ *is contractible to* 0*;*

(v) *if for some $d \in D$ $f(\cdot,d)$ is of class C^1, then $\mathcal{A}_D(0)$ is C^1-diffeomorphic to $\mathbb{R}^n$; and*

(vi) *for every $x \in \partial\mathcal{A}_D(0)$ there exists $d \in \mathcal{D}$ such that $\varphi(t;x,d) \in \partial\mathcal{A}_D(0)$ for all $t \geq 0$.*

Proof:

(i) Clearly we need only show "$\Rightarrow$". Assume that $x \in \mathcal{A}_D(0)$ and there exist sequences $\{d_k\} \subset \mathcal{D}$, $T_k \to \infty$, and $\varepsilon > 0$ such that $\|\varphi(T_k,x,d_k)\| > \varepsilon > 0$ for all $k \in \mathbb{N}$. By uniform stability there exists a $\delta > 0$ such that $\|z\| < \delta$ implies $\|\varphi(t,z,d)\| < \varepsilon$ for all $d \in \mathcal{D}, t \geq 0$. Without loss of generality $d_k \to d \in \mathcal{D}$ in the weak-$*$ topology on $\mathcal{D}$. By assumption there exists a t_0 such that $\|\varphi(t_0,x,d)\| < \delta$. As $\varphi(t_0,x,d_k) \to \varphi(t_0,x,d)$ this means for all k large enough $\|\varphi(t,x,d_k)\| < \varepsilon$ for $t \geq t_0$, a contradiction.

(ii) By assumption there is an open neighborhood V of 0 contained in $\mathcal{A}_D(0)$. By definition from each $x \in \mathcal{A}_D(0)$ there exists a trajectory $\varphi(\cdot,x,d)$ entering V. This shows connectedness. To prove invariance assume that for some $x \in \mathcal{A}_D(0), d_1 \in \mathcal{D}$ there exists a $t > 0$ such that $y := \varphi(t;x,d_1) \notin \mathcal{A}_D(0)$. This implies the existence of a $d_2 \in \mathcal{D}$ such that $\varphi(t,y,d_2) \not\to 0$. But then for the concatenation d given by $d_{|[0,t]} \equiv d_1$, $d_{|(t,\infty)} \equiv d_2(\cdot - t)$ it follows that $\varphi(t;x,d) \not\to 0$ contradicting the choice of x. Finally, to prove that $\mathcal{A}_D(0)$ is open, assume the contrary, let $x \in \mathcal{A}_D(0)$, and assume we are given a sequence $x_k \to x$ with $x_k \notin \mathcal{A}_D(0)$. Then there exist controls d_k such that $\varphi(t;x_k,d_k) \not\to 0$. As in *(i)* this leads to a contradiction.

(iii) If for some $x \in \mathrm{cl}\,\mathcal{A}_D(0)$ and $d \in \mathcal{D}$ we have $\varphi(t;x,d) \notin \mathrm{cl}\,\mathcal{A}_D(0)$ then by continuous dependence on initial conditions we have that $\mathcal{A}_D(0)$ is not invariant, contradicting *(ii)*.

(iv) This follows from Lemma 3.7.

(v) This is a consequence of *(ii)* and Lemma 3.7.

(vi) This follows by definition as $\mathcal{A}_D(0)$ is open and $\mathrm{cl}\,\mathcal{A}_D(0)$ is invariant.

□

The first question to consider is under which conditions $\mathcal{A}_D(0)$ contains a neighborhood of 0, that is, to give a sufficient condition for local uniform stability of system (3.2). To examine this question we use the linearization (3.3) of (3.2) at 0 with maximal Bohl exponent $\beta(A_0, \ldots, A_m, D)$. By [9, Theorem VII.1.3] the Bohl exponent is upper semicontinuous even under nonlinear perturbations. Thus:

Lemma 3.9. *Consider* (3.2) *with linearization* (3.3).

(i) If $\beta(A_0, \ldots, A_m, D) < 0$, *then* $\mathcal{A}_D(0)$ *contains an open neighborhood of* 0.
(ii) If $\beta(A_0, \ldots, A_m, D) > 0$, *then* $0 \in \partial\mathcal{A}_D(0)$.

The following example shows that for the case $\beta(A_0, \ldots, A_m, D) = 0$ both situations are possible.

Example 3.3. Let $A_0, A_1, \ldots, A_m \in \mathbb{R}^{n\times n}$ and $D \subset \mathbb{R}^m$ be such that the maximal Bohl exponent $\beta(A_0, A_1, \ldots, A_m, D) = 0$ and consider the systems

$$\dot{x} = -x(t) < x(t), x(t) > + \left(A_0 + \sum_{i=1}^{m} d_i(t) A_i \right) x(t)\,; \tag{3.17}$$

$$\dot{x} = x(t) < x(t), x(t) > + \left(A_0 + \sum_{i=1}^{m} d_i(t) A_i \right) x(t)\,. \tag{3.18}$$

By [1] there is a norm v on $\mathbb{R}^n$ that is a Lyapunov function for the linearization

$$\dot{x} = \left(A_0 + \sum_{i=1}^{m} d_i(t) A_i \right) x(t)$$

in the following sense. Denote the dual norm by v^*. Then for any x, y with $v(x) = 1, v^*(y) = 1$, and $< x, y >= 1$ it holds that $< A(d)x, y >\leq 0$ for all $d \in D$ and so that for each x there exists a y and $d \in D$ with $< A(d)x, y >= 0$. By homogeneity of the norm we obtain that $< -x(t) < x(t), x(t) > + A(d)x, y >< 0$ for system (3.17) and hence $\mathcal{A}_D(0) = \mathbb{R}^n$. Similarly, for (3.18) one obtains that $\mathcal{A}_D(0) = \{0\}$.

By the results of Section 3.3 the point where the Bohl exponent does not indicate whether $0 \in \mathrm{int}\mathcal{A}_D(0)$ is exactly the perturbation intensity at which the system becomes exponentially unstable.

Furthermore the linearization can be used to obtain a more precise statement on the size of the ball contained in $\mathcal{A}_D(0)$, which is a consequence of [9, Theorem VII.1.3]. To this end denote $L(D) := \max_{d \in D} \|A(d)\|$.

Lemma 3.10. *Let $\beta(A_0, \ldots, A_m, D) < \beta < 0$ and $M_\beta \geq 1$ such that (3.4) is satisfied and fix $\beta < \beta' < 0$ and $M > M_\beta$. Let $h > 0$, $q > 0$ be such that*

$$1 - M_\beta e^{-(\beta - \beta')h} > 0 \quad \textit{and}$$

$$qhe^{(2L(D)+q+\beta')h} = \min\{M - M_\beta, 1 - M_\beta e^{-(\beta-\beta')h}\}.$$

If $\|F(x,d) - A(d)x\| < q$ for all $x \in B(0,\varepsilon), d \in D$ then

$$\|\varphi(t; x, d)\| \leq M e^{\beta' t} \|x\|\,, \quad \forall x \in B(0, \varepsilon/M), d \in \mathcal{D}\,.$$

In particular, it follows that $B(0, \varepsilon/M) \subset \mathcal{A}_D(0)$.

3.6 An Optimal Control Characterization of the Robust Domain of Attraction

In this section we present an optimal strategy for the approximation of the robust domain of attraction . This is motivated as follows. By definition we have

$$x_0 \in \mathcal{A}_D(0) \Leftrightarrow \forall d \in \mathcal{D} \lim_{t\to\infty} \|\varphi(t; x_0, d)\| = 0\,.$$

Assuming that $\beta(A_0, \ldots, A_m, D) < 0$ and applying Lemma 3.9 we obtain immediately that

$$x_0 \in \mathcal{A}_D(0) \Leftrightarrow \forall d \in \mathcal{D}, \quad \limsup_{t\to\infty} \|\varphi(t; x_0, d)\| = 0\,. \tag{3.19}$$

Motivated by Proposition 3.1*(i)*, the domain of attraction can thus be characterized via the following optimization problem. Define

$$J_0(x,d) := \limsup_{t\to\infty} \|\varphi(t; x, d)\|$$

and the corresponding value function

$$v_0(x) = \sup_{d \in D} J_0(x,d)\,;$$

then $\mathcal{A}_D(0) = v_0^{-1}(0)$. In other words, there clearly exists a $c > 0$ such that $x \notin \mathcal{A}_D(0)$ implies that $v_0(x) > c$.

The problem with this value function is obviously its discontinuity at the boundary of $\mathcal{A}_D(0)$. As v_0 is hard to calculate we use the approximation scheme already introduced in Section 3.4. For $\delta > 0$ define

$$J_\delta(x,d) := \int_0^\infty \delta e^{-\delta t} \|\varphi(t;x,d)\| dt$$

with value function $v_\delta(x) = \sup_{d \in \mathfrak{D}} J_\delta(x,d)$. Note that v_δ is continuous w.r.t. x.

Although it is not generally true that v_δ is strictly decreasing w.r.t. δ, it is possible to obtain a convergence result. For $M \geq 1$ and $0 > \beta > \beta(A_0,\dots,A_m,D)$ denote

$$X(M,\beta) := \{x \mid \forall t > 0 \, : \, \sup_{d \in \mathfrak{D}} \|\varphi(t;x_0,d)\| \leq M e^{\beta t}\}.$$

Note that $\mathcal{A}_D(0) \supset X(M,\beta)$. We note the following properties of $X(M,\beta)$.

Proposition 3.2. *Consider* (3.2) *with linearization* (3.3) *and assume that* $\beta(A_0,\dots,A_m,D) < \beta < 0$; *then*

(i) $\mathcal{A}_D(0) = \bigcup_{M \geq 1} \operatorname{int} X(M,\beta)$;

(ii) $x \in X(M,\beta) \Rightarrow v_\delta(x) \leq M\delta/(\delta - \beta)$.

Proof: *(i)* Note that by Lemma 3.10 for some M_0 large enough and some $\varepsilon > 0$ we have $B(0,\varepsilon) \subset \operatorname{int} X(M_0,\beta)$. Let $x \in \mathcal{A}_D(0)$. By Proposition 3.1 *(i)* there exists a $T > 0$ such that for all $d \in \mathfrak{D}$ we have $\varphi(T;x,d) \in B(0,\varepsilon/(2M_0))$. It follows that for a relatively compact neighborhood U of x we have $\varphi(T;y,d) \in B(0,\varepsilon/M_0)$ for all $y \in U, d \in \mathfrak{D}$. Defining

$$M := \max\{M_0, e^{-\beta T} \sup\{ \|\varphi(t;y,d)\| \mid t \in [0,T], d \in \mathfrak{D}, y \in \operatorname{cl} U\}\},$$

we have that $U \subset X(M,\beta)$ as desired.

(ii) This follows from

$$v_\delta(x) = \sup_{d \in \mathfrak{D}} J_\delta(x,d) \leq \sup_{d \in \mathfrak{D}} \int_0^\infty \delta e^{-\delta t} M e^{\beta t} = M \frac{\delta}{\delta - \beta}.$$

□

Corollary 3.1. *Consider* (3.2) *with linearization* (3.3). *It holds that*

$$v_\delta \to v_0 \text{ uniformly on compact subsets of } \mathcal{A}_D(0) \text{ as } \delta \to 0.$$

Proof: Let $K \subset \mathcal{A}_D(0)$ be compact and fix $0 > \beta > \beta(A_0,\dots,A_m,D)$; then by compactness and Proposition 3.2 *(i)* there exists an M, such that $K \subset X(M,\beta)$. Now the assertion follows from Proposition 3.2 *(ii)*. □

The previous statement implies that v_δ converges linearly on compact subsets of $\mathcal{A}_D(0)$ to 0. To obtain an estimate for $\mathcal{A}_D(0)$ define

$$\mathcal{A}(\delta, \varepsilon) := \{x \in \mathbb{R}^n \mid v_\delta(x) < \varepsilon\}.$$

Then we have the following.

Proposition 3.3. *Consider* (3.2) *and assume that* $\beta(A_0, \ldots, A_m, D) < 0$*; then for all* $0 < \varepsilon \leq c_0 := \mathrm{dist}(0, \partial\mathcal{A}_D(0))$,

$$\mathcal{A}_D(0) = \bigcup_{\delta>0} \mathcal{A}(\delta, \varepsilon) = \bigcup_{\delta^*>\delta>0} \mathcal{A}(\delta, \varepsilon), \qquad \forall \delta^* > 0.$$

Proof: This is an immediate consequence of Proposition 3.2 as for any $M \geq 1$ and $\beta(A_0, \ldots, A_m, D) < \beta < 0$ and $\delta > 0$ small enough we have $X(M, \beta) \subset \mathcal{A}(\delta, \varepsilon)$. □

In general, information about c_0 amounts to the solution of the original problem itself, so we need a lower bound on c_0. Using the quantities introduced in Lemma 3.10, assume that $\|F(x, d) - A(d)x\| < q$ for all $x \in B(0, \varepsilon), d \in D$; then $B(0, \varepsilon/M) \subset A_D(0)$ and so $\varepsilon/M \leq c_0$ is the lower bound we require.

In order to use the information provided by Proposition 3.3 we have to obtain estimates for the quantities $\beta(A_0, \ldots, A_m, D)$ and M_β as used in Lemma 3.10. From these the quantities q and ε are determinable and Proposition 3.3 is then applicable.

In order to estimate ε and M, we need some information about the local growth properties of the perturbed system. Via Lemma 3.10 these may be obtained by examining the linearization at 0, defined in Section 3.3. In the following we use the notation of Section 3.4, where one problem has already been discussed, namely, that of approximating the maximal Bohl exponent. Thus if $\beta(A_0, \ldots, A_m, D) < 0$, then choosing $\delta > 0$ small enough in the optimization problem defined by (3.14) we can obtain $0 > \kappa_\delta(D) > \beta(A_0, \ldots, A_m, D)$. It remains to obtain a constant M, such that (3.4) is satisfied.

Let $0 > \kappa > \kappa_\delta(D)$. By (3.13) it is sufficient to find $T > 0$ such that

$$\sup_{||x||=1, d \in \mathcal{D}} \int_0^T q(\psi(s; x, d), d(s)) - \kappa ds < 0.$$

Then it follows that $||\Phi_d(T, 0)|| < e^{\kappa T}$, $\forall d \in \mathcal{D}$, and so

$$||\Phi_d(t, 0)|| < e^{L(D)T} e^{\kappa t}, \ \forall d \in \mathcal{D}, \forall t > 0.$$

Note that in order to find T, the value function v_δ that has already been calculated can be used, and it is sufficient to find T such that

$$\sup_{||x||=1, d \in \mathcal{D}} \int_0^T \delta e^{-\delta s}(q(\psi(s; x, d), d(s)) - \kappa) ds < 0. \qquad (3.20)$$

Also note that solvability of (3.20) depends on the fact that $\kappa > \kappa_\delta(D)$, as for $\kappa_\delta(D)$, the expression on the left is always nonnegative.

With the estimates obtained so far, we are now in a position to describe an algorithm for determining $\mathcal{A}_D(0)$, which is the main contribution of this section.

Algorithm 3.1. *Given $f_0, \ldots, f_m$ and D such that $\beta(A_0, \ldots, A_m, D) < 0$:*

1. *calculate κ_δ for small δ, such that $\kappa_\delta < 0$;*
2. *with the data κ_δ, M satisfying (3.4), determine a ball $B(0, \varepsilon)$ contained in $\mathcal{A}_D(0)$ via Lemma 3.10;*
3. *let $\varepsilon_0 = \varepsilon$, $\hat{\mathcal{A}}_0 = B(0, \varepsilon)$;*
4. *determine the value function $v_{\delta,k}$ associated with the cost functional*
$$J_{\delta,k}(x, d) := \int_0^\infty \delta e^{-\delta t} g_k(\varphi(t; x, d))dt \,, \tag{3.21}$$
where $g_k(x) = \|x\|$ if $\|x\| \notin \hat{\mathcal{A}}_k$, $g_k(x) = 0$, otherwise;
5. *determine ε_{k+1} such that $B(0, \varepsilon_{k+1}) \subset \hat{\mathcal{A}}_{k+1} := v_{\delta,k}^{-1}([0, \varepsilon_k]) \cup B(0, \varepsilon)$. Continue with Step 4.*

Theorem 3.3. *Consider system (3.1) with perturbation structure (3.2). If $\beta(A_0, \ldots, A_m, D) < 0$ then the sets $\hat{\mathcal{A}}_k$, $k = 1, 2, \ldots$ generated by Algorithm 3.1 form a monotonically increasing sequence such that $\cup_{k=0}^\infty \hat{\mathcal{A}}_k = \mathcal{A}_D(0)$.*

Proof: Note that it is clear by definition that $\hat{\mathcal{A}}_0 \subset \hat{\mathcal{A}}_1$ and $g_1 \le g_0$. Thus we may proceed by induction assuming that $\hat{\mathcal{A}}_0 \subset \ldots \subset \hat{\mathcal{A}}_{k-1} \subset \hat{\mathcal{A}}_k$ and $g_0 \ge \ldots \ge g_{k-1} \ge g_k$. With this we obtain for $x \in \hat{\mathcal{A}}_k$,

$$\sup_{d \in \mathcal{D}} \int_0^\infty \delta e^{-\delta t} g_k(\varphi(t; x, d))dt \le \sup_{d \in \mathcal{D}} \int_0^\infty \delta e^{-\delta t} g_{k-1}(\varphi(t; x, d))dt \le \varepsilon_{k-1} \le \varepsilon_k \,.$$

It follows that $x \in \hat{\mathcal{A}}_{k+1}$. So $\hat{\mathcal{A}}_k \subset \hat{\mathcal{A}}_{k+1}$ and consequently, $g_k \ge g_{k+1}$.

Let $x \in \mathcal{A}_D(0)$ and assume $x \notin \hat{\mathcal{A}}_1$, as there is nothing to show otherwise. Define

$$T := \sup\{t \mid \exists d \in \mathcal{D} : \phi(t, x, d) \notin \hat{\mathcal{A}}_1 \}$$

Note that T is finite by Proposition 3.1*(i)* and the fact that $B(0, \varepsilon) \subset \hat{\mathcal{A}}_1$. Let

$$\mathcal{O}^+_{\le t}(x) := \{y \in \mathbb{R}^n \mid \exists d \in \mathcal{D}, 0 \le s \le t : y = \phi(s, x, d) \} \,.$$

Note that $\mathcal{O}^+_{\le T}(x)$ is compact and let R be such that $\mathcal{O}^+_{\le T}(x) \subset B(0, R)$. Let

$$h = -\frac{\log(1 - \varepsilon/R)}{\delta} \,;$$

then it follows for $y \in \mathcal{O}^+_{\leq T}(x) \setminus \mathcal{O}^+_{\leq T-h}(x)$ that

$$h \geq \sup\{t \mid \exists d \in \mathcal{D} : \phi(t,y,d) \notin \hat{A}_1 \},$$

as otherwise we have an immediate contradiction to the definition of T. It follows that for any $d \in \mathcal{D}$ we have

$$\int_0^\infty \delta e^{-\delta t} g_k(\phi(t,y,d))dt \leq \int_0^h \delta e^{-\delta t} R dt = R(1-e^{-\delta h}) = \varepsilon .$$

This implies $\mathcal{O}^+_{\leq T}(x) \setminus \mathcal{O}^+_{\leq T-h}(x) \subset \hat{A}_2$. Continuing this argument we see that $x \in \hat{A}_k$, where k is such that $kh > T$. □

Remark 3.2.

(i) A useful stopping criterion can be applied in Step 5 as follows. If $\varepsilon_k - \varepsilon_{k+1}$ is bigger than some threshold go to 4. Otherwise, determine whether to decrease δ and go to 4 or stop, depending on the size of δ.

(ii) In practice we would suggest stopping the algorithm in Step 1 if $\kappa_\delta \geq 0$ for reasonably small δ. The reason is that although the nonlinear system may be exponentially stable, the Bohl exponent of the linearization is so small that the system is unlikely to be robustly stable in a meaningful sense.

(iii) The reason for choosing the particular form of g_k (3.21) is that once a trajectory enters $\hat{A}_k$, it will robustly converge to 0, and thus there is no longer any need to penalize it in the cost.

(iv) Note that by construction $\hat{A}_k \subset A_D(0)$; thus the algorithm supplies an inner approximation of the robust domain of attraction.

3.7 Conclusion

In this chapter we have studied the robustness of stability of a class of perturbed nonlinear systems, both from a local and a semiglobal perspective.

First we introduced time-varying stability for nonlinear systems. Using linearization techniques and spectral theory for time-varying linear systems it was shown that the nonlinear stability radius equals the linear stability radii provided exponential stability of the unperturbed system holds. A scheme for the calculation of the time-varying stability radius has been proposed, based on discounted optimal control.

Additionally we have discussed robust domains of attraction of singular fixed points. A scheme for the approximation of the robust domain of attraction has been presented. This involves the calculation of approximations of the maximal Bohl exponent of the linearized system and subsequently the solution of an optimal control problem given by the nonlinear system. In this way a sequence of interior approximations to the robust domain of attraction is produced, each approximation being contained in the next.

Acknowledgment: The authors would like to thank Fritz Colonius for pointing out an error in an earlier version of this chapter.

References

[1] N.E. Barabanov, Absolute characteristic exponent of a class of linear nonstationary systems of differential equations, *Sib. Math. J.* 29(4):521–530, 1988.

[2] F. Camilli, L. Grüne, and F. Wirth, A generalization of Zubov's method to perturbed systems, *SIAM J. Control Optimization*, 2000. to appear.

[3] H.-D. Chiang, M.W. Hirsch, and F.F. Wu, Stability regions of nonlinear autonomous dynamical systems, *IEEE Trans. Auto. Control* 33(1):16–27, 1988.

[4] F. Colonius and W. Kliemann, Stability radii and Lyapunov exponents, in *Control of Uncertain Systems*, D. Hinrichsen and B. Mårtensson, eds, Progress in Systems and Control Theory 6, Birkhäuser, Boston, 19–55, 1990.

[5] F. Colonius and W. Kliemann, Maximal and minimal Lyapunov exponents of bilinear control systems, *J. Diff. Equations*, 101:232–275, 1993.

[6] F. Colonius and W. Kliemann, A stability radius for nonlinear differential equations subject to time varying perturbations, *3rd IFAC Symposium on Nonlinear Control Systems Design (NOLCOS'95)*, Lake Tahoe, NV, 44–46, 1995.

[7] F. Colonius and W. Kliemann, The Lyapunov spectrum of families of time varying matrices, *Trans. Amer. Math. Soc.*, 348:4389–4408, 1996.

[8] F. Colonius and W. Kliemann, *The Dynamics of Control*, Birkhäuser, Boston, 2000.

[9] J. Daleckii and M. Krein, *Stability of Solutions of Differential Equations in Banach Spaces*, Translations of Mathematical Monographs 43, American Mathematical Society, Providence, RI, 1974.

[10] L. Grüne, Numerical stabilization of bilinear control systems, *SIAM J. Control Optimization* 34(6):2024–2050, 1996.

[11] L. Grüne, An adaptive grid scheme for the discrete Hamilton–Jacobi–Bellman equation, *Numerische Mathematik* 75:319–373, 1997.

[12] L. Grüne, On the relation between discounted and average optimal control problems, *J. Diff. Equations* 148:65–99, 1998.

[13] L. Grüne and F. Wirth, On the rate of convergence of infinite horizon discounted optimal value functions, *Nonlinear Analysis* 2000, to appear.

[14] W. Hahn, *Stability of Motion*, Springer-Verlag, Berlin, 1967.

[15] D. Hinrichsen, A. Ilchmann, and A. Pritchard, Robustness of stability of time-varying linear systems, *J. Diff. Equations* 82(2):219–250, 1989.

[16] D. Hinrichsen and A. Pritchard, Stability radii of linear systems, *Syst. Control Letters* 7:1–10, 1986.

[17] D. Hinrichsen and A. Pritchard, Real and complex stability radii: A survey, in *Control of Uncertain Systems*, D. Hinrichsen and B. Mårtensson, eds., Progress in Systems and Control Theory 6, Birkhäuser, Basel, 119–162, 1990.

[18] D. Hinrichsen and A. Pritchard, Destabilization by output feedback, *Diff. Integral Equtions* 5(2):357–386, 1992.

[19] Y. Lin, E.D. Sontag, and Y. Wang, A smooth converse Lyapunov theorem for robust stability, *SIAM J. Control Optimization* 34(1):124–160, 1996.

[20] M. Loccufier and E. Noldus, On the estimation of asymptotic stability regions for autonomous nonlinear systems, *IMA J. Math. Control Information* 12:91–109, 1995.

[21] J. Milnor, Differential topology, in *Lectures in Modern Mathematics*, II, T.L. Saaty, ed., Wiley, New York, 165–183, 1964.

[22] A. Packard and J. Doyle, The complex structured singular value, *Automatica* 29(1):71–109, 1993.

[23] A.D. Paice and F.R. Wirth, Robustness of nonlinear systems subject to time-varying perturbations, *Proc. 36th Conference on Decision and Control*, San Diego, CA, December, 4436–4441, 1997.

[24] A.D.B. Paice and F.R. Wirth, Analysis of the local robustness of stability for flows, *Math. Control, Signals, and Syst.* 11(4):289–302, 1998.

[25] A.D.B. Paice and F. Wirth, Robustness analysis of domains of attraction of nonlinear systems, Proc. Mathematical Theory of Networks and Systems MTNS98, Padova, Italy, 353 – 356, 1998.

[26] A. Vannelli and M. Vidyasagar, Maximal Lyapunov functions and domains of attraction for autonomous nonlinear systems, *Automatica* 21(1):69–80, 1985.

[27] F.W. Wilson, The structure of the level surfaces of a Lyapunov function, *J. Diff. Equations* 3:323–329, 1967.

[28] F. Wirth, On the calculation of real time-varying stability radii, *Int. J. Robust Nonlinear Control* 8:1043–1058, 1998.

Chapter 4

On Stability Radii of Slowly Time-Varying Systems

Achim Ilchmann[1]
Iven M.Y. Mareels

ABSTRACT *We consider robustness of exponential stability of time-varying linear systems with respect to structured dynamical nonlinear perturbations. Sufficient conditions in terms of L^2-stability are derived. It is shown that the infimum of the complex stability radii of a family of time-invariant linear systems provides a good estimate for the stability radius of a linear time-varying system if time variations are sufficiently slow.*

Nomenclature

$\mathbb{K}$	$=$	$\mathbb{R}$ or $\mathbb{C}$
A^*	$=$	conjugate complex and transpose of $A \in \mathbb{C}^{p\times m}$
$\|M\|$	$=$	induced Euclidean norm for $M \in \mathbb{K}^{n\times m}$
$L^p(I;\mathbb{K}^{n\times m})$	$=$	the set of p-integrable functions $f : I \to \mathbb{K}^{n\times m}$, $I \subset \mathbb{R}$ an interval, $p \geq 1$
$L^2_{\text{loc}}(\mathbb{R}_{\geq 0};\mathbb{K}^p)$	$=$	the set of locally quadratic integrable functions $f : \mathbb{R}_{\geq 0} \to \mathbb{K}^p$
$\|f\|_{L^p(I;\mathbb{K}^{n\times m})}$	$=$	$\left(\int_I \|f(\tau)\|^p d\tau\right)^{1/p}$
$L^\infty(I;\mathbb{K}^{n\times m})$	$=$	the set of functions $f : I \to \mathbb{K}^{n\times m}$ that are essentially bounded on the interval $I \subset \mathbb{R}$
$\|f\|_{L^\infty(I;\mathbb{K}^{n\times m})}$	$=$	ess $\sup_{t\in I} \|f(t)\|$
$\mathcal{C}_{\text{pw,bdd}}(I;\mathbb{K}^{n\times m})$	$=$	set of piecewise continuous and bounded maps $M(\cdot) : I \to \mathbb{K}^{n\times m}$, $I \subset \mathbb{R}$ an interval
$\pi_t(f)(\tau)$	$=$	$\begin{cases} f(\tau), & \tau \in [0,t] \\ 0, & \tau > t \end{cases}$

[1]This chapter was written while Achim Ilchmann was visiting the Department of Electrical and Electronic Engineering at the University of Melbourne. The support and the hospitality of the Department are hereby gratefully acknowledged.

4.1 Introduction

In this chapter we investigate the robustness of uniformly exponentially stable time-varying systems

$$\dot{x}(t) = A(t)x(t), \qquad t \geq 0, \tag{4.1}$$

where $A(\cdot) \in \mathcal{C}_{\text{pw,bdd}}(\mathbb{R}_{\geq 0}; \mathbb{K}^{n\times n})$ is assumed to be piecewise continuous and bounded.

Definition 4.1. *The system* (4.1) *is called* uniformly exponentially stable *if and only if there exist $L, \lambda > 0$ such that its transition matrix $\Phi(\cdot,\cdot)$ satisfies*

$$\|\Phi(t,t_0)\| \leq Le^{-\lambda(t-t_0)} \qquad \textit{for all } t \geq t_0 \textit{ and all } t_0 \geq 0. \tag{4.2}$$

We investigate the robustness of the stability of systems described by (4.1) with respect to additive nonlinear perturbations:

$$\dot{x}(t) = A(t)x(t) + B(t)\,\mathcal{D}\big(C(\cdot)x(\cdot)\big)(t). \tag{4.3}$$

The structure of the perturbation is represented by piecewise continuous and bounded matrix valued functions

$$B(\cdot) \in \mathcal{C}_{\text{pw,bdd}}(\mathbb{R}_{\geq 0}; \mathbb{K}^{n\times m}), \quad C(\cdot) \in \mathcal{C}_{\text{pw,bdd}}(\mathbb{R}_{\geq 0}; \mathbb{K}^{p\times n}), \tag{4.4}$$

and by the nonlinear causal dynamical perturbation operator

$$\mathcal{D}(\cdot) \,:\, L^2(\mathbb{R}_{\geq 0}; \mathbb{K}^p) \to L^2(\mathbb{R}_{\geq 0}; \mathbb{K}^m).$$

Observe that we allow for infinite-dimensional perturbation systems. Precise definitions of the perturbation classes and global existence of the solution of the possibly nonlinear system (4.3) are given in Section 4.2.

In Section 4.3 we investigate several stability concepts — such as (global uniform) exponential, L^2, L^2-output, and asymptotic stability — for nonlinear systems of the form (4.3). These concepts are nested and the main results in this section are sufficient conditions for (global uniform) exponential stability of the zero solution.

In Section 4.4 we recall the concept of structured stability radius. Loosely speaking, the (complex) stability radius $r_{\mathbb{C}}$ is the sharp bound for the norm of the perturbation operator $\mathcal{D}$ so that the global L^2-stability of the perturbed system (4.3) is preserved as long as $\|\mathcal{D}\| < r_{\mathbb{C}}$ and might be lost if $\|\mathcal{D}\| = r_{\mathbb{C}}$. See Definition 4.5. The main result is that the structured stability radius of the time-varying system is close to the infimum of the structured stability radii of all "frozen" systems; that is, for fixed $\tau \geq 0$,

$$\dot{x}(t = A(\tau)x(t) + B(\tau)\,\mathcal{D}\big(C(\tau)x(\cdot)\big)(t), \qquad t \geq 0, \tag{4.5}$$

if the time-variations of $A(\cdot), B(\cdot), C(\cdot)$ are sufficiently small.

Section 4.5 contains a useful lemma on the convolution of L^p-functions and another lemma collecting smoothness properties of the solution of the algebraic Riccati equation.

4.2 Perturbation Classes

In this section we introduce three different perturbation classes; we then define what we understand as a solution of the perturbed system. For the perturbation classes considered, it turns out that the solutions are well defined into the future and unique.

Definition 4.2. *We consider, for time-varying scaling matrices as in* (4.4)*, the following three classes of the perturbed system* (4.1)*,*

$$\dot{x}(t) = A(t)x(t) + B(t)\,\Delta(t)\,C(t)x(t), \tag{4.6}$$

$$\dot{x}(t) = A(t)x(t) + B(t)\,D\left(t, C(t)x(t)\right), \tag{4.7}$$

$$\dot{x}(t) = A(t)x(t) + B(t)\,\mathcal{D}\big(C(\cdot)x(\cdot)\big)(t), \tag{4.8}$$

where

(i) $\Delta(\cdot) \in L^\infty(\mathbb{R}_{\geq 0}; \mathbb{K}^{m\times p})$ *is a time-varying linear perturbation and*

$$\|\Delta(\cdot)\|_{L^\infty(0,\infty)} := \operatorname{ess\,sup}_{t\geq 0}\|\Delta(t)\|\,;$$

(ii) $D(\cdot,\cdot) : \mathbb{R}_{\geq 0} \times \mathbb{K}^p \to \mathbb{K}^m$ *is a time-varying nonlinear perturbation with:*
$D(t,0) = 0$ *for all* $t \geq 0$,
$t \mapsto D(t,y)$ *is* measurable *for almost all* $y \in \mathbb{K}^p$,
$y \mapsto D(t,y)$ *is* Lipschitz continuous uniformly in t on compact intervals; *that is, for every* $T > 0$ *there exists some* $L_T > 0$ *such that*

$$\|D(t,y) - D(t,\bar{y})\| \leq L_T\|y - \bar{y}\| \quad \textit{for all } y, \bar{y} \in \mathbb{K}^p \textit{ and all } t \in [0,T],$$

and $D(\cdot,\cdot)$ *is of* finite gain*; that is,*

$$\|D(\cdot,\cdot)\|_{\mathrm{nt}} := \inf\left\{\gamma > 0 \mid \forall t \geq 0, \forall y \in \mathbb{K}^p : \|D(t,y)\| \leq \gamma\|y\|\right\} < \infty;$$

(iii) $\mathcal{D}(\cdot) : L^2(\mathbb{R}_{\geq 0}; \mathbb{K}^p) \to L^2(\mathbb{R}_{\geq 0}; \mathbb{K}^m)$ *is a dynamical perturbation satisfying* causality*; that is,*

$$\pi_t \mathcal{D} \pi_t = \pi_t \mathcal{D} \qquad \textit{for all} \quad t \geq 0\,;$$

weakly L^2-Lipschitz continuity*, that is, for every* $(t_0, x_0, \varphi(\cdot)) \in \mathbb{R}_{\geq 0} \times \mathbb{K}^n \times L^2(0, t_0; \mathbb{K}^p)$ *there exist* $r > 0, t_1 > t_0$ *and* $L \geq 0$ *such that for all* $y_1(\cdot), y_2(\cdot) \in L^2(0, t_1; \mathbb{K}^p)$ *satisfying* $y_1(\tau) = y_2(\tau) = \varphi(\tau)$ *for almost all* $\tau \in [0, t_0]$ *and* $\|y_i(\cdot) - \varphi(\cdot)\|_{L^\infty(t_0,t_1)} \leq r$ *for* $i = 1,2$*, we have*

$$\|\mathcal{D}(y_1)(\cdot) - \mathcal{D}(y_2)(\cdot)\|_{L^2(t_0,t_1;\mathbb{K}^m)} \leq L\|y_1(\cdot) - y_2(\cdot)\|_{L^2(t_0,t_1;\mathbb{K}^p)},$$

and $\mathcal{D}$ *is of* finite gain*; that is,*

$$\|\mathcal{D}(\cdot)\|_{\mathrm{dyn}} := \inf\left\{\gamma > 0 \,\middle|\, \begin{array}{l}\forall y(\cdot) \in L^2(\mathbb{R}_{\geq 0}; \mathbb{K}^p) \ : \\ \|\mathcal{D}(y)(\cdot)\|_{L^2(\mathbb{R}_{\geq 0};\mathbb{K}^m)} \leq \gamma\|y(\cdot)\|_{L^2(\mathbb{R}_{\geq 0};\mathbb{K}^p)}\end{array}\right\} < \infty.$$

These three classes of linear time-varying, nonlinear time-varying, and dynamical perturbations are denoted $\mathcal{P}_{\text{lt}}(\mathbb{K})$, $\mathcal{P}_{\text{nt}}(\mathbb{K})$, and $\mathcal{P}_{\text{dyn}}(\mathbb{K})$, respectively.

The above perturbation classes were introduced by Hinrichsen and Pritchard [9] for time-invariant systems; it was shown by them that $D(\cdot,\cdot) \in \mathcal{P}_{\text{nt}}(\mathbb{K})$ can be identified with $\mathcal{D}(\cdot) \in \mathcal{P}_{\text{dyn}}(\mathbb{K})$ by setting

$$\mathcal{D}(y(\cdot))(t) := D(t, y(t)) \qquad \text{for } t \geq 0 \text{ and } y(\cdot) \in L^2(\mathbb{R}_{\geq 0}; \mathbb{K}^p)$$

and the following chain of norm-preserving embeddings holds

$$\mathbb{K}^{m \times p} \subset \mathcal{P}_{\text{lt}}(\mathbb{K}) \subset \mathcal{P}_{\text{nt}}(\mathbb{K}) \subset \mathcal{P}_{\text{dyn}}(\mathbb{K}). \tag{4.9}$$

Since $\mathcal{D}(y(\cdot))(t)$ depends not only on $y(t)$ but on the whole "past" $y(\cdot)|_{[0,t]}$ of $y(\cdot)$, Hinrichsen and Pritchard [9] introduced the following initial value problem, which we extend to the time-varying case in a straightforward manner.

Definition 4.3. *Suppose* (4.1) *is perturbed by $\mathcal{D}(\cdot) \in \mathcal{P}_{\text{dyn}}(\mathbb{K})$ with scaling matrices as in* (4.4) *so that we consider*

$$\dot{x}(t) = A(t)x(t) + B(t)\, \mathcal{D}\big(C(\cdot)x(\cdot)\big)(t). \tag{4.10}$$

Then

$$x(\cdot) = x(\cdot; t_0, x_0, \varphi(\cdot)) \;:\; I \to \mathbb{K}^n, \qquad I = [t_0, t_1),\ t_1 > t_0,$$

is said to be a solution of (4.10) with initial data $(t_0, x_0, \varphi(\cdot)) \in \mathbb{R}_{\geq 0} \times \mathbb{K}^n \times L^2(0, t_0; \mathbb{K}^p)$ *if and only if $x(\cdot)$ is absolutely continuous on I, $x(t_0) = x_0$, and, for almost all $t \in I$,*

$$\dot{x}(t) = A(t)x(t) + B(t)\, \mathcal{D}\big([Cx]^{\varphi}(\cdot)\big)(t), \tag{4.11}$$

where

$$[Cx]^{\varphi}(\tau) \;:=\; \begin{cases} \varphi(\tau), & \tau \in [0, t_0) \\ C(\tau)x(\tau), & \tau \in [t_0, t_1) \\ 0, & \tau \in [t_1, \infty). \end{cases}$$

Note that the perturbed system (4.8), and therefore also the less general ones (4.6) and (4.7), preserve the zero solution.

It turns out that the smoothness assumption on the dynamical perturbation operator $\mathcal{D}$ as defined in Definition 4.2*(iii)* is sufficient to guarantee existence, uniqueness, and that finite escape time does not exist for the initial value problem. One reason for this is, roughly speaking, that the perturbations are assumed to be linearly bounded; see in particular the definition of $\|D(\cdot,\cdot)\|_{\text{nt}}$.

Theorem 4.1. *For any $\mathcal{D}(\cdot) \in \mathcal{P}_{\rm dyn}(\mathbb{K})$ and any initial data $(t_0, x_0, \varphi(\cdot)) \in \mathbb{R}_{\geq 0} \times \mathbb{K}^n \times L^2(0, t_0; \mathbb{K}^p)$, the initial value problem* (4.10) *possesses a solution $x(\cdot; t_0, x_0, \varphi(\cdot))$; and this solution is unique and exists on the whole of $[t_0, \infty)$.*

The proof of Theorem 4.1 is given for constant matrices $(A(\cdot), B(\cdot), C(\cdot)) \equiv (A, B, C) \in \mathbb{K}^{n\times n} \times \mathbb{K}^{n\times m} \times \mathbb{K}^{p\times n}$ by Hinrichsen and Pritchard [9]. The extension to the time-varying case is straightforward and omitted here.

4.3 Stability of Nonlinearly Perturbed Linear Systems

We are interested in seeing to what extent the stability properties of the unperturbed system (4.1) are inherited by the perturbed systems (4.6) to (4.8). To this end we introduce the following different concepts of stability and investigate how they are related.

Definition 4.4. *The origin of the initial value problem* (4.10) *is called*

(i) globally uniformly exponentially stable *if and only if*

$$\|x(t; t_0, x_0, \varphi(\cdot))\| \quad \leq \quad M e^{-\omega(t-t_0)} \left[\|x_0\| + \|\varphi(\cdot)\|_{L^2(0,t_0)}\right],$$

(ii) globally uniformly L^2-stable *if and only if*

$$\begin{aligned} \|x(t; t_0, x_0, \varphi(\cdot))\| &\leq M\left[\|x_0\| + \|\varphi(\cdot)\|_{L^2(0,t_0)}\right] \quad \forall\, t \geq t_0, \\ \|x(\cdot; t_0, x_0, \varphi(\cdot))\|_{L^2(t_0,\infty)} &\leq M\left[\|x_0\| + \|\varphi(\cdot)\|_{L^2(0,t_0)}\right], \end{aligned}$$

(iii) globally uniformly L^2-output stable *if and only if*

$$\begin{aligned} \|C(t)x(t; t_0, x_0, \varphi(\cdot))\| &\leq M\left[\|x_0\| + \|\varphi(\cdot)\|_{L^2(0,t_0)}\right] \quad \forall\, t \geq t_0, \\ \|C(\cdot)x(\cdot; t_0, x_0, \varphi(\cdot))\|_{L^2(t_0,\infty)} &\leq M\left[\|x_0\| + \|\varphi(\cdot)\|_{L^2(0,t_0)}\right], \end{aligned}$$

(iv) globally uniformly asymptotically stable *if and only if*

$$\begin{aligned} \|x(t; t_0, x_0, \varphi(\cdot))\| &\leq M\left[\|x_0\| + \|\varphi(\cdot)\|_{L^2(0,t_0)}\right] \quad \forall\, t \geq t_0, \\ \lim_{t\to\infty} x(t; t_0, x_0, \varphi(\cdot)) &= 0, \end{aligned}$$

holds for some $M, \omega > 0$, and all initial data $(t_0, x_0, \varphi(\cdot)) \in \mathbb{R}_{\geq 0} \times \mathbb{K}^n \times L^2(0, t_0; \mathbb{K}^p)$, respectively; "globally" refers to all $(t_0, x_0, \varphi(\cdot))$ and "uniformly" refers to the independence of M and ω from $(t_0, x_0, \varphi(\cdot))$.

Remark 4.1. *Let $B(\cdot) \equiv 0$ and $C(\cdot) \equiv I_p$ in (4.10); that means we consider the time-varying linear systems (4.1) only. Then all of the stability concepts in Definition 4.4 coincide; the equivalence between (i) and (iv) is well known (see, e.g., Rugh [16]), and the equivalence between (i) and (ii) is proved by Daleckiĭ and Kreĭn [2, Theorem III 6.2]; it even holds true for L^p-stability, where $p \in [1, \infty)$ is arbitrary.*

Before we prove relationships between the different stability concepts, we present the following simple but useful proposition and some formulae and inequalities for the initial value problem (4.10).

Proposition 4.1. *Consider a time-varying uniformly exponentially stable system (4.1) and $B(\cdot), C(\cdot)$ as in (4.4). Then the so-called perturbation operator*

$$\mathfrak{L}_{t_0} : L^2(t_0, \infty; \mathbb{K}^m) \to L^2(t_0, \infty; \mathbb{K}^p), \quad u(\cdot) \mapsto C(\cdot) \int_{t_0}^{\cdot} \Phi(\cdot, \tau) B(\tau) u(\tau)\, d\tau$$

is well defined for any $t_0 \geq 0$. Moreover $t_0 \mapsto \|\mathfrak{L}_{t_0}\|$ is nonincreasing on $\mathbb{R}_{\geq 0}$, and

$$\lim_{t \to \infty} \mathfrak{L}_{t_0}(u(\cdot))(t) = 0 \qquad \text{for every } u(\cdot) \in L^2(t_0, \infty; \mathbb{K}^m).$$

Proof: Since $B(\cdot)$ and $C(\cdot)$ are uniformly bounded and $\Phi(\cdot, \cdot)$ satisfies an inequality of the form (4.2), it follows from the general result of Lemma 4.1 in Section 4.5 that the convolution of two L^2-functions is itself an L^2-function and moreover tends to zero as t tends to infinity. Monotonicity of $t_0 \mapsto \|\mathfrak{L}_{t_0}\|$ is straightforward; see Hinrichsen et al. [10]. This completes the proof.

Applying Variation-of-Constants to (4.11) yields, for every $t \geq t_0$,

$$x(t) = \Phi(t, t_0) x_0 + \int_{t_0}^{t} \Phi(t, \tau) B(\tau) \mathfrak{D}\left([Cx]^{\varphi}\right)(\tau)\, d\tau, \tag{4.12}$$

and if the unperturbed system (4.1) is uniformly exponentially stable, then taking norms in (4.12), invoking (4.2), and applying the Cauchy–Schwarz inequality gives, for all $t \geq t_0$,

$$\begin{aligned} &\|x(t)\| \\ &\leq L e^{-\lambda(t-t_0)} \|x_0\| + L \|B(\cdot)\|_{L^\infty} \int_{t_0}^{t} e^{-\lambda(t-\tau)} \|\mathfrak{D}\left([Cx]^{\varphi}\right)(\tau)\|\, d\tau \qquad (4.13) \\ &\leq L e^{-\lambda(t-t_0)} \|x_0\| + \frac{L}{\sqrt{2\lambda}} \|B(\cdot)\|_{L^\infty} \|\mathfrak{D}([Cx]^{\varphi})(\cdot)\|_{L^2(0,t)}. \qquad (4.14) \end{aligned}$$

Since $\mathcal{D}(\cdot)$ is causal we have, for all $t \geq t_0$,

$$\|\mathcal{D}\left([Cx]^{\varphi}\right)(\cdot)\|^2_{L^2(t_0,t)} \leq \|\mathcal{D}\|^2_{\text{dyn}} \int_0^t \|[Cx]^{\varphi}(\tau)\|^2 \, d\tau \tag{4.15}$$

$$\leq \|\mathcal{D}\|^2_{\text{dyn}} \left[\|\varphi(\cdot)\|^2_{L^2(0,t_0)} + \|C(\cdot)x(\cdot)\|^2_{L^2(t_0,t)}\right] \tag{4.16}$$

and thus

$$\begin{aligned}\|x(t)\| \leq\ & Le^{-\lambda(t-t_0)}\|x_0\| \\ & + \frac{L}{\sqrt{2\lambda}}\|B(\cdot)\|_{L^\infty}\|\mathcal{D}\|_{\text{dyn}}\left[\|\varphi(\cdot)\|_{L^2(0,t_0)} + \|C(\cdot)x(\cdot)\|_{L^2(t_0,t)}\right].\end{aligned} \tag{4.17}$$

Taking L^2-norms in (4.13) and applying Cauchy–Schwarz and (4.16) yields, for all $t \geq t_0$,

$$\begin{aligned}\|x(\cdot)\|_{L^2(t_0,t)} \quad \leq \quad & \frac{L}{\sqrt{2\lambda}}\|x_0\| + \frac{L\|B(\cdot)\|_{L^\infty}\|\mathcal{D}\|_{\text{dyn}}}{\sqrt{2\lambda}}\left[\|\varphi(\cdot)\|_{L^2(0,t_0)}\right] \\ & + \|C(\cdot)x(\cdot)\|_{L^2(t_0,t)}.\end{aligned} \tag{4.18}$$

Multiplying (4.12) by $C(t)$ and using the perturbation operator yields, for all $t \geq t_0$,

$$C(t)x(t) = C(t)\Phi(t,t_0)x_0 + \mathcal{L}_{t_0}\left(\mathcal{D}([Cx]^{\varphi})\right)(t), \tag{4.19}$$

and taking norms and invoking (4.2), Cauchy–Schwarz, and (4.16) again gives, for all $t \geq t_0$,

$$\begin{aligned}\|C(\cdot)x(\cdot)\|_{L^2(t_0,t)} \quad \leq \quad & \|C(\cdot)\|_{L^\infty}\frac{L}{\sqrt{2\lambda}}\|x_0\| + \|\mathcal{D}\|_{\text{dyn}}\|\mathcal{L}_{t_0}\|\left[\|\varphi(\cdot)\|_{L^2(0,t_0)}\right. \\ & \left. + \|C(\cdot)x(\cdot)\|_{L^2(t_0,t)}\right].\end{aligned} \tag{4.20}$$

□

Now we are in a position to prove the following relationships between the stability concepts.

Proposition 4.2. *The stability concepts in Definition 4.4 are related as follows exponentially*

$$\Longrightarrow \quad L^2 \quad \Longleftrightarrow \quad L^2\text{–output} \quad \Longrightarrow \quad \text{asymptotically},$$

where each stability concept holds globally uniformly.

Proof: "exp. $\Rightarrow L^2$" is trivial. "$L^2 \Rightarrow L^2$-output" follows from the boundedness of $C(\cdot)$. "$L^2 \Leftarrow L^2$-output" is a consequence of (4.17) and (4.18). To see "L^2-output $\Rightarrow$ asymptotically", note that $\mathcal{D}\big([Cx]^{\varphi}(\cdot)\big)(\cdot) \in L^2(0,\infty;\mathbb{C}^m)$, and therefore Lemma 4.1 applied to (4.12) yields the result. □

The equivalence between L^2 and L^2-output stability has also been observed by Jacob [13] in a slightly less general context.

Remark 4.2. *It might be worth noticing that the exponentially stable system* (4.10) *can be viewed as an input-to-state stable system in the sense of Sontag. Set* $\omega(\cdot) \equiv C(\cdot)x(\cdot; t_0, x_0, \varphi(\cdot))$ *and consider*

$$\dot{x}(t) = A(t)x(t) + B(t)\,\mathcal{D}\left(\omega(\cdot)\right)(t). \tag{4.21}$$

Then by a little algebraic manipulation (4.18) *yields, for some* $L > 0$,

$$\int_{t_0}^{t} \|x(\tau)\|^2\, d\tau \le L\left[\|x_0\| + \|\varphi(\cdot)\|_{L^2(0,t_0;\mathbb{K}^p)}\right] + \int_{t_0}^{t} L\|\omega(\tau)\|^2\, d\tau,$$

and hence by Theorem 1 in Sontag [18] the system (4.21) *is input-to-state stable.*

In the following theorem we present sufficient conditions for the different stability concepts.

Theorem 4.2. *Let* $0 < \bar{\rho} < \|\mathcal{L}_0\|^{-1}$ *and consider the restricted perturbation class* $\mathcal{P}_{\text{dyn}}(\mathbb{K})$ *with* $\|\mathcal{D}\|_{\text{dyn}} < \bar{\rho}$. *Then the zero solution of the initial value problem* (4.10) *is:*

(i) globally uniformly L^2*-stable;*

(ii) globally uniformly exponentially stable, if $\dot{x}(t) = A(t)x(t), y(t) = C(t)x(t)$ *is uniformly observable; that is, there exist* $\beta_0, \beta_1, \sigma > 0$ *such that*

$$\beta_0 I_n \le \int_{t-\sigma}^{t} \Phi(s, t-\sigma)^*\, C(s)^*\, C(s)\Phi(s, t-\sigma)\, ds \le \beta_1 I_n \qquad \text{for all } t \ge \sigma;$$

(iii) globally uniformly exponentially stable, if there exists some $\varepsilon^* > 0$ *such that for all* $\varepsilon \in (0, \varepsilon^*)$, $t \ge 0$, *and* $\psi(\cdot) \in L^2_{\text{loc}}(\mathbb{R}_{\ge 0}; \mathbb{K}^p)$ *we have*

$$\|e^{\varepsilon\cdot}\mathcal{D}\left(e^{-\varepsilon\cdot}\psi(\cdot)\right)(\cdot)\|^2_{L^2(0,t;\mathbb{K}^m)} \le \|\mathcal{D}\|^2_{\text{dyn}}\|\psi(\cdot)\|^2_{L^2(0,t;\mathbb{K}^p)};$$

(iv) it is globally uniformly exponentially stable if the perturbation class is furthermore restricted to nonlinear time-varying perturbations $\mathcal{P}_{\text{nt}}(\mathbb{K})$.

In the assumption of Theorem 4.2 we could have assumed alternatively that $0 < \bar{\rho} < \|\mathcal{L}_{t'}\|^{-1}$, but then the stability concepts would only be considered for $t_0 \ge t'$.

Proof: (*i*): This statement follows from the well-known small gain theorem, but can readily be established in our context as follows. Since $\|\mathcal{D}\|_{\text{dyn}} <$

$\bar{\rho} < \|\mathcal{L}_0\|^{-1} \leq \|\mathcal{L}_{t_0}\|^{-1}$, there exists $\alpha \in (0,1)$ such that $\|\mathcal{D}\|_{\text{dyn}}\|\mathcal{L}_{t_0}\| < \alpha$ for all $\|\mathcal{D}\|_{\text{dyn}} < \bar{\rho}$ and all $t_0 \geq 0$. Now (4.20) yields, for every $t \geq t_0$,

$$(1-\alpha)\|C(\cdot)x(\cdot)\|_{L^2(t_0,t)} \leq \|C(\cdot)\|_{L^\infty(0,\infty)}\frac{L}{\sqrt{2\lambda}}\|x_0\| + \alpha\|\varphi(\cdot)\|_{L^2(0,t_0)}$$

and hence L^2-stability follows from Proposition 4.2.

(ii): We proceed in several steps.

Step 1: Let $\hat{\rho} \in (\bar{\rho}, \|\mathcal{L}_0\|^{-1})$.
By Lemma 8.1 in Ilchmann [12, p.145] there exists some $\eta > 0$ such that $\hat{\rho} < \|\mathcal{L}_0^\eta\|^{-1}$, where $\mathcal{L}_{t_0}^\eta$ denotes the input-output operator of the "shifted" system

$$\mathcal{L}_{t_0}^\eta : L^2(t_0,\infty;\mathbb{K}^m) \to L^2(t_0,\infty;\mathbb{K}^p),\ u(\cdot) \mapsto C(\cdot)\int_{t_0}^{\cdot} \Phi_{A+\eta I}(\cdot,\tau)B(\tau)u(\tau)\,d\tau,$$

and $\Phi_{A+\eta I}(\cdot,\cdot)$ denotes the transition matrix of the uniformly exponentially stable system $\dot{x}(t) = [A(t)+\eta I]x(t)$. Now by Theorem 5.11 in Hinrichsen et al. [10] there exists a unique stabilizing, positive-definite, continuously differentiable Hermitian solution

$$P(\cdot) = P(\cdot)^* \in L^\infty(0,\infty;\mathbb{C}^{n\times n})$$

of the differential Riccati equation

$$\begin{aligned}&\dot{P}(t) + [A(t)+\eta I_n]^*\ P(t) + P(t)\ [A(t)+\eta I_n]\\ &\qquad = \hat{\rho}^2 C(t)^*C(t) + P(t)B(t)B(t)^*P(t), \quad \text{for all } t \geq 0\end{aligned} \tag{4.22}$$

and stabilizing means that

$$\dot{x}(t) = [A(t) - B(t)B(t)^*P(t)]\ x(t)$$

is uniformly exponentially stable.

Step 2: The nonpositive definite matrix function $P(\cdot)$ serves as a Lyapunov function candidate

$$V(t,x)\ :=\ -x^*P(t)x$$

in the following. Differentiating V along the solution of the perturbed system (4.10) yields, for all $t \geq t_0$, by invoking (4.22) and omitting the argument t for simplicity,

$$\begin{aligned}&\frac{d}{dt}V(t,x(t))\\ &\quad = -\left[Ax + B\mathcal{D}\left([Cx]^\varphi\right)\right]^*\ Px - x^*P\left[Ax + B\mathcal{D}\left([Cx]^\varphi\right)\right] - x^*\dot{P}x\\ &\quad = -\hat{\rho}^2\|Cx\|^2 - \|B^*Px\|^2\\ &\qquad -\langle \mathcal{D}\left([Cx]^\varphi\right), B^*Px\rangle - \langle B^*Px, \mathcal{D}\left([Cx]^\varphi\right)\rangle + 2\eta x^*Px\\ &\quad = -2\eta V(t,x) - \hat{\rho}^2\|Cx\|^2 + \|\mathcal{D}\left([Cx]^\varphi\right)\|^2 - \|B^*Px + \mathcal{D}\left([Cx]^\varphi\right)\|^2\\ &\quad \leq -2\eta V(t,x) - \hat{\rho}^2\|Cx\|^2 + \|\mathcal{D}\left([Cx]^\varphi\right)\|^2.\end{aligned} \tag{4.23}$$

Step 3: Set $\zeta^2 := \hat{\rho}^2 - \|\mathcal{D}\|^2_{\text{dyn}} > 0$.
Then integration of $(d/d\tau)\, V(\tau, x(\tau))$ over $[t_0, t]$ and invoking (4.23) yields, for all $t \geq t_0$,

$$\begin{aligned} V(t,x(t)) \leq & \, V(t_0, x(t_0)) - 2\eta \int_{t_0}^{t} V(\tau, x(\tau))\, d\tau \\ & - \zeta^2 \int_{t_0}^{t} \|C(\tau)x(\tau)\|^2 \, d\tau + \|\mathcal{D}\|^2_{\text{dyn}} \|\varphi(\cdot)\|^2_{L^2(0,t_0)} \\ \leq & \, V(t_0, x(t_0)) - 2\eta \int_{t_0}^{t} V(\tau, x(\tau))\, d\tau + \|\mathcal{D}\|^2_{\text{dyn}} \|\varphi(\cdot)\|^2_{L^2(0,t_0)}. \end{aligned} \tag{4.24}$$

Step 4: Applying the Bellman–Gronwall Lemma to (4.24) gives, for all $t \geq t_0$,

$$V(t, x(t)) \leq \left[V(t_0, x(t_0)) + \|\mathcal{D}\|^2_{\text{dyn}} \|\varphi(\cdot)\|^2_{L^2(0,t_0)} \right] \, e^{-2\eta(t-t_0)}. \tag{4.25}$$

Note that the transition matrix of $\dot{x}(t) = [A(t) + \eta I]x(t)$ is given by $\Phi(t,s)e^{\eta(t-s)}$, and hence it is easy to see that $\dot{x}(t) = [A(t) + \eta I]x(t), y(t) = C(t)x(t)$ is uniformly observable, too. Therefore, by Theorem 6.11*(iv)* in Ilchmann [12, page 137] there exist some $p_1 > p_2 > 0$ such that

$$-p_1 I_n < P(\tau) < -p_2 I_n \quad \text{for all} \quad \tau \geq 0. \tag{4.26}$$

Now substitution of the bounds in (4.26) into (4.25) yields, for all $t \geq t_0$,

$$p_2 \|x(t)\|^2 \leq V(t, x(t)) \leq \left[p_1 \|x(t_0)\|^2 + \|\mathcal{D}\|^2_{\text{dyn}} \|\varphi(\cdot)\|^2_{L^2(0,t_0)} \right] \, e^{-2\eta(t-t_0)}.$$

This proves global uniform exponential stability of (4.10) and completes the proof of *(ii)*.

(iii): First proceed as in Steps 1 and 2 of the proof of *(ii)*. Note that a difference from *(ii)* is that we do not assume uniform observability, and so it is not necessarily true that $P(t)$ satisfies uniform bounds in (4.26). We have to proceed differently. By (4.23) we may conclude that

$$\frac{d}{dt} \left(e^{2\eta t} V(t, x(t)) \right) \leq e^{2\eta t} \left(-\hat{\rho}^2 \|C(t)x(t)\|^2 + \|\mathcal{D}\left([Cx]^{\varphi} \right)(t)\|^2 \right),$$

and hence by integration over $[t_0, t]$,

$$\begin{aligned} e^{2\eta t} V(t, x(t)) \leq & \; e^{2\eta t_0} V(t_0, x(t_0)) \\ & + \int_{t_0}^{t} e^{2\eta \tau} \left[-\hat{\rho}^2 \|C(\tau)x(\tau)\|^2 + \|\mathcal{D}\left([Cx]^{\varphi} \right)(\tau)\|^2 \right] d\tau. \end{aligned} \tag{4.27}$$

Since $\eta > 0$ may be chosen smaller than ε^*, an application of the assumption in *(iii)* yields, for all $t \geq t_0$,

$$\begin{aligned}
&\int_{t_0}^{t} e^{2\eta\tau} \|\mathcal{D}\left([Cx]^\varphi\right)(\tau)\|^2 \, d\tau \\
&\qquad \leq \|\mathcal{D}\|_{\text{dyn}}^2 \int_0^t \|e^{\eta\tau} [Cx]^\varphi(\tau)\|^2 \, d\tau \\
&\qquad \leq \|\mathcal{D}\|_{\text{dyn}}^2 \left[\|e^{\eta\cdot} \varphi(\cdot)\|_{L^2(0,t_0)}^2 + \|e^{\eta\cdot} C(\cdot)x(\cdot)\|_{L^2(t_0,t)}^2\right]
\end{aligned} \tag{4.28}$$

and this applied to (4.27) gives

$$\begin{aligned}
0 \;\leq\; & e^{2\eta t_0} V(t_0, x(t_0)) \\
& - \left[\hat{\rho}^2 - \|\mathcal{D}\|_{\text{dyn}}^2\right] \|e^{\eta\cdot} C(\cdot)x(\cdot)\|_{L^2(t_0,t)}^2 + \|\mathcal{D}\|_{\text{dyn}}^2 \|e^{\eta\cdot} \varphi(\cdot)\|_{L^2(0,t_0)}^2,
\end{aligned}$$

whence, since $\hat{\rho}^2 - \|\mathcal{D}\|_{\text{dyn}}^2 > \hat{\rho}^2 - \bar{\rho}^2$, for all $t \geq t_0$,

$$\begin{aligned}
&\|e^{\eta\cdot} C(\cdot)x(\cdot)\|_{L^2(t_0,t)}^2 \\
&\qquad \leq \frac{1}{\hat{\rho}^2 - \bar{\rho}^2} \left[e^{2\eta t_0} \|P(t_0)\| \, \|x_0\|^2 + \|\mathcal{D}\|_{\text{dyn}}^2 \|e^{\eta\cdot} \varphi(\cdot)\|_{L^2(0,t_0)}^2\right].
\end{aligned} \tag{4.29}$$

Without restriction of generality we may assume that $\eta \in (0, \lambda)$. Applying the Cauchy–Schwarz inequality to (4.13) and applying (4.28) and (4.29) we arrive at, for all $t \geq t_0$,

$$\begin{aligned}
& e^{\eta(t-t_0)} \|x(t)\| \\
& \leq L e^{-(\lambda-\eta)(t-t_0)} \|x_0\| + L\|B(\cdot)\|_{L^\infty} \int_{t_0}^{t} e^{-(\lambda-\eta)(t-\tau)} e^{\eta(\tau-t_0)} \mathcal{D}\left([Cx]^\varphi\right)(\tau) \, d\tau \\
& \leq L e^{-(\lambda-\eta)(t-t_0)} \|x_0\| \\
& \quad + L\|B(\cdot)\|_{L^\infty} \left[\int_{t_0}^{t} e^{-2(\lambda-\eta)(t-\tau)} d\tau\right]^{1/2} \|e^{\eta(\cdot-t_0)} \mathcal{D}\left([Cx]^\varphi\right)(\cdot)\|_{L^2(t_0,t)} \\
& \leq L e^{-(\lambda-\eta)(t-t_0)} \|x_0\| \\
& \quad + \tfrac{L\|B(\cdot)\|_{L^\infty}}{\sqrt{2(\lambda-\eta)}} e^{-\eta t_0} \|\mathcal{D}\|_{\text{dyn}} \left[\|e^{\eta\cdot} \varphi(\cdot)\|_{L^2(0,t_0)} + \|e^{\eta\cdot} C(\cdot)x(\cdot)\|_{L^2(t_0,t)}\right] \\
& \leq L e^{-(\lambda-\eta)(t-t_0)} \|x_0\| \\
& \quad + \tfrac{L\|B(\cdot)\|_{L^\infty}}{\sqrt{2(\lambda-\eta)}} e^{-\eta t_0} \|\mathcal{D}\|_{\text{dyn}} \Big\{ e^{\eta t_0} \|\varphi(\cdot)\|_{L^2(0,t_0)} \\
& \qquad\qquad + \tfrac{e^{\eta t_0}}{\sqrt{\hat{\rho}^2 - \bar{\rho}^2}} \left[\|P(t_0)\|^{1/2} \|x_0\| + \|\mathcal{D}\|_{\text{dyn}} \|\varphi(\cdot)\|_{L^2(0,t_0)}\right] \Big\}.
\end{aligned}$$

Finally multiplication by $e^{-\eta(t-t_0)}$ yields, for all $t \geq t_0$,

$$\|x(t)\| \leq \left[Le^{-\lambda(t-t_0)} + e^{-\eta(t-t_0)} \frac{L\|B(\cdot)\|_{L^\infty}}{\sqrt{2(\lambda-\eta)}} \frac{\|\mathcal{D}\|_{\text{dyn}}}{\sqrt{\hat{\rho}^2 - \bar{\rho}^2}} \|P(t_0)\|^{1/2} \right] \|x_0\|$$

$$+ e^{-\eta(t-t_0)} \frac{L\|B(\cdot)\|_{L^\infty}}{\sqrt{2(\lambda-\eta)}} \|\mathcal{D}\|_{\text{dyn}} \left[1 + \frac{\|\mathcal{D}\|_{\text{dyn}}}{\sqrt{\hat{\rho}^2 - \bar{\rho}^2}} \right] \|\varphi(\cdot)\|_{L^2(0,t_0)},$$

and since $P(\cdot)$ is uniformly bounded from above and $\|\mathcal{D}\|_{\text{dyn}} \leq \bar{\rho}$, the initial value problem (4.10) is globally uniformly exponentially stable. Thus *(iii)* is complete.

(iv): This result has already been proved in Hinrichsen et al. [10, Theorem 6.1]. This completes the proof. □

Remark 4.3. *Note that uniformity in Theorem* 4.2 *(i) to (iv) is due to the assumption* $\|\mathcal{D}\|_{\text{dyn}} < \bar{\rho} < \|\mathcal{L}_0\|^{-1}$. *If we assume* $\|\mathcal{D}\|_{\text{dyn}} < \|\mathcal{L}_{t_0}\|^{-1}$ *instead, then by inspection of the proof we readily see that the statements in Theorem 4.2 (i) to (iv) hold true apart from "uniform," that means* M *and* ω *in Definition* 4.4 *depend on the initial data.*

Related results as in Theorem 4.2 have been achieved for time-varying infinite-dimensional systems by Jacob et al. [14]. However, in their setup they do not use any Lyapunov functions but use operator-theoretic methods. The proof of Theorem 4.2*(ii)* with the use of a Lyapunov function as presented here is crucial for deriving properties of slowly time-varying systems in Section 4.

In the present chapter we are only interested in L^2-stability since we use Riccati equations later on, but the proof of Theorem 4.2 also goes through for L^p-stability, $p \geq 1$, which is defined analogously. As a consequence, the class of L^p-stable systems considered by Jacob [13] is also globally asymptotically stable.

4.4 Stability Radii of Slowly Time-Varying Systems

We are now in a position to formulate the concept of the stability radius for exponentially stable systems (4.1) with respect to the different perturbations (4.6) to (4.8). Loosely speaking, in the case of dynamic perturbations we are interested in a sharp number $r_{\mathbb{K}}$ such that all perturbations of the class $\mathcal{P}_{\text{dyn}}(\mathbb{K})$ preserve global L^2-stability of the origin of the perturbed system (4.11) as long as $\|\mathcal{D}(\cdot)\|_{\text{dyn}} < r_{\mathbb{K}}$ and there exists some $\mathcal{D}(\cdot) \in \mathcal{P}_{\text{dyn}}(\mathbb{K})$ with $\|\mathcal{D}(\cdot)\|_{\text{dyn}} = r_{\mathbb{K}}$ such that the origin of (4.10) is not globally L^2-stable. More formally we define the stability radii as follows.

Definition 4.5. *The (structured)* stability radius *of the exponentially stable system* (4.1) *with respect to dynamic perturbations of the class* $\mathcal{P}_{\mathrm{dyn}}(\mathbb{K})$ *and the structure* $(B(\cdot), C(\cdot))$ *as given in* (4.4) *is*

$$r_{\mathbb{K},dyn}(A(\cdot); B(\cdot), C(\cdot)) := \inf \left\{ \|\mathcal{D}(\cdot)\|_{\mathrm{dyn}} \;\middle|\; \begin{array}{l} \mathcal{D}(\cdot) \in \mathcal{P}_{\mathrm{dyn}}(\mathbb{K}), \textit{ the origin} \\ \textit{of } (4.10) \textit{ is not globally} \\ \textit{uniformly } L^2\textit{-stable} \end{array} \right\}.$$

The stability radii with respect to the perturbation classes $\mathbb{K}^{n\times n}$, $\mathcal{P}_{\mathrm{lt}}(\mathbb{K})$, *and* $\mathcal{P}_{\mathrm{nt}}(\mathbb{K})$ *are defined analogously and are denoted by* $r_{\mathbb{K}}(A(\cdot); B(\cdot), C(\cdot))$, $r_{\mathbb{K},lt}(A(\cdot); B(\cdot), C(\cdot))$, *and* $r_{\mathbb{K},nt}(A(\cdot); B(\cdot), C(\cdot))$, *respectively.*

The stability radius was introduced for time-invariant systems by Hinrichsen and Pritchard [6;7], for time-varying systems by Hinrichsen et al. [10], and for time-varying systems with multiperturbations by Hinrichsen and Pritchard [8].

Theorem 4.3. *Consider the exponentially stable system* (4.1) *and* $(B(\cdot), C(\cdot))$ *as given in* (4.4). *Then for* $\mathbb{K} = \mathbb{R}$ *or* $\mathbb{K} = \mathbb{C}$ *we have*

$$r_{\mathbb{K}} \geq r_{\mathbb{K},lt} \geq r_{\mathbb{K},nt} \geq r_{\mathbb{K},dyn}, \tag{4.30}$$

where for the sake of notation we omitted the arguments $(A(\cdot); B(\cdot), C(\cdot))$, *and*

$$r_{\mathbb{C}}(A(\cdot); B(\cdot), C(\cdot)) \geq r_{\mathbb{C},\mathrm{dyn}}(A(\cdot); B(\cdot), C(\cdot)) = \sup_{t_0 \geq 0} \|\mathcal{L}_{t_0}\|^{-1}, \tag{4.31}$$

and if $(A(\cdot), B(\cdot), C(\cdot)) = (A, B, C)$ *are constant matrices, then*

$$r_{\mathbb{C}}(A; B, C) = r_{\mathbb{C},\mathrm{dyn}}(A; B, C) = \min_{\omega \in \mathbb{R}} \|C(i\omega I_n - A)^{-1} B\|^{-1}. \tag{4.32}$$

The inequality in (4.31) *is strict.*

Proof: (4.30) is immediate from the inclusions (4.9). The inequality in (4.31) is a repetition of (4.30) and strictness is proved by a scalar periodic system in Example 4.4 in Hinrichsen et al. [10]. The left equality in (4.32) is proved in Theorem 3.12 by Hinrichsen and Pritchard [9], while the right equality is proved in Hinrichsen and Pritchard [7]. It remains to prove the equality in (4.31). To this end we need the nice result due to Jacob [13] where she shows that for every $\alpha > \sup_{t_0 \geq 0} \|\mathcal{L}_{t_0}\|^{-1}$ there exists a linear causal operator $\mathcal{D}(\cdot) \in \mathcal{P}_{\mathrm{dyn}}(\mathbb{K})$ such that $\|\mathcal{D}(\cdot)\|_{\mathrm{dyn}} < \alpha$ and the origin of (4.11) is not globally L^2-stable. We therefore have

$$r_{\mathbb{C},\mathrm{dyn}}(A(\cdot); B(\cdot), C(\cdot)) \geq \sup_{t_0 \geq 0} \|\mathcal{L}_{t_0}\|^{-1} \geq r_{\mathbb{C},\mathrm{dyn}}(A(\cdot); B(\cdot), C(\cdot)),$$

where the left inequality follows from Theorem 4.2(i). This completes the proof. □

It is well known (see, e.g., Rugh [16, Theorem 8.7] and the references therein) that a time-varying system $\dot{x}(t) = A(t)x(t)$ with bounded $A(\cdot)$ is uniformly exponentially stable if all real parts of the eigenvalues are uniformly bounded away from the imaginary axis (i.e., $\operatorname{Re}\lambda_i(A(\tau)) \leq -\mu$ for all $\tau \geq 0$ and some $\mu > 0$) and the time-variation of $A(\cdot)$ is sufficiently slow (i.e., $\|\dot{A}(\tau)\|$ is small uniformly in τ).

In the remainder of this section we prove an analogous statement for the stability radius. If the stability radii of the "frozen" systems $r_{\mathbb{C}}(A(\tau) + \eta I_n; B(\tau), C(\tau))$ are uniformly larger than $\rho > 0$ for some arbitrarily small $\eta > 0$, then the stability radius $r_{\mathbb{C},\text{dyn}}(A(\cdot); B(\cdot), C(\cdot))$ of the time-varying system is — provided the time-variation of the matrices $A(\cdot), B(\cdot), C(\cdot)$ is sufficiently small — at least "close" to ρ. More precisely, we have the following theorem.

Theorem 4.4. *Consider* (4.1) *and scaling matrices* (4.4) *with absolutely continuous* $(A(\cdot), B(\cdot), C(\cdot))$. *Assume that there exist* $\rho, \eta > 0$, *such that the stability radii of the "shifted frozen" systems* (4.5) *satisfy*

$$\rho \leq r_{\mathbb{C}}(A(\tau) + \eta I_n; B(\tau), C(\tau)) \qquad \text{for all} \quad \tau \geq 0. \tag{4.33}$$

Suppose that $(A(\tau), C(\tau))$ *is an observable matrix pair for every* $\tau \geq 0$. *Then there exists a* $\delta > 0$ *such that*

$$\|\dot{A}(\tau)\| + \|\dot{B}(\tau)\| + \|\dot{C}(\tau)\| \ < \ \delta \qquad \text{for almost all} \quad \tau \geq 0 \tag{4.34}$$

yields

$$\rho \ \leq \ r_{\mathbb{C},\text{dyn}}(A(\cdot); B(\cdot), C(\cdot)). \tag{4.35}$$

Moreover, the zero solution of the initial value problem (4.10) *is globally uniformly exponentially stable, if we restrict the class of dynamical perturbations* $\mathcal{D}(\cdot) \in \mathcal{P}_{\text{dyn}}(\mathbb{K})$ *to* $\|\mathcal{D}(\cdot)\|_{\text{dyn}} < \rho$.

Certainly, the assumptions (4.33) and (4.34) need only be satisfied "at infinity"; that means $\tau \geq 0$ could be replaced by $\tau \geq \tau_0$ for some $\tau_0 \geq 0$.

Before proving Theorem 4.4, we give some intuition on the "frozen" systems. First note that for any bounded set $\mathcal{K} \subset \mathbb{K}^{n \times n}$ we have

$$\inf_{A \in \mathcal{K}} r_{\mathbb{C}}(A) > 0 \quad \Longleftrightarrow \quad \sup_{A \in \mathcal{K}} \operatorname{Re}\lambda_{\max}(A) < 0. \tag{4.36}$$

Sufficiency follows immediately by contradiction. Necessity also follows by contradiction. Assume there exist sequences $\{A_\tau\}_{\tau \in \mathbb{N}} \subset \mathcal{K}$ and $\{D_\tau\}_{\tau \in \mathbb{N}} \subset \mathbb{K}^{n \times n}$ such that

$$\lim_{\tau \to \infty} \operatorname{Re}\lambda_{\max}(A_\tau + D_\tau) = 0 \qquad \text{and} \qquad \lim_{\tau \to \infty} D_\tau = 0.$$

Then boundedness of $\mathcal{K}$ yields $\lim_{\tau\to\infty} \operatorname{Re}\lambda_{\max}(A_\tau) = 0$, and this is a contradiction.

Note that necessity does not hold true in general if $\mathcal{K}$ is bounded. To this end consider

$$A_\tau \;:=\; -\begin{bmatrix} \tau & \tau^3 \\ 0 & \tau \end{bmatrix}, \qquad D_\tau \;:=\; -\frac{1}{\tau}\begin{bmatrix} -1 & 0 \\ 1 & -1 \end{bmatrix},$$

which satisfy

$$\operatorname{spec}(A_\tau) = \{-\tau\}, \qquad \operatorname{spec}(A_\tau + D_\tau) = \{-2\tau + 1/\tau,\ 1/\tau\}.$$

This shows that the spectrum of $A(\tau)$ might be bounded away from the imaginary axis (in the above example it even tends to $-\infty$), but the stability radius tends to 0.

The following example shows that even if the stability radii of the frozen systems are bounded away from 0, the time-varying systems might not be exponentially stable. Set

$$a(\tau) = \frac{-1}{1+\tau}, \qquad b(\tau) = \frac{1}{1+\tau}, \qquad c(\tau) = 1 \qquad \text{for all } \tau \geq 0.$$

Then a straightforward calculation gives

$$\frac{1}{1+\tau} \;=\; -\lambda_{\max}(a(\tau)) \;\leq\; r_{\mathbb{C}}(a(\tau); b(\tau), 1) \;=\; 1 \qquad \text{for all } \tau \geq 0,$$

and $r_{\mathbb{C}}(a(\cdot)) = 0$ since the solution of $\dot{x}(t) = -(1+t)^{-1}x(t)$ is $x(t) = (1+t)^{-1})$, which is not exponentially decaying.

One might think (as we erroneously did in an earlier version[2]) that (4.33) can be replaced by $\rho \leq r_{\mathbb{C}}(A(\tau); B(\tau), C(\tau))$ for all $\tau \geq 0$, since $\tau \mapsto r_{\mathbb{C}}(A(\tau); B(\tau), C(\tau))$ is continuous and $A(\cdot), B(\cdot), C(\cdot)$ are bounded. However, as the above example shows, the map $\tau \mapsto r_{\mathbb{C}}(A(\tau); B(\tau), C(\tau))$ is not *uniformly* continuous and therefore $\eta > 0$ as in (4.33) does not necessarily exist.

Proof: **(of Theorem 4.4)** We proceed in several steps.

Step 1. By Lemma 4.2 in Section 5, there exists an absolutely continuous map $\tau \mapsto P(\tau)$, where $P(\tau) = P(\tau)^* \leq 0$ is the unique Hermitian solution of the algebraic Riccati equation

$$\begin{aligned} &[A(\tau) + \eta I_n]^*\ P(\tau) + P(\tau)\ [A(\tau) + \eta I_n] \\ &\qquad - \rho^2 C(\tau)^* C(\tau) - P(\tau)B(\tau)B(\tau)^* P(\tau) \;=\; 0 \end{aligned} \qquad (4.37)$$

[2]We are indebted to Fabian Wirth (Bremen) for pointing out the example and error to us.

such that the "frozen" closed-loop systems

$$\dot{x}(t) = \hat{A}(\tau)x(t), \quad \text{where } \hat{\mathrm{A}}(\tau) := \mathrm{A}(\tau) + \eta \mathrm{I_n} - \mathrm{B}(\tau)\mathrm{B}(\tau)^*\mathrm{P}(\tau), \tag{4.38}$$

are uniformly exponentially stable for all $\tau \geq 0$.

Step 2. We show that $\dot{P}(\tau)$ is the unique solution of the Lyapunov equation

$$\hat{A}(\tau)^* \, \dot{P}(\tau) \; + \; \dot{P}(\tau)\hat{A}(\tau) \; = \; R(\tau) \qquad \text{for almost all } \tau \geq 0, \tag{4.39}$$

where

$$R(\tau) := -\Big[\dot{A}(\tau) - B(\tau)\dot{B}(\tau)^* P(\tau)\Big]^* P(\tau) - P(\tau)\Big[\dot{A}(\tau) - B(\tau)\dot{B}(\tau)^* P(\tau)\Big]$$
$$+ \rho^2 \left[\dot{C}(\tau)^* C(\tau) + C(\tau)^* \dot{C}(\tau)\right]. \tag{4.40}$$

Differentiability of $\tau \mapsto P(\tau)$ for almost all $\tau \geq 0$ follows from Lemma 4.2, and a straightforward differentiation of (4.37) yields (4.39).

Step 3. We show that there exists some $r > 0$, independent of $\tau \geq 0$, such that

$$\|R(\tau)\| \; \leq \; r\delta \qquad \text{for almost all } \tau \geq 0. \tag{4.41}$$

$P(\cdot)$ is uniformly bounded by Lemma 4.2. Since $A(\cdot), B(\cdot), C(\cdot)$ are uniformly bounded in τ, the statement readily follows from (4.34) and (4.40).

Step 4. We show that there exists some $p' > 0$, such that

$$\|\dot{P}(\tau)\| \; \leq \; p' \, \delta \qquad \text{for almost all } \tau \geq 0. \tag{4.42}$$

Note that $\hat{A}(\cdot)$ is uniformly bounded and hence, by continuity of $A \mapsto r_{\mathbb{C}}(A;0,0)$, there exists some $\omega > 0$ such that

$$0 < \omega \; := \; \min_{\tau \geq 0} r_{\mathbb{C}}(\hat{A}(\tau);0,0) \leq -\min_{\tau \geq 0} \max_{i=1,\ldots,n} \operatorname{Re} \lambda_i(\hat{\mathrm{A}}(\tau)).$$

Now we are in a position to apply Proposition 3 of Coppel [1, p.4] to conclude that there exists some $M > 0$, independent of τ, such that, for all $\tau \geq 0$,

$$\|e^{\hat{A}(\tau)t}\| \; \leq \; M \, e^{-(\omega/2)t} \qquad \text{for all } t \geq 0. \tag{4.43}$$

Since the solution of (4.39) satisfies

$$\dot{P}(\tau) \; = \; -\int_0^\infty e^{\hat{A}(\tau)^* s} R(\tau) e^{\hat{A}(\tau)s} \, ds \quad \text{for almost all } \tau \geq 0, \tag{4.44}$$

an application of (4.41) and (4.43) to (4.44) yields (4.42).

Step 5. Since $A(\cdot) + \eta I_n$, $B(\cdot)$, and $C(\cdot)$ are uniformly bounded and $(A(\tau) + \eta I_n, C(\tau))$ is observable for every $\tau \geq 0$, there exist by Lemma 4.2 $p_1 > p_2 > 0$ such that

$$0 < p_2 I_n \leq -P(\tau) \leq p_1 I_n \qquad \text{for all} \quad \tau \geq 0. \tag{4.45}$$

We show that the derivative of the time-varying positive-definite Lyapunov function candidate

$$V(t,x) \;:=\; -x^* P(t) x$$

along the solution of the time-varying system (4.10) satisfies, for any $\mathcal{D}(\cdot) \in \mathcal{P}_{\mathrm{dyn}}(\mathbb{K})$ with $\|\mathcal{D}(\cdot)\|_{\mathrm{dyn}} < \rho$ and for almost all $t \geq 0$,

$$\frac{d}{dt} V(t, x(t)) \leq -\eta\, V(t, x(t)) - \rho^2 \|C(t)x(t)\|^2 + \|\mathcal{D}\left([Cx]^{\varphi}\right)(t)\|^2. \tag{4.46}$$

Set δ in (4.34) to $\delta = \eta p_2/p'$. Now differentiation of V along (4.10) and invoking (4.37) and (4.42) yields, where we omit the argument t for simplicity,

$$\begin{aligned}
\tfrac{d}{dt} V(t,x(t)) &= -[Ax + B\mathcal{D}\left([Cx]^{\varphi}\right)]^* \, Px - x^* P[Ax + B\mathcal{D}\left([Cx]^{\varphi}\right)] - x^* \dot{P} x \\
&= 2\eta x^* P x - x^* \dot{P} x - \rho^2 \|Cx\|^2 + \|\mathcal{D}\left([Cx]^{\varphi}\right)\|^2 \\
&\quad - \|B^* P x + \mathcal{D}\left([Cx]^{\varphi}\right)\|^2 \\
&\leq \; -\left[2\eta - \tfrac{p'}{p_2}\delta\right] x^*(-P)x - \rho^2\|Cx\|^2 + \|\mathcal{D}\left([Cx]^{\varphi}\right)\|^2.
\end{aligned}$$

This proves (4.46).

Step 6. (4.46) is of the type (4.23). Now proceeding exactly as in Steps 3 and 4 of the proof of Theorem 4.2 yields global uniform exponential stability of (4.10). As a consequence, we have (4.35) and the proof is complete. □

4.5 Two Lemmas

Although the following lemma on the asymptotic behavior of the convolution of L^p-functions is a consequence of the much more general result of Gripenberg et al. [5, Theorem 2.2, p.39], we present a simple proof for completeness.

Lemma 4.1. *If $f(\cdot) \in L^p(\mathbb{R}_{\geq 0}, \mathbb{R})$ and $g(\cdot) \in L^q(\mathbb{R}_{\geq 0}, \mathbb{R})$ for $p, q \in (1, \infty)$ with $1/p + 1/q = 1$, then*

$$\lim_{t \to \infty} \int_0^t f(t-\tau) g(\tau)\, d\tau \;=\; 0\,.$$

Proof: [3] An application of Hölder's inequality yields

$$\begin{aligned}\int_0^t f(t-\tau)g(\tau)\,d\tau &= \int_0^{t/2} f(t-\tau)g(\tau)\,d\tau + \int_{t/2}^t f(t-\tau)g(\tau)\,d\tau \\ &\leq \|f\|_{L^p(t/2,t)}\|g\|_{L^q(0,t/2)} + \|f\|_{L^p(0,t/2)}\|g\|_{L^q(t/2,t)} \\ &\leq \|f\|_{L^p(t/2,\infty)}\|g\|_{L^q(0,\infty)} + \|f\|_{L^p(0,\infty)}\|g\|_{L^q(t/2,\infty)}\,.\end{aligned}$$

This proves the claim, since by assumption the right-hand side of the last inequality converges to 0 as $t \to \infty$. □

The following lemma is on the existence of Hermitian solutions $P = P^*$ of the *nonstandard algebraic Riccati equation*

$$A^*P + PA - \rho^2 C^*C - PBB^*P = 0. \tag{4.47}$$

The result is known, but we are not aware of a comprehensive presentation, and since it is crucial in our analysis, we present a proof.

Lemma 4.2. *Let $\rho > 0$ be fixed, $\mathbb{K} = \mathbb{R}$ or $\mathbb{K} = \mathbb{C}$, and define*

$$\Sigma = \left\{(A, B, C) \in \mathbb{K}^{n\times(n+m+p)} \;\middle|\; \rho < r_{\mathbb{C}}(A; B, C)\right\}.$$

Then the map

$$\begin{array}{rccl}\varphi : & \Sigma & \longrightarrow & \mathcal{H} = \{P \in \mathbb{K}^{n\times n} \mid P = P^*\} \\ & (A, B, C) & \mapsto & P, \quad \text{where } P \text{ solves (4.47) and } \sigma(A - BB^*P) \subset \mathbb{C}_-\end{array}$$

is well defined; that is, for every (A, B, C) with $\rho < r_{\mathbb{R}}(A; B, C)$ there exists a solution of the algebraic Riccati equation (4.47) which is unique among all exponentially stabilizing solutions, and

(i) *φ is analytic if $\mathbb{K} = \mathbb{C}$, respectively, real analytic if $\mathbb{K} = \mathbb{R}$,*

(ii) *$P = \varphi(A, B, C)$ satisfies*

$$P = -\int_0^\infty e^{A^*s}\left[\rho^2 C^*C + PBB^*P\right] e^{As}\,ds \tag{4.48}$$

and is hence nonpositive. If (A, C) is observable, then P is negative definite.

Proof: Hinrichsen and Pritchard [7] proved the existence and uniqueness of $P = \varphi(A, B, C)$. Their proof is only for $\mathbb{K} = \mathbb{C}$, but carries over if $\mathbb{K} = \mathbb{R}$.

---

[3] We are indebted to Hartmut Logemann (Bath) for pointing out to us this simple proof.

We prove analyticity of φ by using the Implicit Function Theorem; this idea is due to Delchamps [3] who used it for the standard Riccati equation; that is, $-\rho^2 C^* C$ in (4.47) is replaced by $+C^* C$.

The map

$$\begin{array}{rclcl} \psi & : & \mathbb{K}^{n\times(n+m+p)} \times \mathcal{H} & \longrightarrow & \mathcal{H} \\ & & (A,B,C,P) & \mapsto & A^*P + PA - \rho^2 C^*C - PBB^*P \end{array}$$

is certainly analytic and its differential with respect to P is given by

$$\psi_P(A,B,C,P) \,:\, H \mapsto [A - BB^*P]^* H + H[A - BB^*P].$$

Let $(A,B,C) \in \Sigma$ and $P = \varphi(A,B,C)$. Then $\psi(A,B,C,P) = 0$ and since $\sigma(A - BB^*P) \subset \mathbb{C}_-$ it follows that $\psi_P(A,B,C,P)$ is regular. Therefore, *(i)* is a consequence of the Implicit Function Theorem.

It remains to show *(ii)*. (4.48) follows from (4.47) and thus

$$P \leq -\rho^2 \int_0^\infty e^{A^* s} C^* C e^{As}\, ds \leq 0. \tag{4.49}$$

If (A,C) is observable, then an application of the observability Gramian to (4.49) yields that P is negative-definite. □

4.6 Conclusions

This chapter deals with the problem of estimating the stability radius of time-varying linear systems with respect to L^2 finite gain perturbations. Sufficient conditions for uniform exponential stability are given. A conservative but workable bound is presented in terms of the stability radii of the frozen systems, provided the time-variation is sufficiently slow. Thus numerical computations of the stability radii of time-invariant systems (see Hinrichsen et al. [11], Filbir [4], and Sreedhar et al. [19]) might be used to compute a bound of the stability radius of time-varying systems.

The result presented here has implications for the study of gain scheduled control systems. It is generally accepted that for sufficiently slowly scheduled controllers the performance of the frozen system is a good indicator of what to expect in the gain scheduled case. See, for example, Shamma and Athans [17] and Lawrence and Rugh [15]. Quantifying "slow" and the extent to which this folklore is acceptable will be investigated in a separate paper.

References

[1] W. A. Coppel, *Dichotomies in Stability Theory*, Lecture Notes in Mathematics 629, Springer-Verlag, Berlin, 1978.

[2] J. L. Daleckiĭ and M. G. Kreĭn, *Stability of Solutions of Differential Equations in Banach Spaces*, AMS, Providence, RI, 1974.

[3] D. F. Delchamps, Analytic feedback control and the algebraic Riccati equation, *IEEE Trans. Aut. Control* 29:1031–1033, 1984.

[4] F.-D. Filbir, Computation of the structured stability radius via matrix sign function, *Syst. Control Letters* 22:341–349, 1994.

[5] G. Gripenberg, S.-O. Londen, and O. Staffans, *Volterra Integral and Functional Equations*, Cambridge University Press, Cambridge, 1990.

[6] D. Hinrichsen and A.J. Pritchard, Stability radii of linear systems, *Syst. Control Letters* 7:1–10, 1986.

[7] D. Hinrichsen and A.J. Pritchard, Stability radius for structured perturbations and the algebraic Riccati equation, *Syst. Control Letters* 8:105–113, 1986.

[8] D. Hinrichsen and A.J. Pritchard, Robust stability of linear time-varying systems with respect to multi-perturbations, *Proc. European Control Conference*, Grenoble, 1366–1371, 1991.

[9] D. Hinrichsen and A.J. Pritchard, Destabilization by output feedback, *Diff. Integral Equations* 5:357–386, 1992.

[10] D. Hinrichsen, A. Ilchmann, and A.J. Pritchard, Robustness of stability of time-varying linear systems, *J. Diff. Equations* 82:219–250, 1989.

[11] D. Hinrichsen, B. Kelb, and A.J. Pritchard, An algorithm for the computation of the complex stability radius, *Automatica* 25:771–775, 1989.

[12] A. Ilchmann, *Contributions to Time-Varying Linear Systems*, Verlag an der Lottbek, Hamburg, 1989.

[13] B. Jacob, A formula for the stability radius of time-varying systems, *J. Diff. Equations* 142:167–187, 1998.

[14] B. Jacob, V. Dragan, and A.J. Pritchard, Infinite dimensional time-varying systems with nonlinear output feedback, *Integral Equations Oper. Th.* 22:440–462, 1995.

[15] D.A. Lawrence and W.J. Rugh, Gain scheduling dynamic linear controllers for a nonlinear plant, *IEEE Trans. Auto. Control* 31:381–390, 1995.

[16] W.J. Rugh, *Linear System Theory*, 2nd ed., Prentice-Hall, Upper Saddle River, NJ, 1996.

[17] J.S. Shamma and M. Athans, Analysis of gain scheduled control for nonlinear plants, *IEEE Trans. Auto. Control* 35:898–907, 1990.

[18] E. Sontag, Comments on integral variants of ISS, *Syst. Control Letters* 34:93–100, 1999.

[19] J. Sreedhar, P. Van Dooren, and A. L. Tits, A level-set idea to compute the real Hurwitz-stability radius, *Proc. 34th Conference on Dec. and Control*, New Orleans, LA, 126–127, 1995.

Chapter 5

An Invariance Radius for Nonlinear Systems

Fritz Colonius,[1] **Wolfgang Kliemann**

ABSTRACT The stability radius of linear differential equations gives a measure for the robustness of stability with respect to (real, complex, or dynamic) perturbations. In this chapter a generalization for asymptotically stable equilibria of nonlinear systems is proposed and analyzed. It specifies the maximal perturbation range, for which the control set surrounding the equilibrium retains its invariance. It is shown that this value is attained when the invariant control set touches the boundary of its invariant domain of attraction. Then it merges with another (variant) control set and itself becomes variant.

5.1 Introduction

The notion of a stability radius for linear systems which was introduced by D. Hinrichsen and A.J. Pritchard in [5], [6] has proved to be a very efficient and fruitful tool for the analysis of perturbed linear systems. In this chapter we propose a certain nonlinear analogue of this notion.

Consider a nominal nonlinear system in $\mathbb{R}^d$ given by a smooth vector field f_0,

$$\dot{x} = f_0(x)$$

with an asymptotically stable equilibrium x_0; hence $f_0(x_0) = 0$. We are interested in the stability behavior of this equilibrium under persistent, time-varying, and bounded perturbations. Thus we consider a system of the form

$$\dot{x} = f_0(x) + \sum_{i=1}^{m} u_i(t) f_i(x) =: f(x, u(t)), \qquad u(t) = (u_i(t)) \in U,$$

with smooth vector fields $f_1, \ldots, f_m$; the perturbations $u(t)$ take values in a given convex and compact subset U of $\mathbb{R}^m$ with $0 \in \operatorname{int} U$. Also in

[1]F. Colonius was supported by the Nonlinear Control Network, TMR-Project.

this nonlinear situation it is convenient to introduce a parameter $\rho \geq 0$, which indicates the size of the perturbations and to replace the pointwise constraint $u(t) \in U$ by

$$u(t) \in U^{\rho} := \rho \cdot U.$$

In particular, for $\rho = 0$, one recovers the nominal unperturbed system. We are interested in the supremal ρ-value, such that the stability behavior of the nominal system is preserved. In general, however, it is not obvious how to make this precise.

If the equilibrium is singular (i.e., it remains fixed under all perturbations) $f(x_0, u) = 0$ for all $u \in U$, then the stability radius is simply the supremal ρ-value, such that x_0 is asymptotically stable for all $u \in \mathcal{U}^{\rho} := \{u : \mathbb{R} \longrightarrow \mathbb{R}^m,\ u(t) \in \rho \cdot U \text{ for almost all } t\}$. This can be compared to the stability radius of the linearized system

$$\dot{y} = \frac{\partial}{\partial x} f(x_0, u(t))y, \qquad u(t) \in \mathcal{U}^{\rho}.$$

Here Lyapunov or Bohl exponents can be used; see Chapter 3 by A. Paice and F. Wirth in this book.

If the equilibrium does not remain fixed under the disturbances, we have to specify the relevant notion of stability. Considering the perturbation as a control, one finds that for positive $\rho > 0$ there is a control set $D(\rho)$ containing x_0; for ρ small enough, this control set is invariant. We define the corresponding invariance radius as the maximal ρ-value for which the control set $D(\rho)$ containing x_0 remains invariant. As our analysis shows, this ρ-value, where the control set becomes variant, is connected with a merging of control sets and hence with a discontinuity (in the Hausdorff metric) of the map $\rho \longmapsto D(\rho)$. Furthermore, we briefly discuss how the invariance radius is connected with the occurrence of hyperbolic equilibria on the boundary of the invariant control set.

Invariant control sets determine the generic stability behavior of the system (see [3]). They are also of interest in the analysis of Markovian systems, where they are associated with the supports of the invariant measures; see, for example, [2].

In Section 5.2 we give some background material on the behavior of control systems. In Section 5.3, we first present a motivating one-dimensional example. Then the invariance radius is defined and its properties are described. Furthermore, a two-dimensional model for a continuous flow stirred tank reactor is discussed.

5.2 Background on Invariant Control Sets and Chain Control Sets

In this section, we cite the pertinent results from [3] on the global controllability structure of control systems.

We consider the following control system

$$\begin{array}{c} \dot{x}(t) = f_0(x(t)) + \sum_{i=1}^m u_i(t) f_i(x(t)),\ t \in \mathbb{R}, \\ u \in \mathcal{U}^\rho = \{u : \mathbb{R} \to \mathbb{R}^m,\ u(t) \in U^\rho := \rho \cdot U \text{ for all } t \in \mathbb{R}, \text{ measurable}\}, \end{array} \tag{5.1}$$

where for $i = 0, \ldots, m$ the vector fields f_i on $\mathbb{R}^d$ are C^∞. We assume that the control range $U \subset \mathbb{R}^m$ is nonvoid, convex, and compact with $0 \in \operatorname{int} U$, that $\rho \geq 0$, and that for every initial state $x \in \mathbb{R}^d$ and every control function $u \in \mathcal{U}^\rho$ there exists a unique trajectory $\varphi(t, x, u)$, $t \in \mathbb{R}$, satisfying $\varphi(0, x, u) = x$. If the dependence on ρ does not play a role, we simply write $\mathcal{U}$. Throughout we assume that the Lie algebra generated by the vector fields $f_0, f_1, \ldots, f_m$ satisfies for every $x \in \mathbb{R}^d$ the following rank condition

$$\dim \mathcal{LA}\{f_0, f_1, \ldots, f_m\}(x) = d.$$

In particular, this implies local accessibility for all $\rho > 0$; that is, for all $x \in \mathbb{R}^d$ and all $T > 0$ one has

$$\operatorname{int} \mathcal{O}^{\rho,+}_{\leq T}(x) \neq \emptyset \quad \text{and} \quad \operatorname{int} \mathcal{O}^{\rho,-}_{\leq T}(x) \neq \emptyset,$$

where $\mathcal{O}^{\rho,+}_{\leq T}(x) = \{y \in \mathbb{R}^d,\ y = \varphi(t, x, u) \text{ with } 0 \leq t \leq T \text{ and } u \in \mathcal{U}^\rho\}$ and $\mathcal{O}^{\rho,-}_{\leq T}(x) = \{y \in \mathbb{R}^d,\ x = \varphi(t, y, u) \text{ with } 0 \leq t \leq T \text{ and } u \in \mathcal{U}^\rho\}$ are the positive and negative reachable sets (or orbits) from x. We discuss the behavior of these control systems in relation to the behavior of the unperturbed system

$$\dot{x} = f_0(x). \tag{5.2}$$

Fix a compact positively invariant set $K \subset M$ for the unperturbed system and suppose that it has an asymptotically stable equilibrium $x_0 \in \operatorname{int} K$.

In order to describe the global controllability behavior of the control system (5.1), we assume that K is positively invariant for all controls $u \in \mathcal{U}^\rho$, $\rho \geq 0$, and recall the following concepts and results.

Definition 5.1. *A subset D with nonvoid interior of the state space $\mathbb{R}^d$ is* a control set *if it is a maximal subset with the property that for all $x \in D$ the inclusion*

$$D \subset \operatorname{cl}\{\varphi(t, x, u),\ u \in \mathcal{U},\ \textit{and } t \geq 0\}$$

holds. A control set C is an invariant control set if

$$\operatorname{cl} C = \operatorname{cl}\{\varphi(t,x,u),\ u \in \mathcal{U},\ \textit{and}\ t \geq 0\}.$$

If the dependence of D on ρ plays a role, we indicate this by the argument ρ in $D(\rho)$. The same remark applies also to all other notions introduced below. The local accessibility assumption guarantees that the invariant control sets are the closed control sets and that they have nonvoid interior. Recall that every point in K can be steered into an invariant control set.

Definition 5.2. *The* domain of attraction *of a control set D is*

$$\mathbf{A}(D) = \left\{x \in \mathbb{R}^d,\ \operatorname{cl} \mathcal{O}^+(x) \cap D \neq \emptyset\right\},$$

and for an invariant control set C the invariant domain of attraction *is*

$$\mathbf{A}^{inv}(C) = \left\{x \in \mathbb{R}^d,\ \begin{array}{c} \textit{if } \operatorname{cl} \mathcal{O}^+(x) \cap C' \neq \emptyset \textit{ for an invariant} \\ \textit{control set } C',\ \textit{then } C' = C \end{array}\right\}.$$

Note that the domain of attraction of a control set D is open with $D \subset \mathbf{A}(D)$. For an invariant control set C the invariant domain of attraction is the largest invariant subset in the domain of attraction, and clearly, $C \subset \mathbf{A}^{inv}(C)$. However, $\mathbf{A}^{inv}(C)$ is not necessarily open, and trajectories starting in $\mathbf{A}^{inv}(C)$ do not necessarily converge to C. In particular, $\mathbf{A}^{inv}(C)$ may contain variant control sets.

Invariant control sets may be considered as a generalization of asymptotically stable equilibria for the perturbed system. We also need the following weaker concept, chain control sets, which allows for arbitrarily small jumps in the trajectories.

For ε, $T > 0$, a controlled (ε, T)-chain from x to y is given by $n \in \mathbb{N}$, times $T_0, \ldots, T_{n-1} \geq T$, points $x_0 = x, \ldots, x_{n-1}, x_n = y$, and controls $u_0, \ldots, u_{n-1} \in \mathcal{U}$ such that

$$d(\varphi(T_i, x_i, u_i), x_{i+1}) < \varepsilon \qquad \text{for all } i = 0, \ldots, n-1.$$

Definition 5.3. *A* chain control set *$E \subset \mathbb{R}^d$ is a maximal subset such that (i) for all $x, y \in E$ and all ε, $T > 0$ there is a controlled (ε, T)-chain from x to y; and (ii) for all $x \in E$ there is $u \in \mathcal{U}$ with $\varphi(t, x, u) \in E$ for all $t \in \mathbb{R}$.*

Chain control sets are closed and every control set is contained in a chain control set. Furthermore, if two control sets D_1 and D_2 satisfy $\operatorname{cl} D_1 \cap \operatorname{cl} D_2 \neq \emptyset$, then there is a chain control set E with $D_1 \cup D_2 \subset E$.

The following inner pair condition guarantees that an equilibrium x_0 of the nominal system is in the interior of a control set for $\rho > 0$.

$$\text{For every } \rho > 0 \text{ one has } x_0 \in \operatorname{int} \mathcal{O}^{\rho,+}(x_0). \tag{5.3}$$

In this case we call the pair $(0, x_0) \in \mathcal{U}^\rho \times \mathbb{R}^d$ an inner pair. Consider the following maps defined on $[0, \infty)$ with values in the set of compact subsets of $\mathbb{R}^d$,

$$\rho \mapsto \operatorname{cl} D(\rho) \qquad \text{and} \qquad \rho \mapsto E(\rho), \tag{5.4}$$

where $D(\rho)$ and $E(\rho)$ are the control sets and chain control sets containing x_0. Here the existence of $E(\rho)$ is clear; the existence of the control sets $D(\rho)$ follows from the next theorem (see [3, Chapter 4]).

Theorem 5.1. *Assume that the inner pair condition* (5.3) *is satisfied. Then for every* $\rho > 0$ *there is a control set* $D(\rho)$ *with* $x_0 \in \operatorname{int} D(\rho)$ *and*

$$\{x_0\} = \bigcap_{\rho>0} D(\rho) = \bigcap_{\rho>0} E(\rho).$$

It is clear that $\operatorname{cl} D(\rho) \subset E(\rho)$. Under a strengthened inner pair condition, the relation between the control sets and chain control sets above is much closer. Let

$$\mathcal{D}(\rho) := \operatorname{cl}\{(u, x) \in \mathcal{U}^\rho \times \mathbb{R}^d,\ \varphi(t, x, u) \in \operatorname{int} D(\rho) \text{ for all } t \in \mathbb{R}\}.$$

We require that for $\rho' > \rho$ every $(u, x) \in \mathcal{D}(\rho)$ is an inner pair for the ρ'-system.

Theorem 5.2. *Assume that for all* $\rho' > \rho > 0$ *and all* $(u, x) \in \mathcal{D}(\rho)$ *there is* $T > 0$ *with* $\varphi(T, x, u) \in \operatorname{int} \mathcal{O}^{+,\rho'}(x)$. *Then the map* $\rho \mapsto \operatorname{cl} D(\rho)$ *is lower semicontinuous and the map* $\rho \mapsto E(\rho)$ *is upper semicontinuous (with respect to the Hausdorff metric). The sets of continuity points for both maps coincide, and* $\rho^* \in (0, \infty)$ *is a continuity point if and only if* $\operatorname{cl} D(\rho^*) = E(\rho^*)$. *There are at most countably many points of discontinuity.*

Thus the strengthened inner pair condition guarantees that "almost always" the chain control sets coincide with the closures of control sets. The exceptional points are of particular interest.

In what follows we need an existence result for control sets that are invariant relative to a given subset L of the state space; see [3, Section 3.3].

Definition 5.4. *For a subset* $L \subset M$ *a control set* $D \subset L$ *is called* L-invariant, *if* $x \in D$ *and* $\varphi(t, x, u) \notin D$ *for some* $t > 0$ *and* $u \in \mathcal{U}$ *implies* $\varphi(t, x, u) \notin L$.

Hence a trajectory can leave an L-invariant control set only if it also leaves the set L. Control sets with nonvoid interior are closed if and only if they are invariant. For L-invariant control sets one obtains an analogous result, provided that–roughly speaking – a trajectory which starts in L and leaves L cannot return to L and local accessibility in L holds.

Theorem 5.3. *Let L be a subset of the state space M satisfying the no-return condition*

$$\text{if } z \in \operatorname{cl} \mathcal{O}^+(x) \text{ for some } x \in L \text{ and } \mathcal{O}^+(z) \cap L \neq \emptyset, \text{ then } z \in L \tag{5.5}$$

and the L-accessibility condition

$$\text{for all } y \in L \text{ and all } T > 0 \text{ one has } \operatorname{int}\left[\mathcal{O}^+_{\leq T}(y) \cap L\right] \neq \emptyset. \tag{5.6}$$

Assume that there exists a compact set $Q \subset L$ such that for all $y \in L$ one has

$$\operatorname{cl} \mathcal{O}^+(y) \cap Q \neq \emptyset. \tag{5.7}$$

Then there exists an L-invariant control set D.

Remark 5.1. *For an L-invariant control set D the exit boundary*

$$\Gamma^*(D) := \{x \in \partial D, \text{ there are } u \in \mathcal{U} \text{ and } t > 0 \text{ with } \varphi(t, x, u) \notin D\}$$

coincides with $L \cap \partial D$. Also one sees immediately that the no-return condition (5.5) *implies that every control set D with $\operatorname{int} D \cap L \neq \emptyset$ is contained in L.*

Next we discuss the generic (in a topological sense) limit behavior of trajectories. For this purpose we endow the set of control functions $\mathcal{U}$ which is a subset of $L_\infty(\mathbb{R}, \mathbb{R}^m)$ with the corresponding weak* topology. Then $\mathcal{U}$ becomes a compact metric space (cp. [3]) and the domains of attraction in the product space have the following properties.

Theorem 5.4. *Let $C_1, \ldots, C_k$ be the invariant control sets in K. Then for all i,*

$$\mathcal{A}(C_i) := \{(u, x) \in \mathcal{U} \times K, \text{ there is } T > 0 \text{ with } \varphi(T, x, u) \in C_i\}$$

is open in $\mathcal{U} \times K$ and $\bigcup_{i=1}^k \mathcal{A}(C_i)$ is open and dense in $\mathcal{U} \times K$.

5.3 An Invariance Radius for Nonlinear Systems

In order to motivate our definition of an invariance radius for nonlinear systems, we discuss the following simple one-dimensional system.

Example 5.1. Consider the scalar system

$$\dot{x}(t) + V'(x(t)) + u(t) = 0 \qquad \text{in } \mathbb{R},\ u(t) \in [-\rho, \rho],$$

where $\rho \geq 0$, and $V : \mathbb{R} \to \mathbb{R}$ is a smooth function with derivative $V'(x)$. Let the graph of V be as sketched in Figure 5.1. Then for $\rho = 0$ the equilibrium

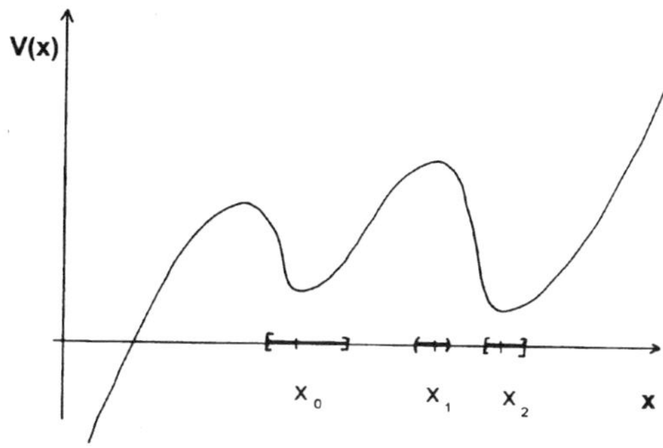

FIGURE 5.1. Equilibria and control sets for the one-dimensional example.

x_0 is asymptotically stable, and x_2 is another asymptotically stable equilibrium, while the equilibrium x_1 is unstable. For small positive ρ there is an invariant control set $D(\rho)$ containing x_0. For increasing ρ the control set $D(\rho)$ loses its invariance: Depending on $V'(x)$, either it is possible to leave it at the left side, and there are unbounded trajectories starting in x_0; or, there are trajectories leaving $D(\rho)$ and entering the invariant control set containing x_2. In general, it is not possible to return to $D(\rho)$ unless ρ is further increased. The only exception is the situation where the invariant control sets around x_0 and x_2 merge directly, which apparently is a nongeneric situation. Observe, furthermore, that there is a (variant) control set containing the unstable equilibrium x_1. The control set $D(\rho)$ loses its invariance when it merges with this control set.

We propose to define the invariance radius r of x_0 as the ρ-value where $D(\rho)$ for the first time loses its invariance.

In the following we show how to generalize the situation in this very special example to general nonlinear systems.

Formally, we define the invariance radius as follows.

Definition 5.5. *Let x_0 be an asymptotically stable equilibrium of the nominal system* (5.2). *Then the* invariance radius *of x_0 for the perturbed system* (5.1) *is*

$$r = \sup\{\rho > 0, \textit{ the control set } D(\rho') \ni x_0 \textit{ is invariant for all } 0 < \rho' < \rho\}.$$

According to Theorem 5.1, this is well defined provided that the inner pair condition (5.3) holds. It is the main purpose of this chapter to describe what happens at $\rho = r$.

First we note that also for $\rho = r$ the control set $D(r)$ is invariant.

Proposition 5.1. *Under the conditions above, the control set $D(r)$ is invariant and*

$$D(r) = \mathrm{cl}\left[\bigcup_{0<\rho<r} D(\rho)\right].$$

Proof: Abbreviate

$$B := \bigcup_{0<\rho<r} D(\rho).$$

For every $x, y \in B$ one has $y \in \mathrm{cl}\,\mathcal{O}^{+,r}(x)$. This follows, since there is a ρ with $0 < \rho < r$ such that $x, y \in D(\rho)$. Hence the assertion follows from controllability in $D(\rho)$. This implies that B is contained in a control set D with nonvoid interior.

We claim that D is contained in $\mathrm{cl}\,B$. Otherwise, note that $\mathrm{cl}\,D = \mathrm{cl}\,\mathrm{int}\,D$. Thus one finds a compact subset A of $\mathrm{int}\,D$ with $x_0 \in A$ and $A \cap (D \setminus \mathrm{cl}\,B) \neq \emptyset$. Then by lower semicontinuity of control sets (see [3, Theorem 3.2.28]) for some $\rho < r$ there is a control set containing A. This control set must be of the form $D(\rho)$ and has nonvoid intersection with $D \setminus \mathrm{cl}\,B$. This is impossible by definition of B. This also proves the equality above with $D(r)$ replaced by $\mathrm{cl}\,D(r)$.

It remains to show that the control set $D(r)$ is invariant. Otherwise, there exist $T > 0$ and $u \in \mathcal{U}^r$ with $\varphi(T, x_0, u) \notin D(r)$. Then continuous dependence on the control and the initial value imply that there are $\rho < r$, and a control $v \in \mathcal{U}^\rho$ with $\varphi(T, x_0, v) \notin D(\rho) \subset D(r)$. This contradicts invariance of $D(\rho)$. □

Let $K = \mathrm{cl}\,\mathrm{int}\,K \subset M$ be a compact, connected, positively invariant (for all $u \in \mathcal{U}$) set containing at least two invariant control sets. Then K contains finitely many invariant control sets $C_1, \ldots, C_k$, $k \geq 2$. In order to study the behavior of a (distinguished) invariant control set, say of C_1, define

$$L = \bigcup_{i=2}^{k} [\mathbf{A}(C_1) \cap \mathbf{A}(C_i)] \cap \mathrm{int}\,K. \tag{5.8}$$

Thus L is a nonvoid open set consisting of the points in $\mathbf{A}(C_1) \cap \mathrm{int}\,K$ that can be steered to at least one other invariant control set C_i. In order to avoid certain degeneracies, we have to be careful about the behavior at the boundary ∂K of K. We require that trajectories in L remain bounded away from ∂K. More precisely, assume that the following strong invariance condition is satisfied.

For all $x \in L$ there is $\varepsilon_x > 0$ with $d(\varphi(t, x, u), \partial K) \geq \varepsilon_x$, $\forall u \in \mathcal{U}$, $t \geq 0$, and there is $\varepsilon_0 > 0$ such that for all $x \in \mathrm{cl}\,L$ and $u \in \mathcal{U}$ we have that $y = \lim_{k\to\infty} \varphi(t_k, x, u) \in L$ for $t_k \to \infty$ implies $d(y, \partial K) \geq \varepsilon_0$.

We need some topological properties of the boundary of L. Let ∂L and $\partial^K L$ denote the boundaries of L in $\mathbb{R}^d$ and K, respectively, and define

$$\partial_j L = \partial^K L \cap \mathbf{A}(C_j), \qquad j = 1, \ldots, l, \quad \text{and} \quad \partial_0 L = \partial L \cap \partial K. \tag{5.9}$$

Note that for each $y \in \partial^K L$ the following alternative holds: if $y \in \mathbf{A}(C_1)$ then $y \notin \mathbf{A}(C_i)$ for every $2 \le i \le k$; or $y \notin \mathbf{A}(C_1)$ and there is (at least one) $i \in \{2, \ldots, k\}$ with $y \in \mathbf{A}(C_i)$. Then

$$\partial^K L = \bigcup_{i=1}^{l} \partial_j L \qquad \text{and} \qquad \partial L = \bigcup_{i=0}^{l} \partial_j L.$$

Using that the domains of attraction are open, one sees that the first union is disjoint; furthermore, the sets $\partial_j L$, $j = 1, \ldots, l$, are open and isolated (i.e., their closures are disjoint) in $\partial^K L$. Here we again use that every point in K can be steered into an invariant control set C_i. In the topology of $\mathbb{R}^d$

$$y \in \operatorname{cl} \partial_1 L \ \cap \operatorname{cl} \bigcup_{i=2}^{l} \partial_j L \ \text{ implies } y \in \partial_0 L \subset \partial K. \tag{5.10}$$

Note further that $\partial^K L$ consists of at least two (different) $\partial_j L$. Analogous definitions can be given for every connected component of L and all properties stated above remain valid.

The following technical lemma is needed.

Lemma 5.1. *For every $x \in L$ there are $J \subset \{2, \ldots, l\}$ and $y \in \mathcal{O}^+(x)$ such that $y \in \mathbf{A}(C_1) \cap \bigcap_{j \in J} \mathbf{A}(C_j)$ and J is a minimal index set in the following sense.*

If $\varphi(t, y, u) \in L$ for some $t > 0$ and $u \in \mathcal{U}$, then $\varphi(t, y, u) \in \bigcap_{j \in J} \mathbf{A}(C_j)$.

Proof: Since $x \in L$, there exists $J_1 \subset \{2, \ldots, l\}$ with $x \in \bigcap_{j \in J_1} \mathbf{A}(C_j)$. If there are $t_1 > 0$ and $v_1 \in \mathcal{U}$ with $y_1 := \varphi(t_1, x, v_1) \in L \setminus \bigcap_{j \in J_1} \mathbf{A}(C_j)$, then there exists a proper subset $\emptyset \neq J_2 \subset J_1$ with $y_1 \in \bigcap_{j \in J_2} \mathbf{A}(C_j)$. Proceeding recursively, one ends up, after finitely many steps, at a point $y \in \mathcal{O}^+(x)$ with a minimal index set J. □

Note that a minimal index set has at least one element. Furthermore, the lemma implies that for each L-invariant control set D there is $J \subset \{2, \ldots, l\}$ such that for each $x \in \operatorname{int} D$ the index set J is minimal.

Next we construct a set Q satisfying condition (5.7).

Lemma 5.2. *Assume that $K \subset M$ with $K = \operatorname{cl} \operatorname{int} K$ is a compact, connected, positively invariant set for the control system (5.1) satisfying the strong positive invariance condition. Assume furthermore that the system is locally accessible from every point in K. Let $x \in \mathbf{A}(C_1) \cap \bigcap_{j \in J} \mathbf{A}(C_j) \cap K$, where $J \subset \{2, \ldots, l\}$*

is some minimal index set for x. Then there exists an L-invariant control set $D \subset \operatorname{cl} \mathcal{O}^+(x)$ with

$$\partial D \cap \partial_j L \neq \emptyset \quad \text{for all } j \in J. \tag{5.11}$$

Proof: Let $x \in L$ and define

$$L_1 := L \cap \operatorname{cl} \mathcal{O}^+(x).$$

The proof is based on Theorem 5.3 applied to L_1 instead of L. It is easily seen that the set L_1 satisfies the no-return condition (5.5) and the L_1-accessibility condition (5.6). In order to construct a compact set Q with (5.7), we first introduce some notations. Denote the connected component of L that contains x by L_x, and let

$$Q_0 := \{y \in L_x,\ d(y, \operatorname{cl} \partial_0 L_x) \geq \varepsilon_x\},$$

where ε_x is chosen according to the strong invariance condition. For $\varepsilon > 0$ and $j \in \{1\} \cup J$ define

$$N_j(\varepsilon) := \{y \in L_x,\ d(y, \operatorname{cl} \partial_j L_x) \leq \varepsilon\},$$

$$Q_j(\varepsilon) := \{y \in Q_0,\ d(y, \operatorname{cl} \partial_j L_x) = \varepsilon\}.$$

Using (5.10), we can choose $\varepsilon > 0$ small enough such that

$$d(\partial_j L_x \cap Q_0, \partial_i L_x \cap Q_0) \geq 5\varepsilon \text{ for all } i,\ j \in J,\ i \neq j.$$

Hence the sets $Q_j(\varepsilon)$ are nonvoid, compact, and pairwise disjoint with distance at least 3ε. Decreasing, if necessary, the number ε further, we may assume that $x \in L_x \setminus \bigcup_{j\in J} N_j(2\varepsilon)$. Every trajectory $\{\varphi(t, y, u),\ t \geq 0\}$ with $y \in \operatorname{cl} \mathcal{O}^+(x) \cap (L_x \setminus N_j(2\varepsilon))$ that approaches C_j for $t \to \infty$ must exit through $\partial_j L_x \cap Q_0$ and must cross $Q_j(\varepsilon)$. For every y in this set there exists a control $u \in \mathcal{U}$ with this property. Furthermore we find that

$$\emptyset \neq Q_j(x, \varepsilon) := \operatorname{cl} \mathcal{O}^+(x) \cap Q_j(\varepsilon) \subset \mathbf{A}(C_1) \cap \bigcap_{j\in J} \mathbf{A}(C_j)$$

for all $j \in J$ and all $\varepsilon > 0$ small enough. Hence by Theorem 5.3 the existence of an L_1-invariant control set $D \subset \operatorname{cl} \mathcal{O}^+(x)$ follows. Obviously, an L_1-invariant control set is also L-invariant. Finally, condition (5.11) holds, because in the construction above ε can be made arbitrarily small and every L-invariant control set D has nonvoid intersection with every $Q_j(x, \varepsilon)$. □

We obtain the following result.

Proposition 5.2. *Assume that $K \subset M$ with $K = \operatorname{cl} \operatorname{int} K$ is a compact, connected, positively invariant set for the control system* (5.1) *satisfying the strong* *invariance condition. Furthermore, let $C_1, \dots, C_l$ be the invariant control sets in K. Consider the set*

$$L = \bigcup_{j\in\{2,\dots,k\}} [\mathbf{A}(C_1) \cap \mathbf{A}(C_j)] \cap \operatorname{int} K.$$

Then there exists at least one and at most finitely many L-invariant control sets D and every point in L can be steered into an L-invariant control set.

Proof: It only remains to show that the number of L-invariant control sets is finite. This follows as [1, Proposition 2.11]. □

The next theorem indicates the controllability situation when an invariant control set touches the boundary of its invariant domain of attraction. In this case, the chain control set containing the invariant control set must contain another control set.

Theorem 5.5. *Assume that $K \subset M$ with $K = \operatorname{cl}\operatorname{int} K$ is a compact, positively invariant set for the control system (5.1) satisfying the strong invariance condition. Let C_1 be an invariant control set with $C_1 \cap \partial \mathbf{A}^{inv}(C_1) \neq \emptyset$. Then there exists a control set D that is invariant relative to the set*

$$L = \bigcup_{j\in\{2,\dots,k\}} [\mathbf{A}(C_1) \cap \mathbf{A}(C_j)] \cap \operatorname{int} K,$$

where $C_2, \dots, C_k$ are the invariant control sets different from C_1 in K, such that the (unique) chain control set E containing C_1 satisfies

$$C_1 \cup D \subset E.$$

Proof: By our assumption, there exists a point $x \in C_1 \cap \partial \mathbf{A}^{inv}(C_1)$. Since the invariant control set C_1 is contained in the interior of its domain of attraction, it follows that there are $x_n \in \mathbf{A}(C_1) \setminus \mathbf{A}^{inv}(C_1)$ with $x_n \to x$. Thus there are invariant control sets different from C_1 that have nonvoid intersection with $\operatorname{cl} \mathcal{O}^+(x_n)$. Since the number of invariant control sets is finite, we may assume that there is a control set $C_j \neq C_1$ with $\operatorname{cl} \mathcal{O}^+(x_n) \cap C_j \neq \emptyset$ for all n.

Hence $x_n \in L$ and by the preceding proposition we find u_n and $t_n > 0$ such that $\varphi(t_n, x_n, u_n) \in \operatorname{int} D_n$ for some L- invariant control set D_n. Since the number of L-invariant control sets is finite and one can steer the system from every point of D_n into C_1, it follows that the chain control set E containing C_1 also contains some D_n. □

Remark 5.2. *The proof above uses arguments similar to those used for the existence of relatively invariant control sets in multistability regions, Colonius et al. [1], see also [3, Section 3.3].*

From the preceding theorem and Theorem 5.2 we immediately obtain the following consequence for systems with increasing control range. It shows that at the invariance radius r the invariant control set merges with another variant control set and itself becomes variant. This is the main result of this chapter.

Theorem 5.6. *Let the inner pair condition from Theorem 5.2 be satisfied and consider an asymptotically stable equilibrium x_0 of the nominal system. Suppose that for an increasing control set family $D(\rho)$ $\rho > 0$, containing x_0, which is invariant for $\rho \leq \rho_0$, one has*

$$D(\rho^0) \cap \partial\mathbf{A}^{inv}(D(\rho^0)) \neq \emptyset.$$

Assume that the chain control set $E(\rho^0)$ containing $D(\rho^0)$ does not contain another invariant control set. Then ρ^0 coincides with the invariance radius r, the map $\rho \mapsto D(\rho)$ is discontinuous at $\rho = r$ and $D(r) \neq E(r)$. Furthermore, there is a control set $D' \neq D(r)$ for the control range U^r such that for $\rho > r$ the control sets $D(\rho)$ satisfy

$$D(\rho) \supset D(r) \cup D'.$$

Finally, we treat the case of unbounded solutions. Instead of the existence of a compact, connected, positively invariant set K we consider a compact connected subset containing all invariant control sets and satisfying a no-return condition.

Theorem 5.7. *Consider an asymptotically stable equilibrium x_0 of the nominal system. Suppose that for an increasing control set family $D(\rho)$, $\rho > 0$, containing x_0, which is invariant for $\rho \leq \rho_0$, one has*

$$D(\rho^0) \cap \partial\mathbf{A}^{inv}(D(\rho^0)) \neq \emptyset.$$

Assume that there is a compact, connected subset $K = \operatorname{cl}\operatorname{int} K$ containing all invariant control sets $C_1, \ldots, C_l$ of the ρ^0-system, satisfying the strong invariance condition and the following no-return condition:

$$\begin{aligned} &\textit{if } \varphi(T, z, u) \notin \operatorname{int} K \textit{ for some } z \in K,\ T > 0, \textit{ and } u \in \mathcal{U}^{\rho^0}, \\ &\textit{then } \varphi(t, z, u) \notin K \textit{ for all } t \geq T. \end{aligned} \tag{5.12}$$

Furthermore, suppose that the chain control set $E(\rho^0)$ containing $D(\rho^0)$ does not contain another invariant control set. Then ρ^0 coincides with the invariance radius r, the map $\rho \mapsto D(\rho)$ is discontinuous at $\rho = r$, and $D(r) \neq E(r)$. Furthermore, there is a control set $D' \neq D(r)$ for the control range U^r such that for $\rho > r$ the control set $D(\rho)$ satisfies

$$D(\rho) \supset D(r) \cup D'.$$

This theorem follows similarly to the preceding one; compare [3, Section 3.3] for analogous arguments in the analysis of multistability regions.

Next we discuss in more detail the question of when an invariant control set C touches the boundary of its invariant domain of attraction. In the one-dimensional system, Example 5.1, this happens when an unstable equilibrium corresponding to some constant control occurs at the boundary of C. A moment's reflection shows that the occurrence of an equilibrium at the boundary of C with nontrivial unstable manifold directed out of C does not necessarily lead to the loss of invariance. It does happen, if the unstable manifold approaches for $t \to \infty$ another invariant control set. If it returns to C, the invariance radius is not attained. If the unstable manifold approaches for $t \to \infty$ a variant control set, then, depending on its (global) properties, the invariance radius may or may not be attained. The following simple example of a chemical reactor shows some of these features (for this model compare Poore [7] or Golubitskii and Schaeffer [4] and also the analysis in [3, Chapter 9]).

Example 5.2. Consider the model of a continuous flow stirred tank reactor given by the equations:

$$\begin{pmatrix} \dot{x}_1 \\ \dot{x}_2 \end{pmatrix} = \begin{pmatrix} -x_1 - a(x_1 - x_c) + B\alpha(1 - x_2)e^{x_1} \\ -x_2 + \alpha(1 - x_2)e^{x_1} \end{pmatrix} + u(t) \begin{pmatrix} x_c - x_1 \\ 0 \end{pmatrix}. \quad (5.13)$$

Here x_1 is the (dimensionless) temperature, x_2 is the product concentration, and a, α, B, x_c are positive constants. The parameter x_c is the coolant temperature, and hence the perturbation affects the heat transfer coefficient. In [7] Poore analyzes the bifurcation behavior of the nominal (i.e., $u(t) \equiv 0$) system. Here we choose parameter values such that for all constant controls $u(t) \equiv u \in [-\rho, \rho]$ the system (5.13) has exactly three fixed points as limit sets. Specifically, we take for our numerical analysis

$$a = 0.15, \qquad \alpha = 0.05, \qquad B = 7.0, \qquad x_c = 1.0. \qquad (5.14)$$

First let $U = [-0.15, 0.15]$. Because of the physical constraints we have to consider the system in the set $M = [0, \infty) \times [0, 1] \subset \mathbb{R}^2$. For each fixed $u \in U$ the equation (5.13) has three fixed points in M. Let $y^i = \alpha e^{z^i}/(1 + \alpha z^i)$, $i = 0, 1, 2$, and let $z^0 < z^1 < z^2$ be the zeros of the transcendental equation

$$-z - (a + u)(z - x_c) + B\alpha[1 - \frac{\alpha e^z}{1 + \alpha e^z}]e^z = 0.$$

Then these fixed points are given as two asymptotically stable ones, $x^0 = (z^0, y^0)$ and $x^2 = (z^2, y^2)$, and a hyperbolic one, $x^1 = (z^1, y^1)$; that is, the linearization about x^1 has one negative and one positive eigenvalue. The phase portrait of the nominal equation is indicated in Figure 5.2. There are exactly three control sets C_1, C_2, and D, containing the fixed points $x^i(u)$, $i = 0, 1, 2$, for $u \in \operatorname{int} U = (-0.15, 0.15)$ in their interior. The control sets C_1 and C_2 are invariant; the control set D is variant. The closures of these control sets are the three chain control sets of the system. Figure 5.2 shows the three control sets.

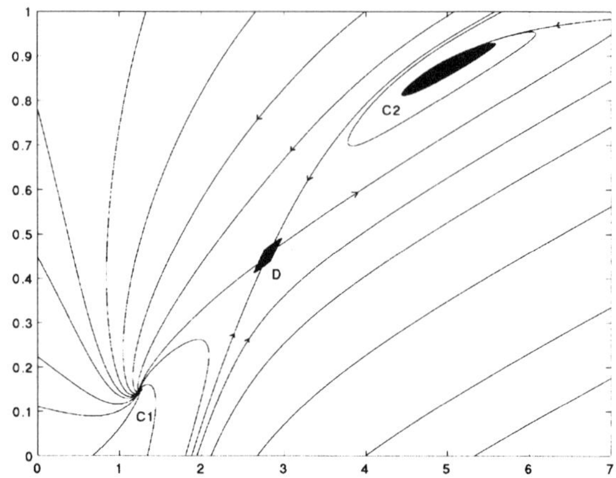

FIGURE 5.2. Phase portrait of the unperturbed ($u(t) \equiv 0$) continuous flow stirred tank reactor and the control sets.

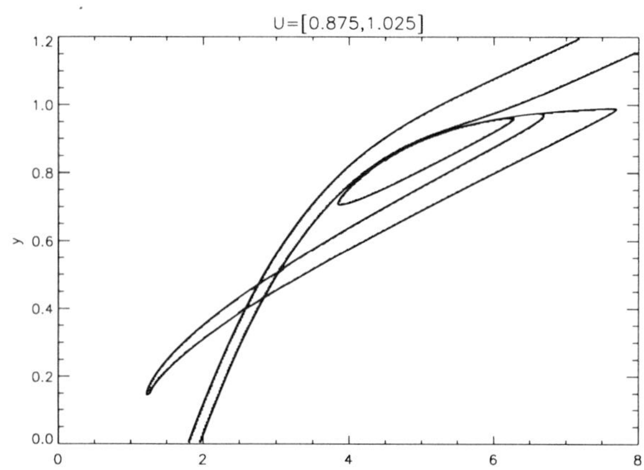

FIGURE 5.3. Control sets of the continuous flow stirred tank reactor and domains of attraction.

For different parameter values, Figure 5.3 shows the invariant control set C_2 as well as the positive and negative orbits from the hyperbolic equilibrium. Their intersection is the variant control set D. Although the numerics seem to indicate that for a slightly larger control range the invariant control set C_2 loses its invariance by intersecting the domain of attraction $\mathbf{A}(D)$ outside of $\operatorname{cl} D$ only, the situation is different: the relevant part of the boundary of $\mathbf{A}(D)$ is the stable manifold of a hyperbolic equilibrium in $\operatorname{cl} D$. Hence, if $\operatorname{cl} \mathbf{A}(D) \cap C_2 \neq \emptyset$, it follows from invariance of C_2 that this hyperbolic equilibrium also is in C_2 and thus $\operatorname{cl} D \cap C_2 \neq \emptyset$. Hence also in this example, the occurrence of an equilibrium on the boundary of the invariant control set with unstable manifold leading out of the invariant control set is responsible for the loss of invariance. If the perturbation range is further increased, the resulting variant control set merges with the remaining invariant control set C_1. This seems to happen when a hyperbolic equilibrium occurs on the boundary of C_1. Here this does not lead to the loss of invariance, since only one invariant control set is present.

References

[1] F. Colonius, F.J. de la Rubia, and W. Kliemann, Stochastic models with multistability and extinction levels, *SIAM J. Appl. Math.* 56:919–945, 1996.

[2] F. Colonius and W. Kliemann, Continuous, smooth, and control techniques for stochastic dynamics, in *Stochastic Dynamics,* H. Crauel and M. Gundlach, eds., Springer-Verlag, New York, 181–208, 1999.

[3] F. Colonius and W. Kliemann, *The Dynamics of Control,* Birkhäuser, Boston, 2000.

[4] M. Golubitsky and D. Schaeffer, *Singularities and Groups in Bifurcation Theory*, Springer-Verlag, New York, 1985.

[5] D. Hinrichsen and A.J. Pritchard, Stability radius for structured perturbations and the algebraic Riccati equation, *Syst. Control Letters* 8:105–113, 1986.

[6] D. Hinrichsen and A.J. Pritchard, Stability radius of linear systems, *Syst. Control Letters* 7:1–10, 1986.

[7] A.B. Poore, A model equation arising from chemical reactor theory, *Arch. Rational Mech. Anal.* 52:358–388, 1974.

Chapter 6

State and Continuity

Jan C. Willems

ABSTRACT We prove that for linear time-invariant differential systems the state consists of the functionals of the system variables that evolve continuously in time.

6.1 Introduction

It is a pleasure to write this chapter in honor of Didi Hinrichsen on the occasion of his 60th birthday. Our professional collaboration goes back almost 25 years, and is due more to intellectual affinity than to geographic proximity, although the latter certainly helped. Didi Hinrichsen was one of the initiators and animators of the Center for Dynamical Systems at the University of Bremen, which grew to be one of the most influential centers for control theory in Europe. The concepts of systems theory and the algebraic structure of linear systems always played a central role in Didi Hinrichsen's research and the notion of state and algorithms associated with state construction and reduction were one of his focal points. The present chapter is devoted to this topic. Specifically, we show that the state of a linear system consists exactly of the system functionals that evolve continuously in time.

My recent work is aimed at foundational aspects of the theory of dynamical systems in interaction with their environment, and in particular with the development of a conceptual framework that smoothly incorporates first principles modeling. Of course, there have been earlier attempts to develop such a setting, the best known being the work of Rosenbrock [4], where the *partial state* is introduced. This concept is actually an intermediary between the classical state and our notion of latent variables. An especially important article, both from the point of view of Rosenbrock systems, and by anticipating some of the developments of the behavioral approach, is [1].

6.2 Linear Differential Systems

We restrict our attention to continuous-time linear time-invariant differential systems with a finite number of real variables, although the basic result (state $\cong$ smoothness) undoubtedly extends to other classes of systems.

A *dynamical system* Σ is triple $\Sigma = (\mathbb{T}, \mathbb{W}, \mathfrak{B})$ with $\mathbb{T} \subset \mathbb{R}$ the time-set, $\mathbb{W}$ the signal space, and $\mathfrak{B} \subset \mathbb{W}^{\mathbb{T}}$ the *behavior*. The intuition behind this definition is that $\mathbb{T}$ is the set of relevant time-instances; $\mathbb{W}$ is the set in which the signals, whose dynamic relation Σ models, take on their values; and the behavior $\mathfrak{B}$ specifies which signals $w : \mathbb{T} \to \mathbb{W}$ obey the laws of the system. The time-set $\mathbb{T}$ equals, for example, $\mathbb{R}$ or $\mathbb{R}_+$ in continuous-time, and $\mathbb{Z}$ or $\mathbb{Z}_+$ in discrete-time systems. Important properties of dynamical systems are linearity and time-invariance; Σ is said to be *linear* if $\mathbb{W}$ is a vector space and $\mathfrak{B}$ a linear subspace of $\mathbb{W}^{\mathbb{T}}$, and *time-invariant* (assuming $\mathbb{T} = \mathbb{R}$ or $\mathbb{Z}$) if $\sigma^t \mathfrak{B} = \mathfrak{B}$ for all $t \in \mathbb{T}$, where σ^t denotes the t-shift (defined by $(\sigma^t f)(t') := f(t' + t)$).

For an explanation of the intuition behind these definitions and for more details, we refer to our earlier writings [2, 5, 6]. The idea underlying the behavioral approach is based on the belief that in a model of a dynamical (physical) phenomenon, it is the behavior $\mathfrak{B}$ (i.e., a set of trajectories $w : \mathbb{T} \to \mathbb{W}$) that is the central object of study. But, this set of trajectories must be specified somehow, and it is here that differential (and difference) equations enter. For $\mathbb{T} = \mathbb{R}$, $\mathfrak{B}$ then consists of the solutions of a system of differential equations as

$$f(w, \frac{d}{dt}w, \ldots, \frac{d^N}{dt^N}w) = 0.$$

We call these *differential* systems. Of particular interest (at least in control, signal processing, circuit theory, etc.) are systems with a signal space that is a finite-dimensional vector space and a behavior that is described by a set of linear constant-coefficient differential equations. A *linear time-invariant differential system* is thus a dynamical system $\Sigma = (\mathbb{R}, \mathbb{W}, \mathfrak{B})$, with $\mathbb{W}$ a finite-dimensional real vector space with dimension is denoted by $\mathtt{w}$, that is, $\mathbb{W} = \mathbb{R}^{\mathtt{w}}$, whose behavior consists of the solutions of

$$R(\frac{d}{dt})w = 0, \tag{6.1}$$

with $R \in \mathbb{R}^{\bullet \times \mathtt{w}}[\xi]$ a real polynomial matrix. Of course, the number of columns of R equals the dimension $\mathtt{w}$ of $\mathbb{W}$. The number of rows of R, which equals the number of equations, is in principle arbitrary. There is much to be said for considering solutions in the space of distributions $\mathfrak{D}'(\mathbb{R}, \mathbb{R}^{\mathtt{w}})$ and interpreting $R(d/dt)w$ as a distribution. Thus (by slight abuse of notation, since formally $\mathfrak{B}$ is not a subset of $\mathbb{W}^{\mathbb{T}}$) in this chapter we consider the behavior corresponding to (6.1) to be

$$\mathfrak{B} := \{w \in \mathfrak{D}'(\mathbb{R}, \mathbb{R}^{\mathtt{w}}) \mid R(\frac{d}{dt})w = 0\}.$$

We denote this behavior as $\ker(R(d/dt))$, the set of linear time-invariant differential systems by $\mathfrak{L}^{\bullet}$, and those with $\dim(\mathbb{W}) = \mathtt{w}$ by $\mathfrak{L}^{\mathtt{w}}$. Whence $\Sigma = (\mathbb{R}, \mathbb{R}^{\mathtt{w}}, \mathfrak{B}) \in \mathfrak{L}^{\mathtt{w}}$ means that there exists an $R \in \mathbb{R}^{\bullet \times \mathtt{w}}[\xi]$ such that $\mathfrak{B} = \ker(R(d/dt))$. We call (6.1) a *kernel representation* of Σ. Note that we may as well write $\mathfrak{B} \in \mathfrak{L}^{\mathtt{w}}$, instead of $\Sigma \in \mathfrak{L}^{\mathtt{w}}$, since the time-axis ($\mathbb{R}$) and the signal space ($\mathbb{R}^{\mathtt{w}}$) are evident from this notation.

6.3 Latent Variables

Mathematical models of complex systems are usually obtained by viewing the system (in a hierarchical fashion) as an interconnection of subsystems. This leads to models that relate the variables whose dynamic relation one wants to model (we call these *manifest* variables) to auxiliary variables (we call these *latent* variables) that have been introduced in the modeling process. For differential systems this leads to equations as

$$f_1(w, \frac{d}{dt}w, \ldots, \frac{d^N}{dt^N}w, \ell, \frac{d}{dt}\ell, \ldots, \frac{d^N}{dt^N}\ell) = f_2(w, \frac{d}{dt}w, \ldots, \frac{d^N}{dt^N}w, \ell, \frac{d}{dt}\ell, \ldots, \frac{d^N}{dt^N}\ell),$$

relating the ($\mathtt{w}$–dimensional vector of) manifest variables w to the ($\mathtt{l}$–dimensional vector of) latent variables ℓ. In the linear time-invariant case this becomes

$$R(\frac{d}{dt})w = M(\frac{d}{dt})\ell, \tag{6.2}$$

with R and M polynomial matrices. Define the *full behavior* of this system as

$$\mathfrak{B}_{\text{full}} := \{(w, \ell) \in \mathfrak{D}'(\mathbb{R}, \mathbb{R}^{\mathtt{w}} \times \mathbb{R}^{\mathtt{l}}) \mid R(\frac{d}{dt})w = M(\frac{d}{dt})\ell\}.$$

This yields the latent variable system $\Sigma_{\text{full}} = (\mathbb{R}, \mathbb{R}^{\mathtt{w}}, \mathbb{R}^{\mathtt{l}}, \mathfrak{B}_{\text{full}})$. This class of systems is denoted by $\mathfrak{L}^{\mathtt{w},\mathtt{l}}$. The corresponding *manifest behavior* is defined as

$$\mathfrak{B} := \{w \in \mathfrak{D}'(\mathbb{R}, \mathbb{R}^{\mathtt{w}}) \mid \exists \ell \in \mathfrak{D}'(\mathbb{R}, \mathbb{R}^{\mathtt{l}}) \text{ such that } R(\frac{d}{dt})w = M(\frac{d}{dt})\ell\}.$$

We call $\Sigma_{\text{full}} = (\mathbb{R}, \mathbb{R}^{\mathtt{w}}, \mathbb{R}^{\mathtt{l}}, \mathfrak{B}_{\text{full}})$ a *latent variable representation* of the manifest system $\Sigma = (\mathbb{R}, \mathbb{R}^{\mathtt{w}}, \mathfrak{B})$. The question occurs whether the manifest system is in $\mathfrak{L}^{\mathtt{w}}$. This is the case indeed: for any real polynomial matrices (R, M) with $\text{rowdim}(R) = \text{rowdim}(M)$, there exists a real polynomial matrix R' such that the manifest behavior of $R(d/dt)w = M(d/dt)\ell$ has the kernel representation $R'(d/dt)w = 0$. This result is called the *elimination theorem.* Of course, the order of the differential equation usually goes up in the elimination process.

In the sequel we need the notion of observability in the context of latent variable systems, pertaining to the possibility of deducing the latent variables from the manifest ones. Let $\Sigma_{\text{full}} = (\mathbb{R}, \mathbb{R}^{\mathtt{w}}, \mathbb{R}^{1}, \mathfrak{B}_{\text{full}})$ be a latent variable system. Then w is said to be *observable* from ℓ (or, simply, the latent variable system is said to be observable) if $(w, \ell'), (w, \ell'') \in \mathfrak{B}_{\text{full}}$ imply $\ell' = \ell''$. Observability thus refers to the possibility of deducing the trajectory ℓ from observation of w and from the laws of the system ($\mathfrak{B}_{\text{full}}$ is assumed to be known). Thus an observable latent variable representation $\Sigma_{\text{full}} \in \mathfrak{L}^{\mathtt{w},1}$ (as we denote $\Sigma_{\text{full}} = (\mathbb{R}, \mathbb{R}^{\mathtt{w}}, \mathbb{R}^{1}, \mathfrak{B}_{\text{full}})$ when $\mathfrak{B}_{\text{full}} \in \mathfrak{L}^{w+n}$) can be expressed by a kernel representation of the form

$$R(\frac{d}{dt})w = 0, \qquad w = M(\frac{d}{dt})\ell.$$

There exist various tests on the polynomial matrix M that characterize observability, and we refer to the references cited for these.

6.4 State Representations

A state model is a special type of latent variable model, one in which the latent variable (called the state, and, in keeping with tradition, denoted by x and with state space denoted $\mathbb{R}^n$) separates the past and the future: given the present value of the state, any past leading to it is compatible (i.e., concatenable) with any future emanating from it. Thus, a latent variable model

$$R(\frac{d}{dt})w = M(\frac{d}{dt})x$$

is said to be a *state model* if whenever (w_1, x_1) and (w_2, x_2) are both elements of its full behavior $\mathfrak{B}_{\text{full}}$, with x_1 and x_2 both continuous at $t = 0$, and if $x_1(0) = x_2(0)$, then the concatenation $(w, x) := (w_1, x_1) \wedge (w_2, x_2)$ also satisfies the differential equation above. Concatenation (at $t = 0$) is defined by $f_1 \wedge f_2(t) := f_1(t)$ for $t < 0$, and $f_1 \wedge f_2(t) := f_2(t)$ for $t \geq 0$.

Let $\mathfrak{B} \in \mathfrak{L}^{w}$. A latent variable representation of $\mathfrak{B}_{\text{full}}$ is called a *state representation* of $\mathfrak{B}$ if it is a state model. Thus, when we have a state model, in order to decide what the future continuations of a trajectory in $\mathfrak{B}$ look like, it is not necessary to analyze the whole past of this trajectory, since only the *present* value of the state is relevant. As such, the state x parameterizes the memory of the system.

The state property can be related to a special differential equation structure, to a special kernel representation. This is explained in the following proposition.

Proposition 6.1. *The latent variable systems $\Sigma_{\text{full}} = (\mathbb{R}, \mathbb{R}^{\mathtt{w}}, \mathbb{R}^{1}, \mathfrak{B}_{\text{full}}) \in \mathfrak{L}^{\mathtt{w},\mathtt{n}}$ is a state system if and only if its full behavior $\mathfrak{B}_{\text{full}}$ admits a kernel representation of the form*

$$R_0 w + M_0 x + M_1 \frac{d}{dt} x = 0, \tag{6.3}$$

with R_0, M_0, M_1 real constant matrices.

Equation (6.3) is an (implicit) first-order differential equation containing derivatives of at most first-order in x and zeroth-order in w. It is easy to see that any latent variable model of the form (6.3) is in fact a state model. Conversely, as stated in the above proposition, every state model (6.3) is equivalent (in the sense that the two have the same full behavior) to a representation of the first-order form (6.3). The number of state variables (i.e., the size of x) is called the *dynamic order* of the state model. If $\Sigma_{\text{full}} \in \mathfrak{L}^{\mathtt{w},\mathtt{n}}$ is a state system with manifest behavior $\mathfrak{B} \in \mathfrak{L}^{\mathtt{w}}$ with dynamic order as small as possible among all state representations of $\mathfrak{B}$, then we call $\Sigma_{\text{full}} \in \mathfrak{L}^{\mathtt{w},\mathtt{n}}$ *minimal.* Precisely, a state system is minimal if whenever $\Sigma'_{\text{full}} \in \mathfrak{L}^{\mathtt{w},\mathtt{n}'}$ is another state representation of $\mathfrak{B} \in \mathfrak{L}^{\mathtt{w}}$, then $n' \geq n$. The dimension of the state space of a minimal state representation of $\mathfrak{B}$ is denoted by $\mathtt{n}(\mathfrak{B})$.

State representations may or may not be observable. It turns out that there always exist observable state representations of $\mathfrak{B} \in \mathfrak{L}^{\mathtt{w}}$. In particular, the minimal ones are always observable. A state representation is observable if and only if there exists $X \in \mathbb{R}^{\bullet \times w}[\xi]$ such that for all $w \in \mathfrak{B}$ we have: $(w, x) \in \mathfrak{B}_{\text{full}} \Rightarrow x = X(d/dt)w$. The operator $X(d/dt)$ is called a *state map* for $\mathfrak{B}$, and a *minimal* one if the resulting state system is minimal. Hence, if (6.1) is a kernel representation of $\mathfrak{B}$, then $X(d/dt)$ is called a state map for $\mathfrak{B}$ if and only if

$$R(\frac{d}{dt})w = 0, \qquad x = X(\frac{d}{dt})w$$

satisfies the axiom of state. A state system $\Sigma_{\text{full}} \in \mathfrak{L}^{\mathtt{w},\mathtt{n}}$ is said to be *state trim* if for all $x_0 \in \mathbb{R}^n$ in the state space, there exists a $(w, x) \in \mathfrak{B}_{\text{full}}$ such that $x(0) = x_0$. We show that minimal state representations are observable and state trim.

In [3], an algorithm is given for constructing a state map for $\mathfrak{B} \in \mathfrak{L}^{\mathtt{w}}$ starting from a kernel representation (6.1) of it. Define the polynomial matrix X_R as follows,

$$X_R := \text{col}(R_1, R_2, \ldots, R_L)$$

with L the highest degree of the elements of R, and the R_ks recursively defined by

$$R_{k+1}(\xi) := \frac{(R_k(\xi) - R_k(0))}{\xi}; \qquad R_0 := R.$$

Then X_R is a state map for $\mathfrak{B}$ and a proper selection of the rows of X_R yield a minimal state map for $\mathfrak{B}$.

For a proof of these results, and details on the state construction, see [3].

6.5 Smoothing Functionals

Any vector polynomial $f \in \mathbb{R}^{1\times\mathtt{w}}[\xi]$ induces the map $f(d/dt) : \mathfrak{D}'(\mathbb{R}, \mathbb{R}^{\mathtt{w}}) \to \mathfrak{D}'(\mathbb{R}, \mathbb{R})$. We call $f(d/dt)$ a *(differential) functional.* Let $\mathfrak{B} \in \mathfrak{L}^{\mathtt{w}}$. A functional is said to be *annihilating* on $\mathfrak{B}$ if $f(d/dt)\mathfrak{B} = 0$, and *smoothing* on $\mathfrak{B}$ if $f(d/dt)(\mathfrak{B} \cap \mathfrak{L}^{\mathrm{loc}}(\mathbb{R}, \mathbb{R}^{\mathtt{w}})) \subset \mathfrak{C}^0(\mathbb{R}, \mathbb{R})$. Smoothness thus means that whenever $\mathtt{w} : \mathbb{R} \to \mathbb{R}^{\mathtt{w}}$ is an "ordinary" vector-valued function (rather than a distribution) in $\mathfrak{B}$, then $f(d/dt)w : \mathbb{R} \to \mathbb{R}$ is a continuous function. Denote by

$$\mathfrak{N}_{\mathfrak{B}} := \{f \in \mathbb{R}^{1\times\mathtt{w}}[\xi] \mid f(\frac{d}{dt}) \text{ is annihilating on } \mathfrak{B}\}$$

the set of annihilating, and by

$$\mathfrak{S}_{\mathfrak{B}} := \{f \in \mathbb{R}^{1\times\mathtt{w}}[\xi] \mid f(\frac{d}{dt}) \text{ is a smoothing on } \mathfrak{B}\}$$

the set of smoothing functionals on $\mathfrak{B}$. More generally, we call the map $F(d/dt) : \mathfrak{B} \to \mathfrak{D}'(\mathbb{R}, \mathbb{R}^{\mathrm{rowdim}(F)})$ induced by the polynomial matrix $F \in \mathbb{R}^{\bullet\times\mathtt{w}}[\xi]$ annihilating (or smoothing) for $\mathfrak{B}$ if each of its rows is.

It is well known and easy to prove that $\mathfrak{N}_{\mathfrak{B}}$ is a submodule of $\mathbb{R}^{1\times\mathtt{w}}[\xi]$. In fact, there is a one-to-one relation between $\mathfrak{L}^{\mathtt{w}}$ and the submodules of $\mathbb{R}^{1\times\mathtt{w}}[\xi]$. Obviously, $\mathfrak{N}_{\mathfrak{B}} \subset \mathfrak{S}_{\mathfrak{B}}$, and $\mathfrak{S}_{\mathfrak{B}}$ is a vector space over $\mathbb{R}$. We show that $\mathfrak{S}_{\mathfrak{B}}(\mathrm{mod}\ \mathfrak{N}_{\mathfrak{B}})$ is finite-dimensional. The aim of the present chapter is to give a meaning to $\mathfrak{S}_{\mathfrak{B}}$ in terms of the state of $\mathfrak{B}$.

6.6 Main Results

The following two theorems are the main results of this chapter. In [3], it is shown that, for discrete-time linear time-invariant systems described by difference equations, a state representation is minimal if and only if it is state trim, and $x(0)$ is observable from the strict past *and* the future of w. This statement can not be generalized to continuous-time systems (contrary to an informal statement that was made in [3]). The main result of this chapter shows that the proper generalization involves in addition to trimness and observability, the smoothness of the state trajectories.

Theorem 6.1. *Let $\Sigma_{\text{full}} = (\mathbb{R}, \mathbb{R}^{\mathtt{w}}, \mathbb{R}^{n}, \mathfrak{B}_{\text{full}}) \in \mathfrak{L}^{\mathtt{w},\mathtt{n}}$ be a state representation of $\mathfrak{B} \in \mathfrak{L}^{\mathtt{w}}$. It is a minimal state representation if and only if it is state trim, observable, and the state map $X(d/dt) : \mathfrak{B} \to \mathfrak{D}'(\mathbb{R}, \mathbb{R}^n)$ is smoothing.*

Theorem 6.2. *The map $X(d/dt) : \mathfrak{B} \to \mathfrak{D}'(\mathbb{R}, \mathbb{R}^n)$ defines a minimal state map for $\mathfrak{B} \in \mathfrak{L}^{\mathtt{w}}$ if and only if its rows form a basis for $\mathfrak{S}_{\mathfrak{B}}$ (mod $\mathfrak{N}_{\mathfrak{B}}$).*

Corollary 6.1. *The real vector space $\mathfrak{S}_{\mathfrak{B}}$ (mod $\mathfrak{N}_{\mathfrak{B}}$) is finite-dimensional.*

6.7 Proofs

Proof: (of Proposition 6.1) See [3]. □

Proof: (of Theorem 6.1)

(*only if*): Assume that $\Sigma_{\text{full}} = (\mathbb{R}, \mathbb{R}^{\mathtt{w}}, \mathbb{R}^{n}, \mathfrak{B}_{\text{full}}) \in \mathfrak{L}^{\mathtt{w},\mathtt{n}}$ is a minimal state system. Then, by Proposition 6.1, there exist constant matrices $R_0, M_0, M_1 \in \mathbb{R}^{\bullet\times\bullet}$ such that (6.3) is a kernel representation of Σ_{full}. Now, by choosing the bases in the state and equation spaces properly, and partitioning x accordingly, this equation becomes

$$\begin{aligned} \tfrac{d}{dt}x_1 &= F_1x_1 + F_2x_2 + G_1w, \\ 0 &= F_3x_1 + F_4x_2 + G_2w. \end{aligned} \tag{6.4}$$

If this representation is minimal, then $F_4 = 0$. Otherwise, x_2 can be further partitioned as $\text{col}(x_2', x_2'')$, while the second equation implies $x_2'' = F_3'x_1 + G_2'w$, which shows that x_2'' can be eliminated, yielding a lower-dimensional state space representation of $\mathfrak{B}$. Next, observe that G_2 must be surjective. Otherwise, x_1 can be further partitioned as $\text{col}(x_1', x_1'')$, while the second equation implies $x_1'' = 0$, which can therefore be eliminated, contradicting minimality. It follows that by choosing the bases in the state, signal, and equation spaces properly and partitioning w and x appropriately, we obtain the kernel representation

$$\begin{aligned} \frac{d}{dt}x &= A_1x_1 + A_2x_2 + B_1w_1, \\ w_2 &= A_3x_1 + A_4x_2 + B_2w_1. \end{aligned}$$

Now, observe that $A_2 = 0$. Otherwise, x_1, x_2 can be further partitioned as $\text{col}(x_1', x_1''), \text{col}(x_2', x_2'')$, while the first equation implies

$$x_2'' = \frac{d}{dt}x_2' - A_1'x_1 - A_2'x_2' - B_1'w,$$

which shows that x_2'' can be eliminated, contradicting minimality. We obtain

$$\begin{aligned} \tfrac{d}{dt}x &= Ax + Bw_1, \\ w_2 &= Cx + Dw_1 \end{aligned} \tag{6.5}$$

as a state representation of Σ_{full}. If this system is minimal, then the pair (A, C) must be observable; otherwise we can choose the basis in the state space and partition x such that the above equation becomes

$$\begin{aligned} \frac{d}{dt}x_1 &= A_1''x_1 + B_1''w_1, \\ \frac{d}{dt}x_2 &= A_2''x_1 + A_3''x_2 + B_2''w_1, \\ w_2 &= C_1''x_1 + Dw_1 \end{aligned}$$

which shows that x_2 can be eliminated, contradicting minimality. We now prove that(6.5)(6.5) is state trim, observable, and that the state map is smoothing. It is obviously state trim. In order to deduce the state map, and hence show observability, write

$$\begin{aligned} w_2 &= Cx + Dw_1, \\ \frac{d}{dt}w_2 &= CAx + CBw_1 + D\frac{d}{dt}w_1, \\ \frac{d^2}{dt^2}w_2 &= CA^2x + CABw_1 + CB\frac{d}{dt}w_1 + D\frac{d^2}{dt^2}w_1, \\ &\vdots \\ \frac{d^{n-1}}{dt^{n-1}}w_2 &= CA^{n-1}x + CA^{n-2}Bw_1 + \cdots + CB\frac{d^{n-2}}{dt^{n-2}}w_1 + D\frac{d^{n-1}}{dt^{n-1}}w_1. \end{aligned}$$

Since the pair (A, C) is observable, $\text{col}(C, CA, \ldots, CA^{n-1})$ is injective. Let L be a left inverse. The above equation then yields the state map, and hence observability of(6.5)(6.5). That the state map is smoothing follows immediately from(6.5)(6.5). Indeed, it shows that any $w_1 \in \mathfrak{L}^{\text{loc}}(\mathbb{R}, \mathbb{R}^{\dim(w_1)})$ yields an x that is absolutely continuous.

(*if*): Assume that Σ_{full} is a state system that is state trim, observable, and that the state map is smoothing. We show that it is then minimal. Again represent Σ_{full} as (6.4). By trimness, G_2 must be surjective, for otherwise the second equation implies a linear relation on the state x. Whence, by choosing the bases properly we obtain

$$\begin{aligned} \tfrac{d}{dt}x_1 &= A_1x_1 + A_2x_2 + B_1w_1, \\ w_2 &= A_3x_1 + A_4x_2 + B_2w_1 \end{aligned}$$

as a kernel representation of Σ_{full}. It follows that both w_1 and x_2 are free. Since this contradicts the smoothing property, we must have $\dim(x_2) = 0$. Therefore Σ_{full} admits a kernel representation as(6.5)(6.5) with (A, C) observable.

We now show that observability of (A, C) implies minimality. Consider therefore $\mathfrak{B}_0 = \{w_2 \in \mathfrak{D}'(\mathbb{R}, \mathbb{R}^{\dim(w_2)}) \mid (0, w_2) \in \mathfrak{B}\}$. It is easily seen that the behavior $\{(w_2, x) \mid (\text{col}(0, w_2), w) \in \mathfrak{B}_{\text{full}}\}$ is a state representation of $\mathfrak{B}_0$. Hence $\dim(x) \geq \mathtt{n}(\mathfrak{B}) \geq \mathtt{n}(\mathfrak{B}_0)$. Note that $\mathfrak{B}_0$ is described by

$$\frac{d}{dt}x = Ax, \qquad w_2 = Cx.$$

It is hence autonomous, and also observable, whence, $\mathtt{n}(\mathfrak{B}_0) = \dim(\mathfrak{B}_0) = \dim(x)$. We conclude that $\mathtt{n}(\mathfrak{B}) = \dim(x)$, showing minimality. □

Proof: (of Theorem 6.2)

Following up the proof of Theorem 6.1, we see that by choosing the basis in the signal space properly, $\mathfrak{B}$ admits a minimal state space representation

$$\frac{d}{dt}x = Ax + Bw_1, \qquad w_2 = Cx,$$

with (A, C) observable. This yields a smoothing minimal state map $X(d/dt) : \mathfrak{B} \to \mathfrak{C}^0(\mathbb{R}, \mathbb{R}^{\mathtt{n}(\mathfrak{B})})$. We need to show that the rows of X span $\mathfrak{S}_{\mathfrak{B}}(\text{mod } \mathfrak{N}_{\mathfrak{B}})$.

Assume that $f \in \mathbb{R}^{1\times \mathtt{w}}[\xi]$ induces a smoothing functional on $\mathfrak{B}$. Write

$$f(\frac{d}{dt})w = f_1(\frac{d}{dt})w_1 + f(\frac{d}{dt})w_2.$$

The above state representation of $\mathfrak{B}$ allows us to eliminate w_2 in terms of w_1 and $X(\frac{d}{dt})w$, and to obtain

$$f(\frac{d}{dt})w = f_1'(\frac{d}{dt})w_1 + LX(\frac{d}{dt})w$$

for all $w \in \mathfrak{B}$, with L a constant matrix. It follows that

$$f(\frac{d}{dt})w - LX(\frac{d}{dt})w = f_1'(\frac{d}{dt})w_1$$

is smoothing on $\mathfrak{B}$. But since w_1 is free, this implies that $f_1' = 0$. Whence, $f(d/dt)w = LX(d/dt)w$, for all $w \in \mathfrak{B}$. Hence the smoothing functionals are exactly those that, up to adding an annihilating functional, are linear combinations of the rows of X. This proves the theorem. □

Proof: (of Corollary 6.1) It is an immediate consequence of Theorem 6.2 that $\mathfrak{S}_{\mathfrak{B}}(\text{mod } \mathfrak{N}_{\mathfrak{B}}) \cong \mathbb{R}^{\mathtt{n}}(\mathfrak{B})$. □

6.8 Remarks

1. It follows from the proofs that $f(d/dt)$ is smoothing for $\mathfrak{B}$ if and only if $f(d/dt)(\mathfrak{B} \cap \mathfrak{C}^0(\mathbb{R}, \mathbb{R}^{\mathtt{w}}) \subset \mathfrak{C}^1(\mathbb{R}, \mathbb{R}^{\mathtt{w}})$. More generally, we call $f(d/dt)$ k–*smoothing* if $f(d/dt)(\mathfrak{B} \cap \mathfrak{C}^0(\mathbb{R}, \mathbb{R}^{\mathtt{w}})) \subset \mathfrak{C}^k(\mathbb{R}, \mathbb{R}^{\mathtt{w}}))$. Define $\mathfrak{S}_{\mathfrak{B}}^k := \{f \in \mathbb{R}^{1\times \mathtt{w}}[\xi] \mid f(d/dt)$ is k–smoothing$\}$. Then $\mathfrak{S}_{\mathfrak{B}}^1$ leads to the state variables,

$\mathfrak{S}^0_{\mathfrak{B}}$ leads to the combination of the free and the state variables, and the dimensions of $\mathfrak{S}^k_{\mathfrak{B}}(\bmod \mathfrak{S}^{k+1}_{\mathfrak{B}})$ lead to the controllability indices.
2. Every functional $f(d/dt)$ is smoothing if and only if $\mathfrak{B}$ is autonomous. In this case the state space of $\mathfrak{B}$ is congruent with $\mathbb{R}^{1\times \mathtt{w}}[\xi](\bmod\ \mathfrak{N}_{\mathfrak{B}})$.
3. The identification of state with smoothness may shed light on the generalization of the state concept to N–D systems described by PDEs.

References

[1] D. Hinrichsen and D. Prätzel-Wolters, Solution modules and system equivalence, *Int. J. Control*, 32: 777–802, 1980.

[2] J.W. Polderman and J.C. Willems, *Introduction to Mathematical Systems Theory: A Behavioral Approach*, 26 Texts in Applied Mathematics, Springer-Verlag, New York, 1998.

[3] P. Rapisarda and J.C. Willems, State maps for linear systems, *SIAM J. Control Optimization*, 35: 1053–1091, 1997.

[4] H.H. Rosenbrock, *State Spaces and Multivariable Theory*, Wiley, New York 1977.

[5] J.C. Willems, Models for dynamics, *Dynamics Reported*, 2: 171–269, 1988.

[6] J. C. Willems, Paradigms and puzzles in the theory of dynamical systems, *IEEE Trans. Auto. Control*, 36: 259–294, 1991.

Chapter 7

Parameterization of Conditioned Invariant Subspaces

Paul A. Fuhrmann[1]
Uwe Helmke[1]

ABSTRACT We study the geometric structure of the set of k–codimensional conditioned invariant subspaces of a fixed observable pair $(\mathcal{C}, \mathcal{A})$.

7.1 Introduction

Conditioned and controlled invariant subspaces play a fundamental role in linear systems theory and provide a unified language in treating topics such as disturbance decoupling, observer design, factorization theory, parameterization problems, and partial realization theory. The study of such invariant subspaces originated in the work of Wonham [20] and Basile and Marro [1] and has led to extensive research in the late 1970s and beginning 1980s. The subsequent development of nonlinear control theory has further deepened the interest in geometric control theory. The situation changed with the advent of H_∞–theory which led to a decline of interest in the geometric approach. Thus, although many important problems in geometric control have been left unsolved, researchers moved in new directions. Today, the pendulum is slowly swinging back, with new applications emerging in factorization theory.

This chapter is to some extent a continuation of Fuhrmann and Helmke [7], where a class of conditioned invariant subspaces was studied in a generic situation. Part of our motivation for studying parameterization problems of conditioned invariant subspaces comes from the resurgence of interest in parameterization problems during the past few years. Moreover, new interest in such problems arose through realizing the crucial role conditioned invariant subspaces play in rectangular spectral factorization problems.

[1]Partially supported by a grant from the German–Israeli Foundation GIF I-526-034.06/97.

We begin with a study of conditioned invariant subspaces in the context of polynomial and rational models, relating these subspaces to kernels of associated Toeplitz operators. An important result shown is a codimension formula for conditioned invariant subspaces in terms of Wiener–Hopf indices. In fact, one thing that becomes clear from the analysis carried out in this chapter is the central importance of Wiener–Hopf factorizations and the corresponding factorization indices in this circle of ideas. To a large extent, they provide the link between the observability indices, the invertibility properties of Toeplitz maps, the surjectivity of the structural map defined by partial reachability matrices, and parameterizations of conditioned invariant subspaces.

Our next central result then is the analysis of uniqueness properties of the representation $\mathcal{V} = X_D \cap T(z)\mathbb{F}^p[z]$ of a conditioned invariant subspace with respect to the shift realization associated with the polynomial matrix D. Again, this question is related to Wiener–Hopf factorization indices, reduced observability indices, modules generated by $\mathcal{V}$, and finally the notion of tightness of $\mathcal{V}$, introduced in Fuhrmann and Helmke [7].

In order to enhance readability we have stated some results both in the polynomial as well as the state space context. A systematic realization theory for conditioned invariant subspaces in terms of similarity orbits of reachable pairs is developed. This leads us to new insight into the topology of conditioned invariant subspaces. It is shown that the set of tight conditioned invariant subspaces of fixed codimension is a smooth manifold and a stratification via reduced observability indices is investigated. Following the work by Ferrer et al. [3, 4] we show that the strata are smooth manifolds. Moreover, a retraction onto a generalized flag manifold is constructed. A cell decomposition via Kronecker invariants is proposed, using state space as well as polynomial methods. We believe that this cell decomposition coincides with the one first introduced by Hinrichsen et al. [15] in a pioneering paper (HMP), but we do not have a rigorous proof for this. The HMP paper was the first one to solve the parameterization problem for conditioned invariant subspaces, a problem that was long regarded as a difficult one. Although topological issues were not really addressed in the paper, subsequent and unpublished work by H.F. Münzner recovered several topological properties of spaces of conditioned invariant subspaces, such as cell decompositions and a combinatorial characterization of the closure relations of the cells. It is perhaps fair to say that the HMP paper is quite hard to read and thus has not received the attention it deserved. Being apparently unaware of the work of the Bremen group, some of the results were rediscovered much later by the Spanish group of Ferrer et al. [3, 4] and Puerta [17].

This is a survey chapter. Owing to space constraints most of the proofs had to be omitted. In many ways our chapter can be seen as an attempt to achieve a better understanding of HMP. We must admit however, that—so far—we have not been completely successful in doing so. Nevertheless we hope that the approach developed here helps to further reveal the fascinating structure of conditioned and controlled invariant subspaces.

Writing about a topic in geometric control theory forces one to look back into the past. Starting in the late 1970s the Bremen group around Didi Hinrichsen has done extensive research on parameterization problems in linear systems theory. Some of the obtained results are still, as we have seen, not fully understood or worked out and continue to give us food for further thought. It is certainly amazing to see that after more than 20 years of research the field of geometric control theory is still alive with open problems and new research directions.

7.2 Preliminaries

We begin by introducing polynomial models. We denote by F^m the space of all column m–vectors with coordinates in a field F. Let π_+ and π_- denote the projections of $F^m((z^{-1}))$, the space of truncated Laurent series, onto $F^m[z]$ and $z^{-1}F^m[[z^{-1}]]$, the space of polynomials and of formal power series vanishing at infinity, respectively. Since

$$F^m((z^{-1})) = F^m[z] \oplus z^{-1}F^m[[z^{-1}]] , \tag{7.1}$$

π_+ and π_- are complementary projections. Given a nonsingular polynomial matrix D in $F^{m\times m}[z]$ we define two projections π_D in $F^m[z]$ and π^D in $z^{-1}F^m[[z^{-1}]]$ by

$$\pi_D f = D\pi_-(D^{-1}f) \qquad \text{for } f \in F^m[z] \tag{7.2}$$

$$\pi^D h = \pi_-(D^{-1}\pi_+(Dh)) \qquad \text{for } h \in z^{-1}F^m[[z^{-1}]] \tag{7.3}$$

and define two linear subspaces of $F^m[z]$ and $z^{-1}F^m[[z^{-1}]]$ by

$$X_D = \operatorname{Im} \pi_D \tag{7.4}$$

and

$$X^D = \operatorname{Im} \pi^D. \tag{7.5}$$

An element f of $F^m[z]$ belongs to X_D if and only if $\pi_+D^{-1}f = 0$; that is, if and only if $D^{-1}f$ is a strictly proper rational vector function. Thus we also have the following description of the polynomial model X_D,

$$X_D = \left\{ f \in F^m[s] \mid f = Dh,\ \ h \in s^{-1}F^m[[s^{-1}]] \right\}. \tag{7.6}$$

The advantage of this characterization is that it makes sense for an arbitrary $p \times m$ polynomial matrix V. Thus we define

$$X_V = \left\{ f \in F^p[s] \mid f = Vh,\ \ h \in s^{-1}F^m[[s^{-1}]] \right\}. \tag{7.7}$$

We refer to X_D as *polynomial models* and to X^D as *rational models*.

We turn X_D into an $F[z]$–module by defining

$$p \cdot f = \pi_D(pf) \qquad \text{for } p \in F[z],\ f \in X_D. \tag{7.8}$$

Since $\operatorname{Ker} \pi_D = DF^m[z]$ it follows that X_D is isomorphic to the quotient module $F^m[z]/DF^m[z]$. Similarly, we introduce in X^D an $F[z]$–module structure by

$$p \cdot h = \pi_-(ph) \qquad \text{for } p \in F[z],\ h \in X^D. \tag{7.9}$$

In X_D we will focus on a special map S_D, a generalization of the classical companion matrix, which corresponds to the shift action of the polynomial z; that is,

$$S_D f = \pi_D z f \qquad \text{for } f \in X_D.$$

Thus the module structure in X_D is identical to the module structure induced by S_D through $p \cdot f = p(S_D)f$. With this definition the study of S_D is identical to the study of the module structure of X_D. In particular, the invariant subspaces of S_D are just the submodules of X_D that are characterized next. They are related to factorization of polynomial matrices.

Similarly, the shift operator acts on X^D via

$$S^D h = \pi_- z h \qquad \text{for } h \in X^D. \tag{7.10}$$

Polynomial and rational models are closely related. In fact, the map $\rho_D : X^D \longrightarrow X_D$ given by $f \mapsto D^{-1}f$ is an intertwining isomorphism; that is, it satisfies $S_D \rho_D = \rho_D S^D$.

In the next theorem we describe the shift realization associated with matrix fraction representations of rational transfer functions.

Theorem 7.1. *Let*

$$G = D^{-1}E$$

be a, not necessarily coprime, matrix fraction representation of a proper $p \times m$ rational function. In the state space X_D a system is defined by

$$\begin{cases} A = S_D \\ B\xi = \pi_D E\xi \\ Cp = (D^{-1}p)_{-1} \\ D_1 = G(\infty). \end{cases} \tag{7.11}$$

Then $G = \left(\begin{array}{c|c} A & B \\ \hline C & D_1 \end{array}\right)$; *this realization is observable and it is reachable if and only if E and D are left coprime.*

Note that in the realization (7.11) the pair (C, A) depends only on D, and we denote it (C_D, A_D).

Polynomial and rational models have convenient representations in terms of realizations. For example, if $(C, A) \in F^{p\times k} \times F^{k\times k}$ is an observable pair and $T(z)^{-1}H(z)$ is a left coprime factorization of $C(zI - A)^{-1}$, then it is well known (see Hautus and Heymann [11] and Wimmer [19]) that

$$X^T = \{C(zI - A)^{-1}\xi \mid \xi \in F^k\}. \tag{7.12}$$

We note that the pair (C, A) determines the nonsingular polynomial matrix T up to a left unimodular factor. Conversely, every nonsingular polynomial matrix T determines, via the shift realization, an observable pair (C, A) that is unique up to a state space isomorphism.

We now quote the characterization of conditioned invariant subspaces, relative to the shift realizations. This is taken from Fuhrmann [6].

Proposition 7.1. *Given a nonsingular $p \times p$ polynomial matrix D, let $(\mathcal{C}, \mathcal{A})$ be the unique pair that minimally realizes D and is defined by (7.11). Then a subspace $\mathcal{V} \subset X_D$ is conditioned invariant if and only if it has a representation of the form*

$$\mathcal{V} = X_D \cap T(z)F^p[z], \tag{7.13}$$

where T is a nonsingular polynomial matrix.

The representation formula (7.13) is at the heart of the chapter and opens up many interesting questions, some of which we describe.

The first difficulty in using this representation arises from the fact that the equivalence class of a polynomial matrix T, modulo right unimodular factors, is not always uniquely determined. So one line of research is to determine the conditions equivalent to the uniqueness of the determination of a nonsingular T by the conditioned invariant subspace $\mathcal{V}$. This leads to the analysis of tight subspaces. Once we are in the situation of uniqueness, up to a right unimodular factor, of the nonsingular polynomial matrix T, we can associate with it an isomorphism class of reachable pairs (A, B) that are associated with the nonsingular polynomial matrix T. The connection is given via the relation

$$(zI - A)^{-1}B = \Phi(z)T(z)^{-1}; \tag{7.14}$$

that is, via the shift realization procedure. Here the matrix fraction representation $\Phi(z)T(z)^{-1}$ is assumed to be right coprime. Thus the concept of tightness introduced formally below allows us to study the set of conditioned invariant subspaces via the study of the set of reachable pairs.

Conversely, one might start with a reachable pair, using equation (7.14) to obtain a nonsingular polynomial matrix T and define a conditioned invariant subspace via (7.13). Exploring these possibilities is one of the main goals of this chapter. It is of interest to characterize the conditions that guarantee that such an obtained subspace is tight.

One of the problems we want to study is, given an observable pair $(\mathcal{C}, \mathcal{A})$, to parameterize the set of all conditioned invariant subspaces, a topic we discuss in detail in Section 7.5. For this the uniqueness of a representation becomes of paramount importance. Without loss of generality we can assume that $(\mathcal{C}, \mathcal{A})$ is in Brunovsky canonical form. This means that, with the coprime factorizations $\mathcal{C}(zI - \mathcal{A})^{-1} = \Delta(z)^{-1}\Theta(z)$, we have

$$\Delta(z) = \begin{pmatrix} z^{\mu_1} & & & \\ & \cdot & & \\ & & \cdot & \\ & & & \cdot \\ & & & & z^{\mu_p} \end{pmatrix}, \tag{7.15}$$

with $\mu_1 \geq \cdots \geq \mu_p \geq 0$. A natural question is how to relate the observability indices of $(\mathcal{C}, \mathcal{A})$ to the controllability indices of the reachable pairs that arise out of a given conditioned invariant subspace. There is a special case worth mentioning and that is when we assume that the observability indices of the observable pair $(\mathcal{C}, \mathcal{A})$ are generic, and in particular the case when they are all equal.

7.3 On Conditioned Invariant Subspaces

Our aim is to parameterize, given an observable pair $(\mathcal{C}, \mathcal{A})$, the set of all conditioned invariant subspaces of a given (co)dimension. Without loss of generality we can assume that $(\mathcal{C}, \mathcal{A})$ is in Brunovsky canonical form. This means that, with the coprime factorizations $\mathcal{C}(zI - \mathcal{A})^{-1} = D(z)^{-1}\Theta(z)$, we have

$$D(z) = \begin{pmatrix} z^{\mu_1} & & & \\ & \cdot & & \\ & & \cdot & \\ & & & \cdot \\ & & & & z^{\mu_p} \end{pmatrix}, \tag{7.16}$$

with $\mu_1 \geq \cdots \geq \mu_p \geq 0$. By Proposition 7.1, given a conditioned invariant subspace $\mathcal{V} \subset X_D$, it has a representation of the form $\mathcal{V} = X_D \cap T(z)F^p[z]$, for some polynomial matrix $T(z)$.

To get a better understanding, we connect the study of conditioned invariant subspaces with that of kernels of Toeplitz operators. We introduce a special case of Toeplitz operators that fits the algebraic context in which we are interested. A convenient general reference for Wiener–Hopf theory is Gohberg and Feldman [10].

Let D and T be $p\times p$ and $p\times m$ polynomial matrices respectively, and assume D is nonsingular. We define the *Toeplitz operator* $\mathcal{T}_{D^{-1}T} : F^m[z] \longrightarrow F^p[z]$, with symbol $D^{-1}T$, by

$$\mathcal{T}_{D^{-1}T} f = \pi_+ D^{-1}T f. \tag{7.17}$$

Lemma 7.1. *Let D and T be $p\times p$ and $m\times p$ polynomial matrices, respectively, and assume D is nonsingular. Let $\mathcal{T}_{D^{-1}T}$ be the Toeplitz operator defined in (7.17). The map $\psi : \operatorname{Ker}\mathcal{T}_{D^{-1}T} \longrightarrow X_D \cap TF^p[z]$ defined by*

$$\psi(p) = Tp$$

is a surjective linear map.

If T has full row rank then ψ is also injective and we have

$$\dim \operatorname{Ker}\mathcal{T}_{D^{-1}T} = \dim X_D \cap TF^p[z]. \tag{7.18}$$

Proof: Let $p \in \operatorname{Ker}\mathcal{T}_{D^{-1}T}$; that is, $\mathcal{T}_{D^{-1}T}p = \pi_+ D^{-1}Tp = 0$. Setting $f = \psi(p) = Tp$, we get $f \in X_D$ as well as $f \in TF^p[z]$; that is, $f \in X_D \cap TF^p[z]$.

Conversely, assume $f \in X_D \cap TF^p[z]$. This implies that for some strictly proper h and polynomial p we have $f = Dh = Tp = \psi(p)$. So $h = D^{-1}Tp$ and hence $p \in \operatorname{Ker}\mathcal{T}_{D^{-1}T}$. This shows that the map ψ is surjective.

If T has full column rank, it is left invertible. Hence $Tp = 0$ implies $p = 0$; that is, ψ is injective. This implies the two spaces are isomorphic, and hence the equality (7.18) follows. □

The analysis of Toeplitz operators is closely related to the study of Wiener–Hopf factorizations. These are introduced next.

Definition 7.1. *Let $G \in F^{p\times m}((z^{-1}))$ be rational. A* left Wiener–Hopf factorization at infinity *is a factorization of G of the form*

$$G = G_- D G_+ \tag{7.19}$$

with $G_+ \in F^{m\times m}[z]$ unimodular, $G_- \in F^{p\times p}[[z^{-1}]]$ biproper, and

$$D(z) = \begin{pmatrix} \Delta(z) & 0 \\ 0 & 0 \end{pmatrix},$$

where $\Delta(z) = \operatorname{diag}(z^{\lambda_1}, \ldots, z^{\lambda_r})$. The integers λ_i, assumed decreasingly ordered, are called the left factorization indices *at infinity. A* right factorization *and the* right factorization indices *are analogously defined with the plus and minus signs in (7.19) reversed.*

The Wiener–Hopf factorization indices give insight into several basic questions. The connection, via coprime factorizations, to the analysis of controllability and observability indices is well known; see Fuhrmann and Willems [9]. In the present analysis we use them to derive some important results. First, they give a necessary and sufficient condition for the computation of the codimension of a conditioned invariant subspace $\mathcal{V} = X_D \cap TF^p[z]$ in terms of T. This is done in Proposition 7.2. Second, we look into the essential uniqueness of a representation of a conditioned invariant subspace in the form $\mathcal{V} = X_D \cap TF^p[z]$. This question is taken up in Theorem 7.2.

Proposition 7.2. *Let D and T be nonsingular $p \times p$ polynomial matrices. Then all left Wiener–Hopf factorization indices of $D(z)^{-1}T(z)$ are nonpositive if and only if we have the codimension formula*

$$\operatorname{codim} X_D \cap TF^p[z] = \deg\det T. \tag{7.20}$$

Proof: Assume all left Wiener–Hopf factorization indices of $D(z)^{-1}T(z)$ are nonpositive. Then there exists a polynomial matrix S for which $D^{-1}TS$ is biproper. We have in this case

$$\mathcal{V} = X_D \cap T(z)F^p[z] = TX_S$$

and $\dim \mathcal{V} = \dim TX_S = \deg\det S$. Now, since $D^{-1}TS$ is biproper, we have

$$\deg\det D = \deg\det(TS) = \deg\det T + \deg\det S.$$

So $\operatorname{codim} \mathcal{V} = \deg\det D - \deg\det S = \deg\det T$.

Conversely, assume the codimension formula (7.20) holds. Thus necessarily T is nonsingular. Assume the left factorization indices of $D^{-1}T$ are $\lambda_1 \geq \cdots \geq \lambda_k > 0 \geq \lambda_{k+1} \geq \cdots \geq \lambda_p$. Set $\Delta_+(z) = \operatorname{diag}(z^{\lambda_1}, \ldots, z^{\lambda_k}, 1, \ldots, 1)$ and $\Delta_-(z) = \operatorname{diag}(1, \ldots, 1, z^{\lambda_{k+1}}, \ldots, z^{\lambda_p})$. The left Wiener–Hopf factorization of $D^{-1}T$ therefore has the form $D^{-1}T = \Gamma\Delta_-^{-1}\Delta_+U$, with U unimodular and Γ biproper. Using our assumption and applying Lemma 7.2 twice, we compute

$$\begin{aligned}
\deg\det D - \deg\det T &= \dim X_D \cap T(z)F^p[z] = \dim \operatorname{Ker} \mathcal{T}_{D^{-1}T} \\
&= \dim \operatorname{Ker} \mathcal{T}_{\Gamma\Delta_-^{-1}\Delta_+U} = \dim \operatorname{Ker} \mathcal{T}_{\Delta_-^{-1}\Delta_+} \\
&= \dim X_{\Delta_-} \cap \Delta_+(z)F^p[z] = \dim X_{\Delta_-} \\
&= -\textstyle\sum_{i=k+1}^p \lambda_i.
\end{aligned}$$

Thus we conclude that

$$\deg\det D - \sum_{i=k+1}^{p} \lambda_i = \deg\det T\,;$$

that is, $\dim \mathcal{V} = -\sum_{i=k+1}^{p} \lambda_i$. On the other hand, we can use the factorization $\Gamma\Delta_-^{-1}\Delta_+U$, and we obtain

$$\deg\det T = \deg\det D + \sum_{i=1}^{k} \lambda_i - \sum_{i=k+1}^{p} \lambda_i.$$

Comparing the two expressions, we conclude that $\sum_{i=1}^{k} \lambda_i = 0$; that is, all the factorization indices are nonpositive. □

Our next aim is to discuss the conditions under which the representation of a conditioned invariant subspace $\mathcal{V}$ in the form $\mathcal{V} = X_D \cap TF^p[z]$ is essentially unique; that is, T is determined up to a left unimodular factor. That such a representation is not unique in general is easily seen. In fact, for every polynomial matrix E, we have $X_D \cap DEF^p[z] = \{0\}$. The answer to this question has to do with reduced observability indices and, in turn, with Wiener–Hopf factorizations.

Given an observable pair $(\mathcal{C}, \mathcal{A})$, let $\mathcal{V}$ be a conditioned invariant subspace. Let J be an output injection map such that $(\mathcal{A} + J\mathcal{C})\mathcal{V} \subset \mathcal{V}$. Then the *restricted pair* $(\overline{A}_{\mathcal{V}}, \overline{C}_{\mathcal{V}})$ acting in the state space $\mathcal{V}$ is defined by

$$\begin{cases} \overline{A}_{\mathcal{V}} &= (\mathcal{A} + J\mathcal{C})|\mathcal{V} \\ \overline{C}_{\mathcal{V}} &= \mathcal{C}|\mathcal{V} \end{cases}$$

The pair $(\overline{C}_{\mathcal{V}}, \overline{A}_{\mathcal{V}})$ is also observable and has a set of observability indices, $\lambda_1 \geq \cdots \geq \lambda_p \geq 0$, associated with it. We refer to the λ_i as the *reduced observability indices.*

If in the representation $\mathcal{V} = X_D \cap T(z)F^p[z]$ of a conditioned invariant subspace the polynomial matrix T is nonsingular, then we can associate with T an isomorphism class of reachable pairs (A, B) acting in F^k, where $k = \deg\det T$. The column indices of T, namely, $\kappa_1 \geq \cdots \geq \kappa_p \geq 0$, are invariant under right multiplication by unimodular matrices. They are equal to the controllability indices of (A_T, B_T). We refer to (A_T, B_T) as the *coreduced system* and to the κ_i as the *coreduced controllability indices.*

It has been shown in Fuhrmann and Helmke [7] that the reduced observability indices $\lambda_1 \geq \cdots \geq \lambda_p$ satisfy

$$\lambda_i \leq \mu_i. \tag{7.21}$$

The factorization indices corresponding to a conditioned invariant subspace have a nice interpretation in terms of reduced observability indices.

Proposition 7.3. *Given a nonsingular $p \times p$ polynomial matrix D, let $\mathcal{V} = X_D \cap TF^p[z]$ be a conditioned invariant subspace. We assume without loss of generality that all left Wiener–Hopf factorization indices of $D^{-1}T$ are nonpositive.*

Then the reduced observability indices are the negatives of the left Wiener–Hopf factorization indices of $D^{-1}T$.

Proof: It can be shown that the Wiener–Hopf indices are nonpositive if and only if there exists a polynomial matrix R such that, with $D_1 := TR$, $D_1^{-1}D$ is biproper. This implies $\mathcal{V} = TX_R$. The reduced system is given by

$$\begin{cases} \overline{A} &= S_{D_1}|TX_R \simeq S_R = A_R \\ \overline{C}(Tg) &= (D_1^{-1}Tg)_{-1} = (R^{-1}g)_{-1} \simeq C_R. \end{cases}$$

So $(\overline{C}, \overline{A})$ is isomorphic to (C_R, A_R). Since the observability indices of (C_R, A_R) are equal to the row indices of R, it follows that the row indices of R are equal to the reduced observability indices.

Now, let $R = U_R \Delta \Gamma_R$ be a right Wiener–Hopf factorization. Since $D_1 = TR$, it follows that $D^{-1}D_1 = D^{-1}TR$ or $D^{-1}T = (D^{-1}D_1)R^{-1} = (\Gamma\Gamma_R^{-1})\Delta^{-1}U_R^{-1}$. This is a left Wiener–Hopf factorization of $D^{-1}T$ and the factorization indices are the negatives of the right factorization indices of R. □

We consider now conditioned invariant subspaces $\mathcal{V} \subset \mathcal{X}$, with respect to an observable pair $(\mathcal{C}, \mathcal{A})$. We recall the definition of tight subspaces; see Fuhrmann and Helmke [7]. Using the dual concept of the notion of a coasting subspace (see Willems [18]) we refer to a conditioned invariant subspace $\mathcal{V}$ as a *tight* conditioned invariant subspace if

$$\mathcal{V} + \operatorname{Ker} \mathcal{C} = \mathcal{X}$$

holds; that is, if $\mathcal{V}$ is transversal to $\operatorname{Ker} \mathcal{C}$.

The following theorem, which generalizes a result from Fuhrmann and Helmke [7], clarifies the various conditions that imply the uniqueness of a representation of a conditioned invariant subspace with respect to the shift realization.

Theorem 7.2. *Let $D(z) = \operatorname{diag}(z^{\mu_1}, \ldots, z^{\mu_p})$ with $\mu_1 \geq \cdots \geq \mu_p > 0$, and let*

$$\mathcal{V} = X_D \cap TF^p[z] \tag{7.22}$$

be a conditioned invariant subspace. Let $\lambda_1 \geq \cdots \geq \lambda_p$ be the reduced observability indices; that is, the observability indices of the system (C_D, A_D) reduced to $\mathcal{V}$. Then the following conditions are equivalent.

(i) All left Wiener–Hopf indices of $D^{-1}T$ are negative.

(ii) The reduced observability indices are all positive, that is, $\lambda_p > 0$.

(iii) $M_{\mathcal{V}}$, the smallest submodule of $F^p[z]$ that includes $\mathcal{V}$, is full, that is, has p generators which are linearly independent over $F[z]$.

(iv) *The factor module* $F^p[z]/M_{\mathcal{V}}$ *is a torsion module.*

(v) $\mathcal{V}$ *is a tight conditioned invariant subspace of* X_D.

(vi) *The corestricted system* (A_T, B_T) *is uniquely determined up to isomorphism.*

(vii) *The representation (7.22) of* $\mathcal{V}$ *is unique up to a right unimodular factor for* T.

As a simple consequence we obtain sufficient conditions for the codimension formula.

Corollary 7.1. *Let* (A, B) *be an arbitrary reachable pair, defined in a* k*–dimensional state space. Let* $(zI - A)^{-1}B = H(z)T(z)^{-1}$ *be coprime factorizations and assume* $D(z) = \operatorname{diag}(z^{\mu_1}, \dots, z^{\mu_p})$ *with* $\mu_1 \geq \cdots \geq \mu_p > 0$. *Then*

(i) *If* $k \leq \mu_p$, *then the codimension formula* $\operatorname{codim} X_D \cap TF^p[z] = \deg \det T$ *holds.*

(ii) *If* $k < \mu_p$, *then the representation of the conditioned invariant subspace* $\mathcal{V} = X_D \cap TF^p[z]$ *is essentially unique; that is,* T *is determined up to a right unimodular factor.*

7.4 The State Space Approach

For the purpose of studying the set of all conditioned invariant subspaces of fixed codimension, we would like to have a description of these subspaces also in state space terms. Indeed, this can be done through the introduction of a new object—the *partial reachability matrix*. A special instance of the partial reachability matrix was introduced in Fuhrmann and Helmke [7]. Here we remove all genericity restrictions on the reachability indices.

Let $\mu = (\mu_1, \dots, \mu_p)$ with $\mu_1 \geq \cdots \geq \mu_p > 0$. Given a reachable pair $(A, B) \in F^{k\times k} \times F^{k\times p}$, we define the *$\mu$-partial reachability matrix* $\mathcal{R}_\mu(A, B) \in F^{k\times n}$ by

$$\mathcal{R}_\mu(A, B) = (b_1, Ab_1, \dots, A^{\mu_1 - 1}b_1, \dots, b_p, \dots, A^{\mu_p - 1}b_p). \tag{7.23}$$

We can consider $\mathcal{R}_\mu$ as a map from F^n into F^k.

Definition 7.2. *Let* $1 \leq k \leq n$.

(i) *A pair* $(A, B) \in F^{k\times k} \times F^{k\times p}$ *is called* μ–regular *if the* $k \times n$ *partial reachability matrix*

$$\mathcal{R}_\mu(A, B) = (b_1, \dots, A^{\mu_1 - 1}b_1, \dots, b_p, \dots, A^{\mu_p - 1}b_p) \tag{7.24}$$

has full rank k.

(ii) *The pair (A, B) is called* tight, *if it is $\mu - 1 = (\mu_1 - 1, \dots, \mu_p - 1)$-regular; that is, if*

$$\operatorname{rank} \mathcal{R}_{\mu-1}(A, B) = \operatorname{rank}(b_1, \dots, A^{\mu_1 - 2} b_1, \dots, b_p, \dots, A^{\mu_p - 2} b_p) = k. \tag{7.25}$$

Obviously, any tight pair is μ–regular and μ–regularity implies reachability. We pass on to the state space characterization of tightness.

Theorem 7.3. *Let $D(z) = \operatorname{diag}(z^{\mu_1}, \dots, z^{\mu_p})$ with $\mu_1 \geq \dots \geq \mu_p > 0$ and $\mu_1 + \dots + \mu_p = n$. Assume $(A, B) \in F^{k \times k} \times F^{k \times p}$ is a reachable pair. Let $H(z)T(z)^{-1}$ be a right coprime factorization of $(zI - A)^{-1}B$. Then the following statements are equivalent.*

(i) *All the left factorization indices of $D^{-1}T$, $-\lambda_i$ are negative, and we assume them decreasingly ordered; that is, $\lambda_1 \geq \dots \geq \lambda_p > 0$.*

(ii) *The pair (A, B) is a tight pair.*

(iii) *The subspace $Ker \mathcal{R}_\mu(A, B)$ is tight; that is,*

$$Ker \mathcal{R}_\mu(A, B) + Ker \mathcal{C} = F^n. \tag{7.26}$$

Proof: The left factorization indices λ_i of $D^{-1}T$ are negative if and only if the restriction $\mathcal{C}|_{\mathcal{V}} : \mathcal{V} \to F^p$ is surjective, or equivalently if and only if $\mathcal{V}$ is tight. Thus the equivalence of the first and last conditions follows from Theorem 7.4 *(b)* (whose proof depends only on the equivalence of *(ii)* and *(iii)*). To complete the proof we note that $Ker\ \mathcal{R}_\mu(A, B)$ is tight if and only if

$$\dim\ (Ker\ \mathcal{R}_\mu + Ker\ \mathcal{C}) = n;$$

that is, if and only if

$$\dim\ Ker \begin{pmatrix} \mathcal{R}_\mu \\ \mathcal{C} \end{pmatrix} = n - rk\ \mathcal{R}_\mu - rk\ \mathcal{C}.$$

Equivalently, this holds if

$$rk \begin{pmatrix} \mathcal{R}_\mu \\ \mathcal{C} \end{pmatrix} = rk\ \mathcal{R}_\mu + rk\ \mathcal{C}.$$

Since $\mathcal{C} = \operatorname{diag}\ (e_{\mu_1}^T, \dots, e_{\mu_p}^T)$ this is equivalent to

$$\begin{bmatrix} b_1 & \dots & A^{\mu_1 - 1} b_1 & \dots & b_p & \dots & A^{\mu_p - 1} b_p \\ 0 & & 1 & & 0 & & 0 \\ \vdots & & \vdots & & \vdots & & \vdots \\ 0 & & 0 & & 0 & & 1 \end{bmatrix}$$

being full row rank. But this just means that

$$rk\ \mathcal{R}_{\mu-1}(A,B) = rk\ \mathcal{R}_\mu(A,B);$$

that is, (A,B) being tight. □

Let $(\mathcal{C},\mathcal{A}) \in F^{p\times n} \times F^{n\times n}$ denote an observable pair in dual Brunovsky canonical form with observability indices $\mu_1 \geq \cdots \geq \mu_p \geq 1$. We now characterize tight conditioned invariant subspaces in terms of similarity orbits of tight pairs.

Theorem 7.4.

(a) For any conditioned invariant subspace $\mathcal{V}$ of $(\mathcal{C},\mathcal{A})$ of codimension k there exists a μ–regular pair $(A,B) \in F^{k\times k} \times F^{k\times p}$ with

$$\mathcal{V} = Ker\ \mathcal{R}_\mu(A,B).$$

Conversely, if $(A,B) \in F^{k\times k} \times F^{k\times p}$ is μ–regular, then $Ker\ \mathcal{R}_\mu(A,B)$ is a conditioned invariant subspace of codimension k.

(b) $Ker\ \mathcal{R}_\mu(A,B)$ is a tight subspace if and only if (A,B) is tight; that is, if and only if $rk\ \mathcal{R}_{\mu-1}(A,B) = k$. For any tight conditioned invariant subspace $\mathcal{V}$ there exists, up to state space similarity, a unique μ–regular (A,B) with $Ker\ \mathcal{R}_\mu(A,B) = V$.

Proof: *(a)* $Ker\ \mathcal{R}_\mu(A,B)$ is conditioned invariant if and only if the subspace $Ker\ \mathcal{R}_\mu(A,B)^\perp = Im\ \mathcal{O}_\mu(B^T,A^T)$ is $(\mathcal{A}^T,\mathcal{C}^T)$–invariant. Here $\mathcal{O}_\mu(H,F)$ denotes the analogously defined partial observability matrix. For

$$\mathcal{A}^T = \left(\begin{array}{cccc|c|cccc}
0 & 1 & & & & & & & \\
 & \ddots & \ddots & & & & & & \\
 & & \ddots & 1 & & & & & \\
 & & & 0 & & & & & \\
\hline
 & & & & \ddots & & & & \\
\hline
 & & & & & 0 & 1 & & \\
 & & & & & & \ddots & \ddots & \\
 & & & & & & & \ddots & 1 \\
 & & & & & & & & 0
\end{array}\right) \qquad \mathcal{C}^T = \left[\begin{array}{c|c|c}
0 & & \\
\vdots & & \\
1 & & \\
\hline
 & \ddots & \\
\hline
 & & 0 \\
 & & \vdots \\
 & & 1
\end{array}\right],$$

and $X \in F^{n\times k}$ partitioned as

$$X = \begin{pmatrix} X_1^0 \\ \vdots \\ X_1^{\mu_1-1} \\ \vdots \\ X_p^0 \\ \vdots \\ X_p^{\mu_p-1} \end{pmatrix},$$

it follows that *span* X is $(\mathcal{A}^T, \mathcal{C}^T)$–invariant if and only if there exists $T \in F^{k\times k}$ with

$$X_j^1 = X_j^0 T, \ldots, X_j^{\mu_j-1} = X_j^{\mu_j-2} T, \qquad j = 1, \ldots, p.$$

Equivalently, for $B := (X_1^{0T}, \ldots, X_p^{0T}), A^T := T$ we obtain that

$$X = \mathcal{O}_\mu(B^T, A^T).$$

This proves *(a)*.

(b) By the previous lemma the subspace $Ker\ \mathcal{R}_\mu(A, B)$ is tight if and only if (A, B) is $(\mu - 1)$–regular (i.e., tight). Suppose $(A_1, B_1), (A_2, B_2)$ are tight pairs with

$$Ker\ \mathcal{R}_\mu(A_1, B_1) = Ker\ \mathcal{R}_\mu(A_2, B_2).$$

Then, up to a similarity transformation on (A_i, B_i), we can assume that

$$\mathcal{R}_\mu(A_1, B_1) = \mathcal{R}_\mu(A_2, B_2).$$

Since $\mu_p \geq 1$ we have $B_1 = B_2$. Moreover it is easily seen that $A_2 = A_1 + X$ for an arbitrary $X \in F^{k\times k}$ with

$$X\ R_{\mu-1}(A_1, B_1) = 0.$$

Since (A_1, B_1) is tight we conclude that $X = 0$ and hence $A_1 = A_2$. This completes the proof. □

Finally we mention an interesting system interpretation of the kernel of the partial reachability matrix as the *module of zero return*, introduced by Münzner and Prätzel–Wolters [16].

Proposition 7.4. *Let* $(A, B) \in F^{k\times k} \times F^{k\times p}$ *and* $u = (u_0^1, \ldots, u_{\mu_1-1}^1, \ldots, u_0^p, \ldots, u_{\mu_p-1}^p)^T \in F^n$; $u_t^j := 0$ *for* $t \geq \mu_j$. *Then* $u \in Ker\ \mathcal{R}_\mu(A, B)$ *if and only if the state trajectory of*

$$x_{t+1} = Ax_t + b_1 u_t^1 + \cdots + b_p u_t^p, \qquad x_0 := 0,$$

satisfies $x_{\mu_1} = 0$.

7.5 On the Parameterization of Conditioned Invariant Subspaces

We proceed now, given an observable pair $(\mathcal{C}, \mathcal{A})$, to parameterize the set of all conditioned invariant subspaces. This problem was first solved in Hinrichsen et al. [15]. Unfortunately, this pathbreaking paper did not get the attention it deserved and for a long time the study of this problem was effectively abandoned. A recurrence of interest in this parameterization problem occurred in Fuhrmann and Helmke [7] and in the papers arising out of X. Puerta's thesis [17], namely Ferrer et al. [3, 4].

In Hinrichsen et al. [15] a module– theoretic approach was used based on the characterization of conditioned invariant subspaces obtained in Fuhrmann [5], [6], but put in the context of rational models. Our approach is inspired by that of Hinrichsen et al. as well as the thesis of Eckberg [2]. We give a full proof of the reduction of a polynomial matrix to (modified) Kronecker–Hermite canonical form which, we believe, is more transparent than those given in Eckberg [2] and Hinrichsen and Prätzel–Wolters [14]. The addition is a new count of the number of free parameters in the Kronecker–Hermite canonical form.

The first instinct might be, given a conditioned invariant subspace $\mathcal{V} \subset X_D$, to choose a basis for $< \mathcal{V} >$, the submodule of $F^p[z]$ generated by it and reduce it to some, say column proper, canonical form. This turns out to be unfruitful. The main reason for this is the fact that in such a reduction the basis elements may not stay in X_D. Moreover the corresponding column indices have nothing to do with D, so they can not have a system–theoretic significance. This shows that the significant object should be $D^{-1}T$.

In order not to make the notation overly cumbersome, we assume that $\mathcal{C}$ has full row rank. This is equivalent to the positivity of all of the observability indices; that is,

$$\mu_1 \geq \cdots \geq \mu_p > 0. \tag{7.27}$$

Since the lattice of conditioned invariant subspace is invariant, up to isomorphism, under the full output injection group, it entails no loss of generality to assume that $(\mathcal{C}, \mathcal{A})$ is in dual Brunovsky form; that is, that it corresponds to the polynomial matrix

$$D(z) = \begin{pmatrix} z^{\mu_1} & & & \\ & \cdot & & \\ & & \cdot & \\ & & & \cdot \\ & & & & z^{\mu_p} \end{pmatrix}. \tag{7.28}$$

In order to gain some intuition, we consider the simplest case, namely, when the submodule $< \mathcal{V} >$ has a single generator, say $h \in \mathcal{V}$. This generator is uniquely determined up to a nonzero scalar factor. We can use this freedom to make one of the polynomials h_i monic. Now, there are two clear invariants for the subspace $\mathcal{V}$. The first one is $d = \dim \mathcal{V}$ which we know, by Proposition 7.3, is equal to the negative of the only one left Wiener–Hopf factorization index δ of $g = D^{-1}h$. Since $D(z) = \operatorname{diag}(z^{\mu_1}, \ldots, z^{\mu_p})$, we have $-d = \max_k \deg z^{-\mu_k} h_k$. So there exists at least one index for which this equality holds and we assume i is the largest index for which this holds. Thus with each singly generated conditioned invariant subspace, two integers δ, i exist, defined as above. The set of all conditioned invariant subspaces having those invariants is denoted $\mathcal{M}(\delta, i)$. Each singly generated conditioned invariant subspace is contained in exactly one of these sets.

Clearly, $g = D^{-1}h \in \mathcal{M}(\delta, i)$ if and only if with $g = \begin{pmatrix} g_1 \\ \cdot \\ \cdot \\ \cdot \\ g_p \end{pmatrix}$, we have

$$g_k(z) = \begin{cases} \sum_{j=0}^{\mu_k-\delta} h_{kj} z^{-\mu_k+j} & k < i \\ \sum_{j=0}^{\mu_i-\delta} h_{ij} z^{-\mu_i+j} & k = i \ h_{ii} \text{ is monic} \\ \sum_{j=0}^{\mu_k-\delta-1} h_{kj} z^{-\mu_k+j} & k > i \ \& \ \mu_k > \delta \\ 0 & k > i \ \& \ \mu_k \leq \delta\,. \end{cases} \tag{7.29}$$

The number of free parameters we need to parameterize $\mathcal{M}(\delta, i)$ is the number $N(\delta, i)$ of coefficients, not counting h_{ii}. This is given by

$$N(\delta, i) = \sum_{k=1}^{i-1} (\mu_k - \delta + 1) + (\mu_i - \delta) + \sum_{k>i} (\mu_k - \delta)^+$$

$$= \sum_{k=1}^{p} (\mu_k - \delta)^+ + (i - 1).$$

Let G be any generator matrix. Two generating matrices for $< \mathcal{W} >$ differ by a right unimodular factor. Let ν be any integer for which $z^\nu G = K$ is a polynomial matrix. Recall that D is in Brunovsky form (7.28) and hence such an integer exists. One should note that, starting from a generating matrix H for the submodule $< \mathcal{V} >$, we have $K = H$, for some integer ν, if and only if all the left Wiener–Hopf factorization indices are equal. This corresponds to our observation in Section 7.3 that the main object of interest is not the generator H of $< \mathcal{V} >$, but rather the generator $G = D^{-1}H$ of $< \mathcal{W} > = < D^{-1}\mathcal{V} >$. Clearly, if U is any unimodular polynomial matrix such that GU is column or row proper, the same holds for K. So, without loss of generality, we can look for a canonical form, under right unimodular equivalence, for rectangular polynomial matrices. We want to exhibit the minimum number of nonintegral parameters. This means reducing the degrees of elements of K as much as possible. So reduction to column proper form is a natural step. However, after such a reduction, there is still some freedom of applying elementary column operation. This would allow us to choose a pivot element in each column and reduce elements in its row. This leads to a canonical form, introduced in Eckberg [2] and Hinrichsen et al. [15], where it is called the Kronecker–Hermite form. We recall the definition.

Definition 7.3. *Let $H(z)$ be a $p \times l$, full column rank polynomial matrix. Let $h_1, \dots, h_l$ denote the columns of H and let h_{ij} denote the ith element of h_j. We say that H is in* Kronecker–Hermite canonical form *if there exists uniquely determined indices $1 \le i_1 < \cdots < i_l \le p$, such that*

(i) $h_{i_j j}$ is monic with $\delta_j := \deg h_{i_j j} = \deg h_j$;

(ii) we have $\deg h_{i_j k} < \delta_j$ for $1 \le k \le l, k \ne j$;

(iii) we have if $i > i_j$ then $\deg h_{ij} < \delta_j$.

Note that a matrix H in the Kronecker–Hermite canonical form is column proper and the $l \times l$ submatrix $\overline{H}$ consisting of the $i_1, \dots, i_l$ rows is row proper with $[H]_l$, the matrix of leading row coefficients of H, equal to I_l.

It is useful for our purposes to modify somewhat the definition of the Kronecker–Hermite canonical form.

Definition 7.4. *Let $H(z)$ be a $p \times l$, full column rank polynomial matrix. We say that H is in* modified Kronecker–Hermite canonical form *if there exist two uniquely determined lists of indices $\nu_1 < \cdots < \nu_s$ and $0 < k_1, \dots, k_s$ with $\sum_{i=1}^{s} k_i = l$, and disjoint sets of row indices $R_i = \{1 \le \rho_1^{(i)} < \cdots < \rho_{k_i}^{(i)} \le p\}$ such that*

(i) We have the partitioning

$$H(z) = (H_1(z); \dots ; H_s(z)); \tag{7.30}$$

(ii) $H^{(i)}$ is a $p \times k_i$ column proper matrix with all column degrees equal to ν_i;

(iii) denoting the λ, μ entry of $H^{(i)}$ by $h_{\lambda,\mu}^{(i)}$, we have

(a) $h_{\rho_j,j}^{(i)}$ is monic of degree ν_i;

(b) for $\lambda > \rho_j$, we have $\deg h_{\lambda,j}^{(i)} < \deg h_{\rho_j,j}^{(i)}$, and

(c) for all $i = 1, \dots, s$, we have

$$\deg h_{\rho_j,\mu}^{(i)} < \deg h_{\rho_j,j}^{(i)} \quad \text{for} \begin{cases} j = 1, \dots, k_i \\ \mu \ne j; \end{cases} \tag{7.31}$$

(d) for all $i = 1, \ldots, s$, we have

$$\deg h^{(t)}_{\rho_j,\mu} < \deg h^{(i)}_{\rho_j,j} \quad \textit{for} \begin{cases} t \neq i \\ j = 1, \ldots, k_i \\ \mu = 1, \ldots, k_t\,. \end{cases} \tag{7.32}$$

The partition (7.30) *is referred to as the* canonical partition *of H.*

Note that the Kronecker–Hermite canonical form is obtained from the modified Kronecker–Hermite canonical form by a permutation of columns that arranges the set indices $\bigcup_{i=1}^{s} R_i$ in increasing order.

Before proceeding with the statement of the principal reduction result, we prove the following simple lemmas we use in the sequel.

Lemma 7.2. *Let $T(z) = T_0 + \cdots + T_\nu z^\nu \in F^{m\times m}[z]$ with T_ν nonsingular. Let $f \in F^m[z]$ with $\deg f = n$. The f can be uniquely represented as*

$$f = Tg + r \tag{7.33}$$

with $r \in F^m[z]$ and $\deg r < \nu$. Moreover, we have $\deg g = \deg f - \nu$.

Proof: Clearly, the nonsingularity of T_ν implies the nonsingularity of T. We apply the projection π_T to f and write $r = \pi_T f$. Now $h = T^{-1}r$ is strictly proper, so

$$\begin{aligned} r(z) &= T(z)h(z) = (T_0 + \cdots + T_\nu z^\nu)(\frac{h_1}{z} + \frac{h_2}{z^2} + \cdots) \\ &= T_\nu h_1 z^{\nu-1} + \cdots, \end{aligned}$$

which shows that $\deg r \leq \nu - 1$. Assume $\deg g = \mu$. If $\deg f < \nu$, we have $g = 0$. Otherwise $\deg f = \deg T + \deg g$ for the highest term in Tg is $T_\nu g_\mu z^{\nu+\mu}$ and the coefficient is nonzero by the assumed nonsingularity of T_ν. Hence $\deg g = \deg f - \nu$. □

The previous lemma can be extended to the case that $T(z)$ is a row proper matrix; see Eckberg [2].

Lemma 7.3. *Let $T(z) \in F^{m\times m}[z]$ be row proper with row degrees $\nu_1, \ldots, \nu_m$. Let $f \in F^m[z]$. The f can be uniquely represented in the form*

$$f = Tg + r \tag{7.34}$$

with $r = \begin{pmatrix} r_1 \\ \cdot \\ \cdot \\ \cdot \\ r_m \end{pmatrix} \in F^m[z]$ *and* $\deg r_i < \nu_i$.

Moreover, if $\gamma = \max_{1\le i\le m}(\deg f_i - \nu_i)$ *then* $\deg g = \gamma$ *when* $\gamma \le 0$*; otherwise* $g = 0$.

We have the following theorem due to Eckberg [2] and Hinrichsen and Prätzel–Wolters [14].

Theorem 7.5. *Every full column rank,* $p \times l$ *polynomial matrix* $H(z)$ *can be reduced to a unique modified Kronecker–Hermite canonical form by right multiplication by a unimodular matrix.*

Proof: The basic idea of the proof is the reduction of the number of free parameters, using elementary column operations. The technical tool is a repeated application of the division rule for polynomials.

We prove the theorem by induction. We show that in this case H is reducible, by elementary column operations, to the Modified Kronecker–Hermite canonical form. To this end we assume without loss of generality that H is in column proper form with s distinct column indices $\nu_1 < \cdots < \nu_s$ and with k_i columns of degree ν_i. Thus $\sum_{i=1}^{s} k_i = l$ and $\sum_{i=1}^{s} k_i\nu_i = n$. Let us write

$$H(z) = (H_1(z); \ldots ; H_s(z));$$

H_i are $p \times k_i$ column proper polynomial matrices. Our proof goes by induction on the number s of distinct minimal column degrees.

If $s = 1$, all column degrees are equal to ν_1 and we can reduce the highest coefficient column matrix to reverse echelon form, using multiplication on the right by constant elementary matrices. The resulting matrix is obviously in the modified Kronecker–Hermite canonical form and this form is uniquely determined. We assume the pivot elements to be in the rows $i_1^{(1)} < \cdots < i_{k_1}^{(1)}$.

The proof is by induction on the number of distinct column indices. So our induction hypothesis is that the $p\times(k_1+\cdots+k_{i-1})$ matrix $(H^{(1)}(z); \ldots ; H^{(i-1)}(z))$ is in modified Kronecker–Hermite form. In particular the $p\times k_j$ matrices $H^{(j)}, j = 1, \ldots , i-1$ are column proper with the columns of $H^{(j)}$ of degree ν_j and the leading column coefficient matrix in reverse echelon form. Furthermore, we assume that $\deg h_\sigma^{(j)} < \nu_m$ for $\sigma \in \{i_1^{(m)}, \ldots , i_{k_m}^{(m)}\}$.

We consider now the $p \times k_i$ matrix $H^{(i)}$. $(H^{(1)}(z); \ldots ; H^{(i-1)}(z))$, the $k_1 + \cdots + k_{i-1}$ submatrix of $H(z)$ which is based on the rows $\bigcup_{m=1}^{i-1}\{i_1^{(m)}, \ldots , i_{k_m}^{(m)}\}$ is, by the induction hypothesis, nonsingular column proper as well as row proper with its leading row coefficient matrix a permutation matrix.

We reduce now, using Lemma 7.3, the $(k_1 + \cdots + k_{i-1}) \times k_i$ submatrix of $H^{(i)}$, which is based on the rows $\bigcup_{m=1}^{i-1}\{i_1^{(m)}, \ldots , i_{k_m}^{(m)}\}$ of $(H^{(1)}(z); \ldots ; H^{(i-1)}(z))$ with respect to the submatrix of $(H^{(1)}(z); \ldots ; H^{(i-1)}(z))$ of size $(k_1 + \cdots + k_{i-1}) \times (k_1 + \cdots + k_{i-1})$ which is based on the rows $\bigcup_{m=1}^{i-1}\{i_1^{(m)}, \ldots , i_{k_m}^{(m)}\}$. We retain the notation $H^{(i)}$ for the reduced matrix and we note that this reduction does not increase the degrees of the columns. The reduction process did not affect the columns of the submatrices $H^{(j)}$, $j \neq i$. Thus it follows that $H^{(i)}$ is still column proper with column degrees ν_i. Moreover the elements in the rows $i_1^{(j)}, \ldots, i_{k_j}^{(j)}$, for $j = 1, \ldots, i-1$, have degrees $< \nu_j$. So the degree ν_i elements occur in the complementary rows. By constant elementary column operations on the columns of $H^{(i)}$ we reduce the leading column coefficient matrix to reverse echelon form. These elementary operations keep the degrees of the elements in rows $i_1^{(j)}, \ldots, i_{k_j}^{(j)}$, for $j = 1, \ldots, i-1$, below ν_j. On the other hand the elements of $(H^{(1)}(z), \ldots, H^{(i-1)}(z))$ in rows $i_1^{(i)}, \ldots, i_{k_i}^{(i)}$, have degrees less than ν_i. Thus we have reduced $(H^{(1)}(z); \ldots ; H^{(i)}(z))$ to canonical form and this holds also for $i = s$.

We note that, by construction, the $l \times l$ submatrix based on the rows $i_1^{(j)}, \ldots, i_{k_j}^{(j)}$, $j = 1, \ldots, s$, is both row and column proper and its determinant has degree ν.

We proceed now to count the number of free continuous parameters in the canonical form. Obviously, the canonical form is completely determined by the integral parameters $R_i = \{\rho_1^{(i)}, \ldots, \rho_{k_i}^{(i)}\}, \nu_i$, $i = 1, \ldots, s$, where the sets of row indices R_i are disjoint and the degrees ν_i are distinct. Note that we have $\sum_{i=1}^{s} k_i = l$.

We first count the number of free parameters in $H^{(i)}$, disregarding the degree reduction due to the constraints (7.31) and (7.32). Consider first the jth column, $j = 1, \ldots, k_i$, of the submatrix $H^{(i)}$. It has a monic degree ν_i polynomial in row $\rho_j^{(i)}$. This polynomial has ν_i free parameters as do all polynomials in the lower rows of the same column. The polynomials in the rows above it have degree ν_i and therefore have $\nu_i + 1$ free parameters. Altogether the jth column has

$$(\nu_i + 1)(\rho_j^{(i)} - 1) + \nu_i(p - \rho_j^{(i)} + 1) = (\nu_i + 1)p - (p - \rho_j^{(i)} + 1)$$

free parameters. Adding over all columns of $H^{(i)}$ and over all submatrices, we get

$$\begin{aligned}
\sum_{i=1}^{s} \sum_{j=1}^{k_i} [(\nu_i + 1)p - (p - \rho_j^{(i)} + 1)] &= \sum_{i=1}^{s} k_i(\nu_i + 1)p - \\
&\quad - \sum_{i=1}^{s} \sum_{j=1}^{k_i} (p - \rho_j^{(i)} + 1) \\
&= p(n + l) - \sum_{i=1}^{s} \sum_{j=1}^{k_i} (p - \rho_j^{(i)} + 1)\,.
\end{aligned}$$

From this number we have to subtract the number of parameters cancelled by the constraints (7.31) and (7.32). We treat the constraint (7.31) first.
In this case each pivot element reduces the degrees of the terms to its right by one. Thus the total number of parameters in $H^{(i)}$ reduced due to this constraint is

$$(k_i - 1) + (k_i - 2) + \cdots + 1 + 0 = \frac{(k_i - 1)k_i}{2}.$$

Summing up over all indices $i = 1, \ldots, s$, we get

$$\sum_{i=1}^{s} \frac{(k_i - 1)k_i}{2}.$$

Clearly, the constraint (7.32) affects the entries of $H^{(i)}$ only by the submatrices to its left. Now the pivot row indices of $H^{(i)}$ are

$$\rho_1^{(i)} < \cdots < \rho_{k_i}^{(i)},$$

whereas the pivot row indices of $H^{(j)}$, $1 \leq j < i$, are

$$\rho_1^{(j)} < \cdots < \rho_{k_j}^{(j)}$$

and the sets of indices are disjoint. The relative position of the second set with respect to the first determines the extra reduction in degrees. This is a combinatorial problem and for this purpose we need to count the number $\omega_{j,\mu}^{(i)}$ of $\rho_t^{(j)}$ that satisfy

$$\rho_\mu^{(i)} < \rho_t^{(j)} < \rho_{\mu+1}^{(i)}, \qquad \mu = 0, \ldots, k_i. \tag{7.35}$$

To account for the boundary cases, we take $\rho_0^{(i)} = 0$ and $\rho_{k_i+1}^{(i)} = p + 1$; that is, in the extremal cases there is only one nontrivial inequality to satisfy.

Taking, for $k \in \mathbf{Z}$, $k^+ = (k + |k|)/2$, we clearly have that the number of $\rho_t^{(j)}$ satisfying (7.35) is given by

$$\omega_{j,\mu}^{(i)} = \sum_{t=1}^{k_j} \frac{\operatorname{sign}(\rho_{\mu+1}^{(i)} - \rho_t^{(j)})^+ + \sum_{t=1}^{k_j} \operatorname{sign}(\rho_t^{(j)} - \rho_\mu^{(i)})^+}{2}. \tag{7.36}$$

For each index $\rho_t^{(j)}$ satisfying (7.35), the extra reduction in the number of parameters of $H^{(i)}$ is

$$\mu(\nu_i - \nu_j) + (k_i - \mu)(\nu_i - \nu_j + 1) = k_i(\nu_i - \nu_j + 1) - \mu.$$

So, the total number of parameters of $H^{(i)}$ that is reduced by satisfying (7.35) is

$$\sum_{j=1}^{i-1} \sum_{\mu=0}^{k_i} \omega_{j,\mu}^{(i)} [k_i(\nu_i - \nu_j + 1) - \mu].$$

Finally, the number of free continuous parameters in H is

$$\begin{aligned} N &= p(n+l) - \textstyle\sum_{i=1}^{s} \sum_{j=1}^{k_i} (p - \rho_j^{(i)} + 1) - \sum_{i=1}^{s} \dfrac{(k_i - 1)k_i}{2} \\ &\quad - \textstyle\sum_{i=2}^{s} \sum_{j=1}^{i-1} \sum_{\mu=0}^{k_i} \omega_{j,\mu}^{(i)} [k_i(\nu_i - \nu_j + 1) - \mu]. \end{aligned} \tag{7.37}$$

To show uniqueness of the Kronecker–Hermite canonical form, assume that H, H' are two column equivalent matrices, both in canonical form. Clearly, by the definition, they have the same row indices $i_1, \ldots, i_l$ and column degrees $\delta_1, \ldots, \delta_l$. Let $\overline{H}, \overline{H}'$ be the two submatrices consisting of the $i_1, \ldots, i_l$ rows of H. By column equivalence, there exists a unimodular polynomial matrix such that $H = H'U$ and hence also $\overline{H} = \overline{H}'U$. Since the row degrees are the same, we can write $\overline{H} = \Delta\Gamma$ and $\overline{H}' = \Delta\Gamma'$, where $\Delta(z) = \operatorname{diag}(z^{\delta_1}, \ldots, z^{\delta_l})$. Here Γ and Γ' are biproper with leading term equal to I_l. This implies that $(\Gamma')^{-1}\Gamma = U$ and hence both terms are constant matrices, necessarily equal to I_l. Thus $H = H'$. □

7.6 Topology of Tight Conditioned Invariant Subspaces

We now turn to an analysis of the topology of spaces of condition invariant subspaces. For simplicity we focus on the tight situation. Let $(\mathcal{C}, \mathcal{A}) \in F^{p\times n} \times F^{n\times n}$ denote an observable pair in dual Brunovsky canonical form with observability indices $\mu_1 \geq \cdots \geq \mu_p \geq 1$. We denote by

$$Inv_k(\mathcal{C}, \mathcal{A}) := \{\mathcal{V} \in Grass_{n-k}(F^n) \mid \mathcal{A}(\mathcal{V} \cap Ker\mathcal{C}) \subset \mathcal{V}\}$$

the set of all conditioned invariant subspaces of codimension k, viewed as a subset of the Grassmann manifold of all codimension k linear subspaces of F^n. Likewise, we consider the open subset

$$Tight_k(\mathcal{C}, \mathcal{A}) := \{\mathcal{V} \in Inv_k(\mathcal{C}, \mathcal{A}) \mid \mathcal{V} \text{ tight}\}$$

of *tight conditioned invariant* subspaces; that is, conditioned invariant subspaces $\mathcal{V} \subset F^n$ that satisfy the transversality condition

$$\mathcal{V} + Ker\ \mathcal{C} = F^n.$$

Given any controllable pair $(A, B) \in F^{k\times k} \times F^{k\times p}$ let

$$\mathcal{R}_\mu(A, B) = (b_1, \ldots, A^{\mu_1 - 1}b_1, \ldots, b_p, \ldots, A^{\mu_p - 1}b_p)$$

denote the $k \times n$ partial reachability matrix.

To state a geometric reformulation of our previous parameterization results we recall some basic facts about the geometry of controllable pairs. We refer to Helmke [12] for further results. Let $\mathbb{F} = \mathbb{R}, \mathbb{C}$. Let $\sum_{k,p}(\mathbb{F})$ denote the orbit space of controllable pairs $(A, B) \in \mathbb{F}^{k\times k} \times \mathbb{F}^{k\times p}$ with respect to state space similarity. Thus $\sum_{k,p}(\mathbb{F})$ is the set of all orbits

$$[A, B] = \{(SAS^{-1}, SB) \mid S \in GL_k(\mathbb{F})\}$$

of controllable pairs. It is well known that $\sum_{k,p}(\mathbb{F})$ is a smooth connected manifold of dimension (over $\mathbb{F}$) kp. Let

$$\sum_{k,p}(\mu) := \{[A,B] \in \sum_{k,p} \mid rk\ \mathcal{R}_\mu(A,B) = k\}$$

denote the subset of all similarity orbits of μ–regular pairs and

$$\sum_{k,p}^t(\mu) := \{[A,B] \in \sum_{k,p} \mid rk\ \mathcal{R}_{\mu-1}(A,B) = k\}$$

the subset consisting of similarity orbits of tight pairs (A, B). Then $\sum_{k,p}(\mu)$ and $\sum_{k,p}^t(\mu)$ are *open* subsets of $\sum_{k,p}(\mathbb{F})$ and $\sum_{k,p}^t(\mu)$ is nonempty if and only if $k \leq n-p$. Moreover, $\sum_{k,p}(\mu)$ and $\sum_{k,p}^t(\mu)$ are dense subsets (the latter only if $k \leq n-p$).

Theorem 7.6. *The map*

$$\begin{aligned} \rho_\mu \quad : \quad & \sum_{k,p}(\mu) \to Inv_k(\mathcal{C},\mathcal{A}) \\ & [A,B] \mapsto Ker\ \mathcal{R}_\mu(A,B) \end{aligned}$$

is a surjective algebraic map. It restricts to a bijection

$$\rho_\mu : \sum_{k,p}^t(\mu) \to Tight_k(\mathcal{C},\mathcal{A})$$

onto the tight subspaces. In particular, $\sum_{k,p}^t(\mu)$ and $Tight_k(\mathcal{C},\mathcal{A})$ are homeomorphic. $Inv_k(\mathcal{C},\mathcal{A})$ has nonempty interior in $Grass_{n-k}(\mathbb{F}^k)$ if and only if $p = n-k$, provided $k \leq n-p$.

Corollary 7.2. *Let $k \leq n-p$.*

(a) $Inv_k(\mathcal{C},\mathcal{A})$ is connected and contains $Tight_k(\mathcal{C},\mathcal{A})$ as an open and dense subset.

(b) $Tight_k(\mathcal{C},\mathcal{A})$ is a smooth connected manifold of dimension kp.

Conjecture 7.1. *$Inv_k(\mathcal{C},\mathcal{A})$ is homotopy equivalent to $Inv_k(\mathcal{A})$.*

Generalizing the work in Fuhrmann and Helmke [7] we inspect the generic situation. Recall, that the generic observability indices of $(\mathcal{C},\mathcal{A})$ are "as equal as possible."

Theorem 7.7. *Let $\mu = (n,\ldots,n)$ or $\mu = (n,\ldots,n,n-1)$. Then*

(a)

$$\sum_{k,p}(\mu) = \{[A,B] \in \sum_{k,p} \mid \kappa_1(A,B) \leq n\},$$

where κ_1 denotes the largest controllability index of (A,B);

(b) for $\mu = (n, \ldots, n)$,

$$\sum_{k,p}^{t}(\mu) = \{[A, B] \in \sum_{k,p} \mid \kappa_1(A, B) \leq n - 1\}$$

and for $\mu = (n, \ldots, n, n-1)$,

$$\sum_{k,p}^{t}(\mu) = \{[A, B] \in \sum_{k,p} \mid \kappa_1(A, B) \leq n - 1 \text{ and } \kappa_p(A, B) \leq n - 2\}.$$

Proof: For any observability indices $\mu_1 \geq \cdots \geq \mu_p \geq 1$ it is easily seen that a μ–tight pair (A, B) satisfies $\kappa_1 \leq \mu_1 - 1$. Conversely, if $\mu = (n, \ldots, n)$ or $\mu = (n, \ldots, n, n-1)$ is generic, then $\kappa_1 \leq n$ and controllability of (A, B) implies μ–regularity (similar for tightness). Thus the result follows. □

Consequently, in the generic situation, $\sum_{k,p}(\mu)$ is a union of Brunovsky strata. For any partition $\kappa_1 \geq \cdots \geq \kappa_p$ of k let

$$\sum(\kappa) := \{[A, B] \in \sum_{k,p} \mid (A, B) \text{ has Brunovsky indices } \kappa\}$$

denote the Brunovsky stratum for κ. The topology of Brunovsky strata is well understood and it is in principle easy to determine, for example, explicit formulas for the mod 2 Betti numbers of $\sum(\kappa)$. However we do not do so here. Rather we show how $\Sigma_{k,p}(\mu)$ decomposes as the union of Brunovsky strata by means of examples.

Example 7.1.
$\sum_{1,2}(2, 2) = \sum_{1,2} = \sum(1, 0)$.

$\sum_{2,2}(2, 2) = \sum_{2,2}$, $\sum_{3,2}(2, 2) = \sum(2, 1)$.

$\sum_{3,2}(3, 3) = \sum_{3,2}$, $\sum_{4,2}(3, 3) = \sum(3, 1) \cup \sum(2, 2)$.

7.7 Brunovsky Strata for Conditioned Invariant Subspaces

We investigate the geometry of tight conditioned invariant subspaces with fixed restriction indices. Here we follow and extend the work of Ferrer et al. [3],

Let $\mathcal{V} \subset F^n$ be a conditioned invariant subspace of codimension k and let $J \in F^{n \times p}$ be such that $(\mathcal{A} + J\mathcal{C})\mathcal{V} \subset \mathcal{V}$. The *restricted system* to $\mathcal{V}$ is the observable pair $(\mathcal{C}|_{\mathcal{V}}, (\mathcal{A} + J\mathcal{C})|_{\mathcal{V}})$ with observability indices $\lambda_1 \geq \cdots \geq \lambda_p$, $\lambda_1 + \cdots + \lambda_p = n - k$. The observability indices $\lambda_1, \ldots, \lambda_p$ are called the *restriction indices* of $\mathcal{V}$. They satisfy the inequalities

$$\lambda_1 \leq \mu_1, \ldots, \lambda_p \leq \mu_p.$$

Since observability indices are output injection invariants, the restriction indices do not depend on the choice of J.

Lemma 7.4.

(a) $\mathcal{V} \in Inv_k(\mathcal{C}, \mathcal{A})$ *is tight if and only if* $\lambda_p \geq 1$.

(b) Let $\mathcal{V} \in Inv_k(\mathcal{C}, \mathcal{A})$ *be tight and* $J_0 \in F^{n\times p}$ *such that* $(\mathcal{A} + J_0\ \mathcal{C})\mathcal{V} \subset \mathcal{V}$. *Then the set of all output injection matrices* $J \in F^{n\times p}$ *with* $(\mathcal{A}+J\mathcal{C})\mathcal{V} \subset \mathcal{V}$ *is given as* $\{J \in F^{n\times p} \mid Im(J - J_0) \subset \mathcal{V}\}$.

Proof: *(a)* $\mathcal{V} \subset F^n$ is tight if and only if $\mathcal{V} + Ker\mathcal{C} = F^n$; that is, if and only if the restriction $\mathcal{C}|_{\mathcal{V}} : \mathcal{V} \to F^p$ is surjective. This is equivalent to $\lambda_p \geq 1$.

(b) $(\mathcal{A} + J\mathcal{C})\mathcal{V} \subset \mathcal{V}$ is equivalent to $(J - J_0)\mathcal{C}\mathcal{V} \subset \mathcal{V}$. By tightness of $\mathcal{V}$ this is equivalent to $Im(J - J_0) \subset \mathcal{V}$. □

We are now deriving explicit characterizations of conditioned invariant subspaces in terms of basis representatives. For simplicity we focus on the tight case. Thus let $Z \in F^{n\times(n-k)}$ be a full rank matrix with $\mathcal{V} = Im\ Z$. Then $\mathcal{V}$ is a tight conditioned invariant subspace if and only if there exist $J \in F^{n\times p}$, $\overline{A} \in F^{(n-k)\times(n-k)}$, and $\overline{C} \in F^{p\times(n-k)}$, $rk\ \overline{C} = p$, such that

$$\begin{aligned}(\mathcal{A} + J\mathcal{C})Z &= Z\overline{A}\\ \mathcal{C}Z &= \overline{C}.\end{aligned}$$

The pair $(\overline{C}, \overline{A})$ is observable and isomorphic by similarity to the restriction $(\mathcal{C}|_{\mathcal{V}},\ (\mathcal{A} + J\mathcal{C})|_{\mathcal{V}})$. The observability indices of $(\overline{C}, \overline{A})$ are the restriction indices of $Im\ Z$.

Lemma 7.5. *Let* $Z \in F^{n\times(n-k)}$ *be full rank and* $J, \overline{A}, \overline{C}$ *be such that*

(i)

$$(\mathcal{A} + J\mathcal{C})Z = Z\overline{A}, \qquad \mathcal{C}Z = \overline{C}$$

holds. Assume $Im\ Z$ *is tight; that is,* $rk\ \overline{C} = p$ *holds. If* $J', \overline{A}', \overline{C}'$ *also satisfy (i) then there exists a unique* $L \in F^{(n-k)\times p}$ *such that*

$$\overline{A}' = \overline{A} + L\overline{C}, \qquad \overline{C}' = \overline{C}.$$

The set

$$(\mathcal{C}, \mathcal{A})|_{\mathcal{V}} := \{(S(\overline{A} + L\overline{C})S^{-1}, \overline{C}S^{-1}) | S \in GL_{n-k}, L \in F^{(n-k)\times p}\}$$

is the set of all possible restrictions of $(\mathcal{C}, \mathcal{A})$ *on* $\mathcal{V}$.

Proof: We have

$$(\mathcal{A} + J'\mathcal{C})Z = Z\overline{A}', \qquad \mathcal{C}Z = \overline{C}'.$$

Thus $\overline{C} = \overline{C}'$ and

$$\begin{aligned} Z\overline{A}' &= (\mathcal{A} + J\mathcal{C})Z + (J' - J)\mathcal{C}Z \\ &= Z\overline{A} + (J' - J)\overline{C}. \end{aligned}$$

Since $Im(J' - J) \subset Im\ Z$, there exists a unique $L \in F^{(n-k)\times p}$ such that $J' - J = ZL$. Hence

$$Z\overline{A}' = Z(\overline{A} + L\overline{C});$$

that is, $\overline{A}' = \overline{A} + L\overline{C}$. The result follows. □

Using Lemma 7.5 we see that every tight conditioned invariant subspace $\mathcal{V} \subset F^n$ has a representation $\mathcal{V} = Im\ Z$ with Z satisfying

$$(\mathcal{A} + J\mathcal{C})Z = Z\overline{A}, \qquad \mathcal{C}Z = W\overline{C}$$

with $(\overline{A}, \overline{C})$ in Brunovsky canonical form and $W \in GL_p$ invertible. It is possible to eliminate J from the above equation and obtain the following characterization.

Theorem 7.8. *(Ferrer et al. [3]). A subspace $\mathcal{V} \subset F^n$ of codimension k is a tight conditioned invariant subspace if and only if there exists $Z \in F^{n\times(n-k)}$ full rank with $\mathcal{V} = Im\ Z$ and an observable pair $(\overline{C}, \overline{A}) \in F^{p\times(n-k)} \times F^{(n-k)\times(n-k)}$ in Brunovsky canonical form with $rk\ \overline{C} = p$ and*

(i)

$$\mathcal{A}Z = Z\overline{A} + \mathcal{A}Z\overline{C}^T\overline{C},$$

(ii)

$$\mathcal{C}Z = \mathcal{C}Z\overline{C}^T\overline{C},$$

(iii)

$$\mathcal{C}Z\overline{C}^T \in F^{p\times p} \textit{ is invertible }.$$

Proof: From the above we see that every tight conditioned invariant subspace $\mathcal{V} \subset F^n$ has a representation $\mathcal{V} = ImZ$ with

$$(\mathcal{A} + J\mathcal{C})Z = Z\overline{A}, \qquad \mathcal{C}Z = W\overline{\mathcal{C}},$$

and $(\overline{C}, \overline{A})$ in Brunovsky canonical form, $W \in GL_p$. By tightness we have $rk\overline{C} = p$ and thus $W = \mathcal{C}Z\overline{C}^T$, as $\overline{C}\overline{C}^T = I_p$. Therefore

$$\begin{aligned} \mathcal{A}Z &= Z\overline{A} - J\mathcal{C}Z\overline{C}^T\overline{C} \\ &= Z\overline{A} - (\mathcal{A} + J\mathcal{C})Z\overline{C}^T\overline{C} + \mathcal{A}Z\overline{C}^T\overline{C} \\ &= Z\overline{A} - Z\overline{A}\ \overline{C}^T\overline{C} + \mathcal{A}Z\overline{C}^T\overline{C} \\ &= Z\overline{A} + \mathcal{A}Z\overline{C}^T\overline{C}, \end{aligned}$$

as $\overline{A}\,\overline{C}^T = 0$ holds for any system in Brunovsky canonical form. Since $W = \mathcal{C}Z\overline{C}^T$ is invertible this shows *(i)* through *(iii)*. Conversely, suppose *(i)* through *(iii)* hold. Then by *(ii)*,

$$\begin{aligned} \mathcal{A}Z &= Z\overline{A} + \mathcal{A}Z\overline{C}^T\overline{C} \\ &= Z\overline{A} + \mathcal{A}Z\overline{C}^T(\mathcal{C}Z\overline{C}^T)^{-1}\mathcal{C}Z. \end{aligned}$$

Thus for $J := \mathcal{A}Z\overline{C}^T(\mathcal{C}Z\overline{C}^T)^{-1}$ we obtain

$$(\mathcal{A} - J\mathcal{C})Z = Z\overline{A}, \qquad \mathcal{C}Z = \mathcal{C}Z\overline{C}^T\overline{C}.$$

Therefore $\mathcal{V} = Im\ Z$ is a conditioned invariant subspace. It is tight, as $rk(\mathcal{C}\ Z\overline{C}^T\overline{C}) = rk\ \overline{C} = p$. □

Using the well–known structure of the stabilizer subgroup of Brunovsky transformations we derive a uniqueness result for the canonical representation $\mathcal{V} = ImZ$ of conditioned invariant subspaces, for solutions Z of *(i)* through *(iii)*.

Let $(\overline{C}, \overline{A})$ denote the observable pair in Brunovsky canonical form with observability indices $\lambda_1 \geq \cdots \geq \lambda_p \geq 1$. Then

$$\Gamma(\lambda) := \{S \in GL_{n-k} \mid (S(\overline{A} + L\overline{C})S^{-1}, W\overline{C}S^{-1}) = (\overline{A}, \overline{C}) \text{ for suitable } L, W \in GL_p\}$$

measures the nonuniqueness of state space similarity transformations occurring in the Brunovsky canonical form. Obviously, $\Gamma(\lambda)$ is a closed subgroup of GL_{n-k} whose structure has been determined, for example, by Münzer and Prätzel–Wolters [16] and Fuhrmann and Willems [8]. The proof of the next result is similar to the previous one and is therefore omitted.

Theorem 7.9. *Let $(\overline{C}, \overline{A})$ be in Brunovsky canonical form, $rk\ \overline{C} = p$, with observability indices $\lambda_1 \geq \cdots \geq \lambda_p \geq 1$.*

(a) Two full rank matrices $Z_1, Z_2 \in F^{n\times(n-k)}$ with $Im\ Z_1 = Im\ Z_2$ satisfy (i) through (iii) if and only if there exists a unique $S \in \Gamma(\lambda)$ with $Z_2 = Z_1 S^{-1}$.

(b) $\Gamma(\lambda)$ is the set of all $S \in F^{(n-k)\times(n-k)}$ satisfying

(i) $\overline{A}S = S\overline{A} + \overline{A}S\overline{C}^T\overline{C}$;

(ii) $\overline{C}S = \overline{C}S\overline{C}^T\overline{C}$;

(iii) $\overline{C}S\overline{C}^T$ is invertible.

We are now in a position to study the geometry of tight conditioned invariant subspaces with fixed restriction indices. Here we assume $F =$

$\mathbb{R}, \mathbb{C}$. Let $\lambda_1 \geq \cdots \geq \lambda_p \geq 1$ be a partition of $n-k$ compatible with $\mu_1 \geq \cdots \geq \mu_p \geq 1$; that is,

$$\lambda_1 \leq \mu_1, \ \ldots, \ \lambda_p \leq \mu_p.$$

The set

$$Inv_k(\lambda, \mu) := \{\mathcal{V} \in Inv_k(\mathcal{C}, \mathcal{A}) \mid \mathcal{V} \text{ has restriction indices } \lambda\}$$

is called a *Brunovsky stratum* of $Tight_k(\mathcal{C}, \mathcal{A})$. Since $\lambda_p \geq 1$ each Brunovsky stratum is actually a subset of $Tight_k(\mathcal{C}, \mathcal{A})$ and $Tight_k(\mathcal{C}, \mathcal{A})$ is in fact the union of pairwise disjoint Brunovsky strata $Inv_k(\lambda, \mu)$, with λ compatible with μ. In Ferrer et al. [3] it is shown that each Brunovsky stratum of $Tight_k(\mathcal{C}, \mathcal{A})$ is a smooth manifold. To formulate their result in more precise terms, let

$$M(\lambda, \mu) := \{Z \in F^{n \times (n-k)} \mid Z \text{ satisfies } (i) \text{ through } (iii)\ \}.$$

Since *(i)* and *(ii)* are linear equations and *(iii)* defines an open class of matrices Z (which actually with *(i)* and *(ii)* implies Z being full rank) we see that $M(\lambda, \mu)$ is a smooth (algebraic) submanifold of $F^{n \times (n-k)}$ on which $\Gamma(\lambda)$ acts freely via $Z \mapsto ZS^{-1}$. From the above theorems the map

$$\begin{aligned} \rho : M(\lambda, \mu)/\Gamma(\lambda) &\rightarrow Tight_k(\mathcal{C}, \mathcal{A}) \\ [Z] &\mapsto Im\ Z \end{aligned}$$

is an embedding. Thus we obtain the following.

Theorem 7.10. *(Ferrer et al. [3]) Each Brunovsky stratum $Inv_k(\lambda, \mu)$ is a smooth submanifold of $Grass_{n-k}(\mathbb{F}^n)$, which is diffeomorphic to the orbit space $M(\lambda, \mu)/\Gamma(\lambda)$.*

One can say a bit more about the geometry of Brunovsky strata. In particular, it is possible to show that $Inv_k(\lambda, \mu)$ retracts onto a generalized flag manifold. The details will appear in a future joint paper with X. Puerta; here we briefly describe the construction of the retraction.

For any $Z \in M(\lambda, \mu)$ and $S \in \Gamma(\lambda)$ we have $p \times p$ matrices

$$\gamma(Z) := \mathcal{C}Z\overline{C}^T, \qquad \bar{\gamma}(S) := \overline{C}S\overline{C}^T.$$

Let

$$\begin{aligned} St(\lambda, \mu) &:= \{\mathcal{C}Z\overline{C}^T \mid Z \in M(\lambda, \mu)\} \\ G(\lambda) &:= \{\overline{C}S\overline{C}^T \mid S \in \Gamma(\lambda)\}. \end{aligned}$$

Then $\gamma : M(\lambda, \mu) \rightarrow St(\lambda, \mu), \gamma(Z) = \mathcal{C}Z\overline{C}^T$, is a smooth surjective map and it is easily verified that $\overline{\gamma} : \Gamma(\lambda) \rightarrow G(\lambda)$ is a group homomorphism. In particular, $St(\lambda, \mu) \subset GL_p(F)$ is seen as a smooth submanifold of $GL_p(F)$ and $G(\lambda) \subset GL_p(F)$ is a Lie subgroup of $GL_p(F)$ that acts freely on $St(\lambda, \mu)$ via the action

$$\begin{array}{ccc} G(\lambda) \times St(\lambda, \mu) & \to & St(\lambda, \mu) \\ (\overline{C}S\overline{C}^T, \, \mathcal{C}Z\overline{C}^T) & \mapsto & \mathcal{C}Z\overline{C}^T\overline{C}S^{-1}\overline{C}^T = \mathcal{C}Z \, S^{-1}\overline{C}^T. \end{array}$$

The topology of the orbit space $St(\lambda, \mu) \, / \, G(\lambda)$ has been studied in Helmke and Shayman [13] where it is shown that $St(\lambda, \mu) \, / \, G(\lambda)$ is homeomorphic to a generalized flag manifold and explicit formulas for the Betti numbers of $St(\lambda, \mu) \, / \, G(\lambda)$ are obtained. Moreover, from the previous theorems 7.1 and 7.2 we conclude that $\gamma : M(\lambda, \mu) \to St(\lambda, \mu)$ induces a smooth map on orbit spaces

$$\hat{\gamma} : M(\lambda, \mu)/\Gamma(\lambda) \to St(\lambda, \mu)/G(\lambda).$$

Theorem 7.11. *The map $\hat{\gamma}$ given above defines a homotopy equivalence between $Inv_k(\lambda, \mu)$ the generalized flag manifold $St(\lambda, \mu)/G(\lambda)$.*

In particular, $Inv_k(\lambda, \mu)$ is connected with explicit (although somewhat complicated) formulae for the Betti numbers. Moreover, the Schubert cell decomposition of the flag manifold $St(\lambda, \mu)/G(\lambda)$ induces via the above retraction a cell decomposition of $Inv_k(\lambda, \mu)$. We believe that this cell decomposition is equivalent to the following one, and also to that first described in Hinrichsen et al. [15] (as well as to the one described in Section 7.5). For any combination $K := (K_1, \ldots, K_p)$ $K_i \geq 1$, of $n - k$ let

$$Kro(K, \mu) := \{\mathcal{V} \in Tight_k(\mathcal{C}, \mathcal{A}) \mid (\mathcal{C}, \mathcal{A})|_{\mathcal{V}} \text{ has Kronecker indices } K\}.$$

It is easily seen that $Kro(K, \mu)$ is a well-defined (possibly empty) subset of $Tight_k(\mathcal{C}, \mathcal{A})$.

Conjecture 7.2. *Each Kronecker cell $Kro(K, \mu)$ defines a cell of the set $Tight_k(\mathcal{C}, \mathcal{A})$. The corresponding partition of $Tight_k(\mathcal{C}, \mathcal{A})$ into Kronecker cells defines an analytic cell decomposition of $Tight_k(\mathcal{C}, \mathcal{A})$.*

References

[1] G. Basile and G. Marro, Controlled and conditioned invariant subspaces in linear system theory, *J. Optimization Theory Appl.* 3:306–316, 1973.

[2] A.E. Eckberg, A characterization of linear systems via polynomial matrices and module theory, Ph.D. thesis, M.I.T., Cambridge, MA, 1974.

[3] J. Ferrer, F. Puerta, and X. Puerta, Differentiable structure of the set of controllable $(A, B)^t$–invariant subspaces, *Linear Algebra Appl.* 275/276:161–177, 1998.

[4] J. Ferrer, F. Puerta, and X. Puerta, Stratification of the set of general (A, B)–invariant subspaces, preprint, 1999.

[5] P.A. Fuhrmann, *Linear Operators and Systems in Hilbert Space*, McGraw–Hill, New York, 1981.

[6] P.A. Fuhrmann, Duality in polynomial models with some applications to geometric control theory, *IEEE Trans. Auto. Control* AC–26:284–295, 1981.

[7] P.A. Fuhrmann and U. Helmke, A homeomorphism between observable pairs and conditioned invariant subspaces, *Syst. Control Letters* 30:217–223, 1997.

[8] P.A. Fuhrmann and J.C. Willems, Factorization indices at infinity for rational matrix functions, *Integral Equations Oper. Th.* 2:287–301, 1979.

[9] P.A. Fuhrmann and J.C. Willems, A study of (A, B)–invariant subspaces via polynomial models, *Int. J. Control* 31:467–494, 1980.

[10] I. Gohberg and I.C. Feldman, *Convolution Equations and Projection Methods for their Solution*, Translations of Mathematical Monographs 41, American Mathematical Society, Providence, RI, 1974.

[11] M.L.J. Hautus and M. Heymann, Linear feedback—an algebraic approach, *SIAM J. Control* 16:83–105, 1978.

[12] U. Helmke, *The Cohomology of Moduli Spaces of Linear Dynamical Systems*, Regensburger Mathematische Schriften 24, Department of Mathematics, University of Regensburg, 1993.

[13] U. Helmke and M. Shayman, The biflag manifold and the fixed point set of a nilpotent transformation on the flag manifold, *Linear Algebra Appl.* 92:125–159, 1987.

[14] D. Hinrichsen and D. Prätzel–Wolters, Generalized Hermite matrices and complete invariants of strict system equivalence, *SIAM J. Control & Optimization* 21:289–305, 1983.

[15] D. Hinrichsen, H.F. Münzner, and D. Prätzel–Wolters, Parameterization of (C, A)–invariant subspaces, *Syst. Control Letters* 1:192–199, 1981.

[16] H.F. Münzner and D. Prätzel–Wolters, Minimal bases of polynomial modules, structural indices and Brunovsky–transformations, *Int. J. Control* 30:291–318, 1979.

[17] X. Puerta, Contributions to the geometric study of multivariable linear systems, Ph.D. thesis, Universitat Politècnica de Catalunya, Barcelona, 1998.

[18] J.C. Willems, Almost A mod B–invariant subspaces, *Asterisque* 75/76: 239–248, 1980.

[19] H.K.Wimmer, The structure of nonsingular polynomial matrices, *Math. Syst. Theory* 14:367–379, 1979.

[20] W.M. Wonham, *Linear Multivariable Control*, 2nd ed., Springer Verlag, New York, 1979.

Chapter 8

Duality Between Multidimensional Convolutional Codes and Systems

Heide Gluesing-Luerssen[1]
Joachim Rosenthal[1]
Paul A. Weiner[2]

ABSTRACT *Multidimensional convolutional codes arise as a generalization of "classical" one-dimensional codes. We introduce m-dimensional convolutional codes of length n as submodules of $\mathcal{D}^n$ where $\mathcal{D}$ is the polynomial ring in m variables over a finite field. Besides their coding theoretic significance, they can also be regarded as the annihilating modules of systems of partial difference equations, the latter being studied in much detail in discrete-time multidimensional systems theory. We apply the duality theorem of Oberst [5] to this particular case and employ the duality to investigate certain first-order representations of one-dimensional convolutional codes.*

8.1 Introduction

Data transmission over noisy channels requires implementation of good coding devices. Convolutional codes are among the most widely implemented codes. These codes represent in essence discrete-time linear systems over a fixed finite field $\mathbb{F}$. Because of this, a study of convolutional codes requires a good understanding of techniques from linear systems theory.

Multidimensional convolutional codes generalize (one-dimensional) convolutional codes, and they correspond to multidimensional systems widely studied in the systems literature. (See [5] and its references.) These codes are very suitable if, for example, the data transmission requires the encoding of a sequence of pictures; we explain this at the end of this section.

[1]Supported in part by NSF grant DMS-96-10389.

[2]Paul Weiner would like to thank the Center for Applied Mathematics at Notre Dame for a fellowship that financially supported the presented research.

In the sequel we assume that a certain message source is already encoded through a sequence of vectors $m_i \in \mathbb{F}^k$, $i = 1, \ldots, \gamma$. If every vector in $\mathbb{F}^k$ is a valid message word, then the change of one coordinate of a vector $m \in \mathbb{F}^k$ will result in another valid message vector $\tilde{m} \in \mathbb{F}^k$ and the error can neither be detected nor corrected. In order to overcome this difficulty one can add some redundancy by constructing an injective linear map

$$\varphi : \mathbb{F}^k \longrightarrow \mathbb{F}^n$$

having the property that the Hamming distance $\operatorname{dist}(\varphi(m_1), \varphi(m_2))$, that is, the number of different entries in the vectors $\varphi(m_1)$ and $\varphi(m_2)$ is at least d whenever $m_1 \neq m_2$. If one transmits the n-vector $\varphi(m_1)$ instead of the k-vector m_1, then it is possible to correct up to $\lfloor (d-1)/2 \rfloor$ errors for every transmitted n-vector; for details see Lemma 8.1.

The linear transformation φ defines an encoder and $\operatorname{im}(\varphi) \subset \mathbb{F}^n$ is called a linear block code. In order to describe the encoding of a whole sequence of message words $m_0, m_1, \ldots, m_\gamma \in \mathbb{F}^k$ it is convenient to introduce the polynomial vector $m(z) := \sum_{i=0}^{\gamma} m_i z^i \in \mathbb{F}^k[z]$. The encoding procedure is then compactly written by:

$$\hat{\varphi} : \mathbb{F}^k[z] \longrightarrow \mathbb{F}^n[z], \quad m(z) \longmapsto \hat{\varphi}(m(z)) = \sum_{i=0}^{\gamma} \varphi(m_i) z^i.$$

If $\mathcal{D}$ denotes the polynomial ring $\mathcal{D} = \mathbb{F}[z]$, then one immediately verifies that $\hat{\varphi}$ describes an injective module homomorphism between the free modules $\mathcal{D}^k$ and $\mathcal{D}^n$ and $\operatorname{im}\left(\hat{\varphi}(\mathcal{D}^k)\right) \subset \mathcal{D}^n$ is a submodule.

In general not every injective module homomorphism between $\mathcal{D}^k$ and $\mathcal{D}^n$ is of this form. Indeed $\hat{\varphi}$ has the peculiar property that the ith term of $\hat{\varphi}(m(z))$ only depends on the ith term of $m(z)$. In other words the encoder $\hat{\varphi}$ has "no memory". In general it is highly desirable to invoke encoding schemes where $\hat{\varphi} : \mathcal{D}^k \to \mathcal{D}^n$ is an arbitrary injective module homomorphism. The image of such a module homomorphism is then called a one-dimensional (1-D) convolutional code.

1-D convolutional codes are very well suited for the encoding of sequences of message blocks. Sometimes it might be desirable that the data be represented through polynomial rings in several variables. This leads us then to the definition of an m-dimensional convolutional code whose basic properties we study in this chapter.

The following example illustrates the usefulness of multidimensional convolutional codes.

Example 8.1. Let $\mathcal{D} = \mathbb{F}[z_1, z_2, z_3]$ be the polynomial ring in the indeterminates z_1, z_2, z_3. A whole motion picture (without sound) can be described by one element of $\mathcal{D}^k$. Indeed if $f \in \mathcal{D}^k$,

$$f(z_1, z_2, z_3) = \sum_{x=0}^{\xi} \sum_{y=0}^{\rho} \sum_{t=0}^{\tau} f_{(x,y,t)} z_1^x z_2^y z_3^t \in \mathbb{F}^k[z_1, z_2, z_3],$$

then we can view the vector $f_{(x,y,t)} \in \mathbb{F}^k$ as describing the color and the intensity of a pixel point with coordinates (x, y) at time t.

In practice the encoding of the element $f(z_1, z_2, z_3) \in \mathcal{D}^k$ is done in the following way. At a particular time instance t all the data vectors $f_{(x,y,t)}$ are combined into a large vector $\hat{f}_t \in \mathbb{F}^K$, where K depends on the size of k and the number of pixel points on the screen. In this way we can identify each element $f(z_1, z_2, z_3) \in \mathcal{D}^k$ of the above type with a polynomial vector $\hat{f}(z_3) \in \mathbb{F}^K[z_3]$. The vector $\hat{f}(z_3)$ is then encoded with a usual 1-D encoding scheme. This encoding scheme is shift invariant with respect to time but it is in general not shift invariant with respect to the z_1 and z_2 directions on the screen.

In order to achieve an encoding scheme that is also shift-invariant with respect to the coordinate axes of the screen one can do the following. Construct an injective module homomorphism $\hat{\varphi} : \mathcal{D}^k \to \mathcal{D}^n$. The image then describes a three-dimensional convolutional code that is invariant with respect to time and both coordinate axes. The transmission of an element $\mathcal{D}^n$ is then done by choosing a term order among the monomials of the form $z_1^x z_2^y z_3^t$.

8.2 Multidimensional Convolutional Codes

In this section we introduce multidimensional convolutional codes as submodules of $\mathcal{D}^n$, where $\mathcal{D}$ denotes a polynomial ring in m variables. Our presentation in this section closely follows [12, Chapter 2].

We begin by setting some notations. Let $\mathbb{F}$ be any finite field and define $\mathcal{D} = \mathbb{F}[z_1, \ldots, z_m]$ to be the polynomial ring in m indeterminates over $\mathbb{F}$. We mainly use the shorter form

$$\mathcal{D} = \mathbb{F}[z] = \Big\{ \sum_{\alpha \in \mathbb{N}^m}{}' f_\alpha z^\alpha \;\Big|\; f_\alpha \in \mathbb{F} \Big\},$$

where for $\alpha = (\alpha_1, \ldots, \alpha_m) \in \mathbb{N}^m$ the notation z^α stands for $z_1^{\alpha_1} \cdot \ldots \cdot z_m^{\alpha_m}$ and where $\sum'$ means this sum being finite. Note that $\mathcal{D}$ is $\mathbb{F}$-isomorphic to the m-dimensional finite sequence space

$$\mathcal{S} = \{ f : \mathbb{N}^m \longrightarrow \mathbb{F} \mid f \text{ has finite support} \},$$

the isomorphism given by

$$\begin{aligned} \psi : \mathcal{S} &\longrightarrow \mathcal{D} \\ f &\longmapsto \sum_{\alpha \in \mathbb{N}^m} f(\alpha) z^\alpha. \end{aligned}$$

One can visualize the elements of $\mathcal{S}$ by using the integer lattice of the first quadrant of $\mathbb{R}^m$ and attaching the element $f(\alpha_1, \ldots, \alpha_m) \in \mathbb{F}$ to the point with coordinates $(\alpha_1, \ldots, \alpha_m)$. It is convenient to omit the attachment if $f(\alpha_1, \ldots, \alpha_m) = 0$.

Example 8.2. We visualize the polynomial $f(z_1, z_2) = 1 + 2z_1^2 + 2z_1 z_2 \in \mathbb{F}_3[z_1, z_2]$ as well as $z_1 f(z_1, z_2)$.

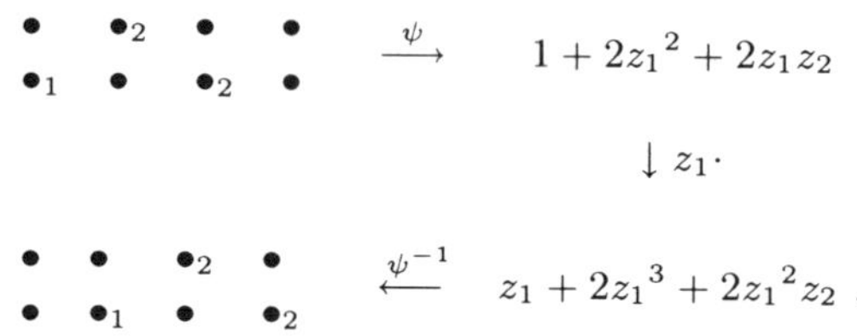

As the example indicates, multiplication with z_i in the ring $\mathcal{D}$ corresponds to the forward shift along the ith axis in $\mathcal{S}$. This can be verified with the help of the following commutative diagram

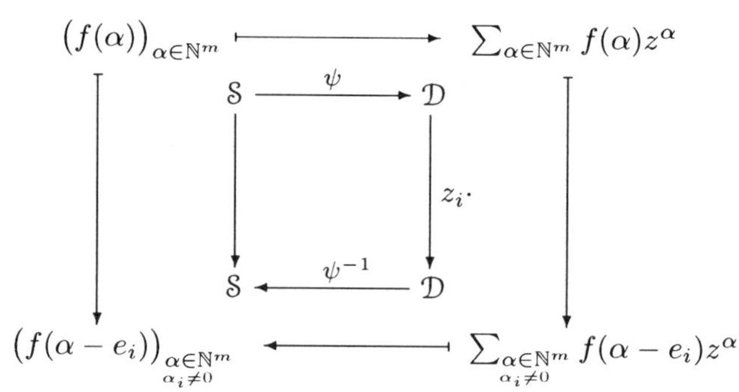

Here $e_i \in \mathbb{N}^m$ denotes the ith standard basis vector.

Throughout this chapter a code is defined to be an $\mathbb{F}$-linear subspace of some $\mathcal{S}^n$ that is invariant under the forward shifts along all axes. By virtue of the above diagram this can simply be phrased as the following.

Definition 8.1. A linear m-dimensional convolutional code *(for short,* m-D code*) of length n over $\mathbb{F}$ is a $\mathcal{D}$-submodule of $\mathcal{D}^n$. An element of a code is said to be a* codeword.

Remark 8.1. *In the coding literature (see, e.g.,* [6]*) convolutional codes are usually not restricted to sequence spaces whose elements have finite support. There is however no engineering reason behind this. After all, every transmitted message created by mankind did have finite length. Convolutional codes with finite support were first studied by Fornasini and Valcher* [1, 2, 11]. *These authors defined a convolutional code as a submodule of $\tilde{\mathcal{D}}^n$, where $\tilde{\mathcal{D}}$ represents the ring of Laurent polynomials $\mathbb{F}[z_1, \ldots, z_m, z_1^{-1}, \ldots, z_m^{-1}]$. In doing so, a convolutional code then corresponds to an $\mathbb{F}$-linear subspace of some $\tilde{\mathcal{S}}^n$, where $\tilde{\mathcal{S}} = \{f : \mathbb{Z}^m \longrightarrow \mathbb{F} \mid f \text{ has finite support}\}$.*

Since $\mathcal{D}$ is a Noetherian ring, each code $\mathcal{C} \subseteq \mathcal{D}^n$ is finitely generated. In other words, there exists some $l \in \mathbb{N}$ and a matrix $G \in \mathcal{D}^{n\times l}$ such that $\mathcal{C} = \mathrm{im}_{\mathcal{D}}G$. We call such a matrix G a generator matrix of $\mathcal{C}$. Note that we do not use the row vector notation common in coding theory. It would force us to use the same notation also for the dual system-theoretic version, which is very unusual. The notation $\mathrm{im}_{\mathcal{D}}G$ means of course the set of all Gp with $p \in \mathcal{D}^l$. This notation instead of only $\mathrm{im}G$ is necessary later when interpreting G as a different type of operator. Analogously, we might also use the notation $\ker_{\mathcal{D}} G = \{p \in \mathcal{D}^l \mid Gp = 0\}$.

As a finitely generated $\mathcal{D}$-module each code has a well-defined rank, say $\mathrm{rank}\,\mathcal{C} = k$. It can simply be calculated as $\mathrm{rank}\,G$, where one may use any generator matrix G of $\mathcal{C}$, considered as a matrix over the quotient field $\mathbb{F}(z_1, \dots, z_m)$. The rate of $\mathcal{C}$ is defined to be the quotient k/n.

The code $\mathcal{C}$ is called free if $\mathcal{C}$ is a free $\mathcal{D}$-module, that is, if $\mathcal{C}$ has a $\mathcal{D}$-basis. This is the case if and only if $\mathcal{C}$ has a generator matrix $G \in \mathcal{D}^{n\times k}$ with $\mathrm{rank}\,G = k = \mathrm{rank}\,\mathcal{C}$. Such a generator matrix is called an encoder. If $\mathcal{C}$ has an encoder, say $G = [G_1, \dots, G_k] \in \mathcal{D}^{n\times k}$, then each codeword can be written in a unique way as a $\mathcal{D}$-linear combination of $G_1, \dots, G_k$. This is certainly a very desirable property for a code. It is a well-known fact that each one-dimensional code, that is, each $\mathbb{F}[z_1]$-module is free. However, for higher dimensions (i.e., for $m > 1$) this is no longer true.

Example 8.3. Let $\mathcal{D} = \mathbb{F}[z_1, z_2]$ and

$$G(z_1, z_2) = \begin{bmatrix} {z_1}^2 & z_1 + z_1 z_2 \\ z_1 & 1 + z_2 \\ z_1 z_2 & z_2 + {z_2}^2 \end{bmatrix}.$$

It can easily be shown that $\mathrm{im}_{\mathcal{D}}(G)$ is rate 1/3 but not free. That is, $\mathrm{im}_{\mathcal{D}}(G)$ has no 3×1 encoder. The code $\mathrm{im}_{\mathcal{D}} \begin{bmatrix} z_1 & 1 & z_2 \end{bmatrix}^{\mathsf{T}}$ is free of rate 1/3 and properly contains $\mathrm{im}_{\mathcal{D}}(G)$.

It is easy to see that encoder matrices for a given code are unique up to unimodular right multiplication; that is, for $G_i \in \mathcal{D}^{n\times k}$ with $\mathrm{rank}\,G_i = k$ it is

$$\mathrm{im}_{\mathcal{D}}G_1 = \mathrm{im}_{\mathcal{D}}G_2 \Longleftrightarrow G_2 = G_1 U \quad \text{for some } U \in Gl_k(\mathcal{D}). \tag{8.1}$$

An important measure for the "goodness" of a (convolutional) code is its distance. In the remainder of this section we introduce this parameter.

Definition 8.2. *Let $a \in \mathbb{F}^n$. The* weight *of a is given by the number of nonzero entries of a. It is denoted by $wt(a)$.*

For $w = \sum_{\alpha\in\mathbb{N}^m} b_\alpha z^\alpha \in \mathcal{D}^n$ with $b_\alpha \in \mathbb{F}^n$ the weight of w is defined as

$$wt(w) = \sum_{\alpha\in\mathbb{N}^m} wt(b_\alpha)\,.$$

Hence the weight of a vector in $\mathcal{D}^n$ measures the distance to the all-zero vector by counting all nonzero terms in the vector. The weight has the characteristic of a discrete norm. In particular the weight induces a metric on $\mathcal{D}^n$.

Definition 8.3. *Given two elements $w, \tilde{w} \in \mathcal{D}^n$, the* (Hamming) distance *between w and $\tilde{w}$ is given by $dist(w, \tilde{w}) = wt(w - \tilde{w})$. Given any m-D code $\mathcal{C}$ of length n, the distance of $\mathcal{C}$ is defined as*

$$\operatorname{dist}(\mathcal{C}) = \min\{\operatorname{dist}(w, \tilde{w}) : w, \tilde{w} \in \mathcal{C}, w \neq \tilde{w}\}.$$

Remark 8.2.

(i) The Hamming distance defines a metric on $\mathcal{D}^n$ called the Hamming metric.

(ii) For a code $\mathcal{C}$ we have that $dist(\mathcal{C}) = \min\{\operatorname{wt}(w) : w \in \mathcal{C}, w \neq 0\}$. This is because $dist(w_1, w_2) = \operatorname{wt}(w_1 - w_2)$ and $w_1 - w_2 \in \mathcal{C}$ whenever $w_1, w_2 \in \mathcal{C}$.

The following result is standard in coding theory. It follows immediately from the definition of Hamming distance and from the properties of a metric space.

Lemma 8.1. *Let $\mathcal{C}$ be a convolutional code with $d = \operatorname{dist}(\mathcal{C})$. Let furthermore $t = \lfloor (d-1)/2 \rfloor$ where $\lfloor x \rfloor$ denotes the greatest integer that is less than or equal to x. Let $y \in \mathcal{D}^n$. If $w \in \mathcal{C}$ is a codeword such that $\operatorname{dist}(w, y) \leq t$, then w is the unique codeword nearest (with respect to the Hamming metric) to y.*

We say that $\mathcal{C}$ can correct up to t errors. In practice it is not a simple task to compute the transmitted vector $w \in \mathcal{D}^n$ from the received vector $y \in \mathcal{D}^n$. One way to do this is by syndrome decoding. In order to explain this we first state the following.

Proposition 8.1. *Suppose $\mathcal{C}$ is a free convolutional code of rate k/n with encoder $G \in \mathcal{D}^{n \times k}$. Then*

$$0 \longrightarrow \mathcal{D}^k \xrightarrow{G\cdot} \mathcal{D}^n \xrightarrow{\pi} \mathcal{D}^n/\mathcal{C} \longrightarrow 0$$

is a short exact sequence.

The proof is left to the reader.

In the above short exact sequence π is often called a syndrome former. Syndrome decoding works as follows. If $y \in \mathcal{D}^n$ is received, one seeks the vector $e \in \pi^{-1}(\pi(y)) = y + \mathcal{C}$ of smallest possible weight. The vector y is then decoded as $w := y - e$.

8.3 Duality Between Codes and Behaviors

There have been several instances in the recent literature about coding theory, in which certain types of duality between convolutional codes and behaviors in the system theoretical sense of [13] have been mentioned or used; see, for example, [3, 9, 11, 12].

In this section we are going to make this duality precise by introducing the appropriate bilinear form. Exploiting the very comprehensive and powerful paper of Oberst [5], many results about this duality are available even in the multidimensional case. However, as it seems to us, most interesting is the duality in the one-dimensional case, where various minimal first-order representations exist and have been studied systematically and exhaustively by Kuijper [4]. They can be translated into corresponding descriptions for codes. This is studied in Section 8.4.

We introduce the notations and the setting along the lines of [5]. Only those results needed for our purposes are cited afterwards.

Throughout this section the field $\mathbb{F}$ need not be finite; the results hold for any field. First we have to define the underlying setting for the behaviors. Let

$$\mathcal{A} := \mathbb{F}[\![z]\!] = \Big\{ \sum_{\alpha \in \mathbb{N}^m} f_\alpha z^\alpha \,\Big|\, f_\alpha \in \mathbb{F} \Big\}$$

be the set of power series in the m variables $z_1, \ldots, z_m$ over $\mathbb{F}$. On $\mathcal{A}$ we consider the backward shifts along the ith axis followed by truncation; that is, for each $i = 1, \ldots, m$ define

$$\begin{aligned} L_i : \quad \mathcal{A} \quad &\longrightarrow \quad \mathcal{A} \\ \sum_{\alpha \in \mathbb{N}^m} f_\alpha z^\alpha &\longmapsto \sum_{\alpha \in \mathbb{N}^m} f_{\alpha + e_i} z^\alpha \end{aligned} \tag{8.2}$$

(see also [5, p. 15]). This action can also be expressed in the following ways.

$$L_i\Big(\sum_{\alpha \in \mathbb{N}^m} f_\alpha z^\alpha \Big) = z_i^{-1} \Big(\sum_{\substack{\alpha \in \mathbb{N}^m \\ \alpha_i \neq 0}} f_\alpha z^\alpha \Big) = \Pi_+ \Big(z_i^{-1} \sum_{\alpha \in \mathbb{N}^m} f_\alpha z^\alpha \Big), \tag{8.3}$$

where Π_+ denotes the projection that cuts off the terms with negative exponents. Clearly, the operators L_i are $\mathbb{F}$-linear. Moreover, $\mathcal{A}$ gets the structure of a $\mathcal{D}$-module via the scalar multiplication

$$p(z_1, \ldots, z_m) \cdot f := p(L_1, \ldots, L_m)(f) \in \mathcal{A} \quad \text{for } p \in \mathcal{D},\ f \in \mathcal{A}.$$

Example 8.4. Let $m = 2$ and $p = 1 + z_1^2 + z_1 z_2 \in \mathcal{D} = \mathbb{F}_5[z_1, z_2]$. Then $p \cdot (1 + 3z_1 z_2^3 + 2z_1^2 + 4z_2) = 3 + 3z_1 z_2^3 + 2z_1^2 + 4z_2 + 3z_2^2$.

The example shows that the notation $p \cdot f$ has to be read with care. It is not the usual convolutional product in $\mathcal{A}$. Instead from (8.3) one can derive the formula

$$\sum_{\alpha\in\mathbb{N}^m}{}' p_\alpha z^\alpha \cdot \sum_{\beta\in\mathbb{N}^m} f_\beta z^\beta = \sum_{\beta\in\mathbb{N}^m}\Big(\sum_{\alpha\in\mathbb{N}^m}{}' p_\alpha f_{\alpha+\beta}\Big) z^\beta. \tag{8.4}$$

Since we never use ordinary convolution in $\mathcal{A}$, this should not cause confusion.

Remark 8.3.

(a) *Obviously, $\mathcal{D}$ is a $\mathcal{D}$-submodule of $\mathcal{A}$. However, it is worth mentioning that the canonical injection $\iota : \mathcal{D} \longrightarrow \mathcal{A}$, $p \mapsto p$ is not $\mathcal{D}$-linear. In fact, for example, $z_1 = \iota(z_1) \neq z_1 \cdot \iota(1) = z_1 \cdot 1 = 0$ in $\mathcal{A}$. This is not really an issue as the inclusion $\mathcal{D} \subset \mathcal{A}$ is never considered in this setting. While $\mathcal{D}$ is the set of operators, either generator matrices for codes or shift operators, $\mathcal{A}$ serves as the space of trajectories for the behaviors.*

(b) *$\mathcal{A}$ is not finitely generated as a $\mathcal{D}$-module; see* [5, p. 55].

Each polynomial matrix $G \in \mathcal{D}^{k\times n}$ gives rise to a linear partial difference operator $G(L_1, \ldots, L_m)$ which we denote for short G; thus

$$\begin{aligned} G :\ \mathcal{A}^n &\longrightarrow \mathcal{A}^k \\ a &\longmapsto G\cdot a := G(L_1,\ldots,L_m)(a). \end{aligned}$$

These operators are going to be the objects dual to generator matrices for codes. The following notations are useful in the sequel. For $G \in \mathcal{D}^{k\times n}$ define

$$\operatorname{im}_{\mathcal{A}} G = \{G\cdot a \mid a \in \mathcal{A}^n\} \subseteq \mathcal{A}^k \quad\text{and}\quad \ker_{\mathcal{A}} G = \{a \in \mathcal{A}^n \mid G\cdot a = 0\} \subseteq \mathcal{A}^n.$$

Definition 8.4. *An m-dimensional* behavior *$\mathcal{B}$ in $\mathcal{A}^n$ is defined to be a $\mathcal{D}$-submodule $\mathcal{B} \subseteq \mathcal{A}^n$ of the form $\mathcal{B} = \ker_{\mathcal{A}} G$ for some $G \in \mathcal{D}^{k\times n}$ (not necessarily of full row rank).*

This setting is identical to the study of $m - D$-discrete-time systems in the behavioral context; see, for example, [7].

We observe that, while each $\mathcal{D}$-submodule of $\mathcal{D}^n$ is a code, not every $\mathcal{D}$-submodule of $\mathcal{A}^n$ is a behavior. Characterizations for an $\mathbb{F}$-subspace of $\mathcal{A}^n$ being a behavior are given in the one-dimensional case in [13, III.1] and for the general case in [5, pp. 61,62].

Now the bilinear form to be used for the duality is obvious. For each

$n \geq 1$ a $\mathcal{D}$-bilinear nondegenerate form is given by (cf. [5, p. 22])

$$\begin{aligned} \mathcal{D}^n \times \mathcal{A}^n &\longrightarrow \quad \mathcal{A} \\ (p, a) &\longmapsto \langle p, a \rangle := p^\mathsf{T} \cdot a = \sum_{i=1}^{n} p_i \cdot a_i = \sum_{i=1}^{n} p_i(L_1, \ldots, L_m)(a_i), \end{aligned} \tag{8.5}$$

where $p = (p_1, \ldots, p_n)^\mathsf{T}$, $a = (a_1, \ldots, a_n)^\mathsf{T}$. In the literature related to codes and behaviors a certain $\mathbb{F}$-bilinear form also has been used, see [9] and [12, p. 20]. We clarify the relationship between this one and (8.5) at the end of this section.

Using the above bilinear form we define the duals in the obvious way.

Definition 8.5.

(a) *The* dual *of a subset* $\mathcal{B} \subseteq \mathcal{A}^n$ *is defined to be* $\mathcal{B}^\perp := \{p \in \mathcal{D}^n \mid \langle p, a \rangle = 0$ *for all* $a \in \mathcal{B}\}$.

(b) *The* dual *of a subset* $\mathcal{C} \subseteq \mathcal{D}^n$ *is given by* $\mathcal{C}^\perp := \{a \in \mathcal{A}^n \mid \langle p, a \rangle = 0$ *for all* $p \in \mathcal{C}\}$.

Obviously, duals are $\mathcal{D}$-modules and one has $\mathcal{B} \subseteq \mathcal{B}^{\perp\perp}$ as well as $\mathcal{C} \subseteq \mathcal{C}^{\perp\perp}$.

Now we are in the position to state the results given in [5]. Essentially, they amount to the fact that $\mathcal{A}$ is a large injective cogenerator in the category of $\mathcal{D}$-modules. Instead of going into an explanation of this statement, we simply extract from [5] the following consequences of this very strong result. Statements (*iv*), (*v*), and (*vii*) of the next theorem are exactly the duality between codes and behaviors for which we were looking.

Theorem 8.1. *Let* $P \in \mathcal{D}^{l\times n}$, $Q \in \mathcal{D}^{k\times l}$, $R \in \mathcal{D}^{r\times n}$. *Then*

(i) *If the sequence* $\mathcal{D}^k \xrightarrow{Q^\mathsf{T}} \mathcal{D}^l \xrightarrow{P^\mathsf{T}} \mathcal{D}^n$ *is exact, then so is the sequence* $\mathcal{A}^n \xrightarrow{P} \mathcal{A}^l \xrightarrow{Q} \mathcal{A}^k$;

(ii) $\ker_\mathcal{A} P \subseteq \ker_\mathcal{A} R$ *if and only if* $R = XP$ *for some* $X \in \mathcal{D}^{r\times l}$;

(iii) *if* $\operatorname{rank} P = l$, *then the operator* $P : \mathcal{A}^n \to \mathcal{A}^l$ *is surjective;*

(iv) $(\operatorname{im}_\mathcal{D} Q^\mathsf{T})^\perp = \ker_\mathcal{A} Q$;

(v) $(\ker_\mathcal{A} Q)^\perp = \operatorname{im}_\mathcal{D} Q^\mathsf{T}$;

(vi) $(\operatorname{im}_\mathcal{A} Q)^\perp = \ker_\mathcal{D} Q^\mathsf{T}$;

(vii) $\mathcal{C} = \mathcal{C}^{\perp\perp}$ *and* $\mathcal{B} = \mathcal{B}^{\perp\perp}$ *for each code* $\mathcal{C} \in \mathcal{D}^n$ *and each behavior* $\mathcal{B} \in \mathcal{A}^n$.

(*iii*) means in other words, for each P with full row rank and for each $g \in \mathcal{A}^l$ the associated linear partial difference equation $P \cdot f = g$ has a solution in $\mathcal{A}^n$. This is a well-known fact in the one-dimensional case; that is, $\mathcal{D} = \mathbb{F}[z_1]$. Even more, one can also prescribe initial conditions up to a certain order. In the m-dimensional case this is more involved. Statement (iv) shows especially that the dual of a code is not only a $\mathcal{D}$-module but even a behavior.

As for the proof, all the above results go back to [5, p. 33], which is just the large injective cogenerator property. However, we give some more detailed references and arguments from the paper to show how things are related to each other, although this might be a bit different from the order they have been proven.

(*i*) is exactly the injectivity of the module $\mathcal{A}$ which is defined in [5, p. 24]. (*ii*) is in [5, p. 36]. (*iii*) is a consequence of (*i*). (*iv*) and (*v*) are in [5, pp. 30,31], but they can also be derived directly from the above as follows. (*iv*) and also (*vi*) follow immediately from

$$\langle Q^{\mathsf{T}} p, a\rangle = \langle p, Q \cdot a\rangle \text{ for each } p \in \mathcal{D}^k \text{ and } a \in \mathcal{A}^l \tag{8.6}$$

together with the nondegeneracy of the bilinear form (8.5). (*v*) can be shown with the help of (*ii*) via

$$\begin{aligned} p \in (\ker_{\mathcal{A}} Q)^{\perp} \iff \ker_{\mathcal{A}} Q \subseteq \ker_{\mathcal{A}} p^{\mathsf{T}} &\iff p^{\mathsf{T}} = v^{\mathsf{T}} Q \text{ for some } v \in \mathcal{D}^k \\ &\iff p \in \operatorname{im}_{\mathcal{D}} Q^{\mathsf{T}}. \end{aligned}$$

(*vii*) is a consequence of (*iv*) and (*v*).

Remark 8.4. *(Compare with Remark 8.1). If convolutional codes are defined as submodules of $\tilde{\mathcal{D}}^n$, where $\tilde{\mathcal{D}}$ represents the ring of Laurent polynomials $\mathbb{F}[z, z^{-1}]$ then this results in a duality between codes and linear behaviors defined on $\tilde{\mathcal{A}}^n$, where $\tilde{\mathcal{A}} := \mathbb{F}[\![z, z^{-1}]\!]$ is the ring of formal power series in the variables $z_1, \ldots, z_m, z_1^{-1}, \ldots, z_m^{-1}$.*

Next we want to concentrate on two specific descriptions of behaviors that are significant for one-dimensional first-order representations in the next section. In fact, the following two types of representations, applicable to both codes and behaviors, are dual to each other as is proven next. They specialize to the so-called (P, Q, R)- and (K, L, M)-representations in the one-dimensional case.

Theorem 8.2. *Let $R \in \mathcal{D}^{n \times l}$, $N \in \mathcal{D}^{k \times l}$, and $M \in \mathcal{D}^{k \times n}$. Then the following are true.*

(a) *The module $R \cdot (\ker_{\mathcal{A}} N) := \{R \cdot \zeta \mid \zeta \in \mathcal{A}^l,\ N \cdot \zeta = 0\} \subseteq \mathcal{A}^n$ is a behavior and its dual is given by $\big(R \cdot (\ker_{\mathcal{A}} N)\big)^{\perp} = \{p \in \mathcal{D}^n \mid R^{\mathsf{T}} p \in \operatorname{im}_{\mathcal{D}} N^{\mathsf{T}}\}$.*

(b) *The module $\{a \in \mathcal{A}^n \mid M \cdot a \in \operatorname{im}_{\mathcal{A}} N\} \subseteq \mathcal{A}^n$ is a behavior. Its dual is $\{a \in \mathcal{A}^n \mid M \cdot a \in \operatorname{im}_{\mathcal{A}} N\}^{\perp} = M^{\mathsf{T}}(\ker_{\mathcal{D}} N^{\mathsf{T}})$.*

(c) $\big(R(\ker_{\mathcal{D}} N)\big)^{\perp} = \{a \in \mathcal{A}^n \mid R^{\mathsf{T}} \cdot a \in \operatorname{im}_{\mathcal{A}} N^{\mathsf{T}}\}$.

(d) $\{p \in \mathcal{D}^n \mid Mp \in \operatorname{im}_{\mathcal{D}} N\}^{\perp} = M^{\mathsf{T}} \cdot (\ker_{\mathcal{A}} N^{\mathsf{T}})$.

Proof: *(a)* The first part is proven in [5, p. 26]. As for the second part, note the following equivalences, which hold for each $p \in \mathcal{D}^n$ using equation (8.6)

$$\begin{aligned}\langle p, R \cdot a\rangle = 0 \,\forall\, a \in \ker_{\mathcal{A}} N &\Longleftrightarrow \langle R^{\mathsf{T}} p, a\rangle = 0 \,\forall\, a \in \ker_{\mathcal{A}} N \\ &\Longleftrightarrow R^{\mathsf{T}} p \in (\ker_{\mathcal{A}} N)^{\perp} = \operatorname{im}_{\mathcal{D}} N^{\mathsf{T}}.\end{aligned}$$

(b) Using $N = 0$ in *(a)* we obtain especially that a $\mathcal{D}$-submodule of the form $\operatorname{im}_{\mathcal{A}} R$ is a behavior. Thus, write $\operatorname{im}_{\mathcal{A}} N = \ker_{\mathcal{A}} Q$ with some appropriate $Q \in \mathcal{D}^{q \times k}$. Then $\{a \in \mathcal{A}^n \mid M \cdot a \in \operatorname{im}_{\mathcal{A}} N\} = \ker_{\mathcal{A}} QM$ is a behavior (see also [5, p. 27]) and moreover

$$\begin{aligned}\{a \in \mathcal{A}^n \mid M \cdot a \in \operatorname{im}_{\mathcal{A}} N\}^{\perp} &= (\ker_{\mathcal{A}} QM)^{\perp} = \operatorname{im}_{\mathcal{D}} (QM)^{\mathsf{T}} = M^{\mathsf{T}}(\operatorname{im}_{\mathcal{D}} Q^{\mathsf{T}}) \\ &= M^{\mathsf{T}} \cdot \big((\ker_{\mathcal{A}} Q)^{\perp}\big) = M^{\mathsf{T}}\big((\operatorname{im}_{\mathcal{A}} N)^{\perp}\big) \\ &= M^{\mathsf{T}}(\ker_{\mathcal{D}} N^{\mathsf{T}}).\end{aligned}$$

(c) and *(d)* follow now from *(a)* and *(b)* with Theorem 8.1 *(vii)*. □

In the following we briefly discuss parity check matrices for multidimensional codes.

Definition 8.6. *Let $\mathcal{C} \subseteq \mathcal{D}^n$ be a code. A matrix $H \in \mathcal{D}^{l \times n}$ is called a* parity check matrix *of $\mathcal{C}$ if $\mathcal{C} = \ker_{\mathcal{D}} H$.*

Not every code has a parity check matrix; for example, for $\mathcal{D} = \mathbb{F}[z_1]$ the code $\operatorname{im}_{\mathcal{D}} \begin{bmatrix} z_1 \\ z_1 \end{bmatrix}$ has no parity check matrix, since each matrix $H \in \mathbb{F}[z_1]^{l \times 2}$ having $(z_1, z_1)^{\mathsf{T}}$ in its kernel would also have $(1,1)^{\mathsf{T}} \in \ker_{\mathcal{D}} H$.

The following result about the existence of parity check matrices can be found in [12, 3.3.8].

Theorem 8.3. *Let $\mathcal{C} = \operatorname{im}_{\mathcal{D}} G$ with $G \in \mathcal{D}^{n \times k}$ be a free code; thus* $\operatorname{rank} G = k$. *Then $\mathcal{C}$ has a parity check matrix if and only if G is minor-prime, that is, if the greatest common divisor of all full-size minors of G is a unit in $\mathcal{D}$. If a parity check matrix exists, then one also has a parity check matrix $H \in \mathcal{D}^{(n-k) \times n}$ with* $\operatorname{rank} H = n - k$.

This result can be dualized by using Theorem 8.1.

Theorem 8.4. *Let $\mathcal{C} \subseteq \mathcal{D}^n$ be a free code. Then $\mathcal{C}$ has a parity check matrix if and only if the behavior $\mathcal{C}^{\perp} \subseteq \mathcal{A}^n$ has an image representation; that is,*

$$\mathcal{C} = \ker_{\mathcal{D}} H \text{ for some } H \in \mathcal{D}^{l \times n} \Longleftrightarrow \mathcal{C}^{\perp} = \operatorname{im}_{\mathcal{A}} H^{\mathsf{T}} \text{ for some } H \in \mathcal{D}^{l \times n}.$$

Hence a behavior $\ker_{\mathcal{A}} G \subseteq \mathcal{A}^n$ *has an image representation if and only if* G *is minor-prime.*

Proof: This follows from Theorem 8.1*(i)*, *(vi)*, and *(vii)*. □

Recall that for one-dimensional behaviors the existence of image representations is equivalent to controllability; see [13]. For $m > 1$, at least one direction is true; namely, behaviors with image representations are always controllable; see [15, Theorem 4.2]. Equivalence can be established for $m = 2$ or for $m \geq 2$ if certain directions of the time-space axes are two-sided; see [7] and [14, Theorem 6].

At the end of this section we discuss the relationship of the above bilinear form with an $\mathbb{F}$-bilinear form that also has been used in the literature within this context. Let

$$\begin{array}{ccc} \mathcal{D}^n \times \mathcal{A}^n & \longrightarrow & \mathbb{F} \\ \Big(\sum'_{\alpha\in\mathbb{N}^m} p_\alpha z^\alpha, \sum_{\alpha\in\mathbb{N}^m} f_\alpha z^\alpha\Big) & \longmapsto & \langle\!\langle p, f\rangle\!\rangle := \sum'_{\alpha\in\mathbb{N}^m} p_\alpha^{\mathsf{T}} f_\alpha, \end{array}$$

where $p_\alpha^{\mathsf{T}} f_\alpha \in \mathbb{F}$ denotes the usual scalar product in $\mathbb{F}^n$. Observe that the sum on the right-hand side is indeed finite.

Example 8.5. Let $\mathcal{D} = \mathbb{F}_2[z_1]$ and $n = 2$. For $p = (1,0)^{\mathsf{T}} \in \mathcal{D}^2$ and $f = (z_1, 1)^{\mathsf{T}} \in \mathcal{A}^2$ we obtain $\langle\!\langle p, f\rangle\!\rangle = (1,0)\begin{pmatrix}0\\1\end{pmatrix} = 0$, whereas the previously used $\mathcal{D}$-bilinear form yields $\langle p, f\rangle = (1,0)\begin{pmatrix}z_1\\1\end{pmatrix} = z_1$. Hence p and f are orthogonal with respect to $\langle\!\langle\ ,\ \rangle\!\rangle$ but not with respect to $\langle\ ,\ \rangle$.

However, there is a close relationship between these two forms as we derive next. In order to do so, we use the notation $L^\alpha := L_1^{\alpha_1} \circ \ldots \circ L_m^{\alpha_m}$ for $\alpha = (\alpha_1, \ldots, \alpha_m) \in \mathbb{N}^m$ and the shifts L_i defined in (8.2). Let $p = \sum'_{\alpha\in\mathbb{N}^m} p_\alpha z^\alpha \in \mathcal{D}^n$ and $f = \sum_{\alpha\in\mathbb{N}^m} f_\alpha z^\alpha \in \mathcal{A}^n$. First, using Definition (8.5) and equation (8.4) one obtains

$$\begin{aligned} \langle p, f\rangle = 0 &\iff \sum'_{\alpha\in\mathbb{N}^m} p_\alpha^{\mathsf{T}} L^\alpha(f) = 0 \iff \sum'_{\alpha\in\mathbb{N}^m} p_\alpha^{\mathsf{T}} \sum_{\beta\in\mathbb{N}^m} f_{\beta+\alpha} z^\beta = 0 \\ &\iff \sum'_{\alpha\in\mathbb{N}^m} p_\alpha^{\mathsf{T}} f_{\beta+\alpha} = 0\ \forall\, \beta \in \mathbb{N}^m \iff \langle\!\langle p, z^\beta \cdot f\rangle\!\rangle = 0\ \forall\, \beta \in \mathbb{N}^m. \end{aligned}$$

Second, it is

$$\langle\!\langle z^\beta p, f\rangle\!\rangle = \langle\!\langle \sum'_{\alpha\in\mathbb{N}^m} p_\alpha z^{\alpha+\beta}, \sum_{\alpha\in\mathbb{N}^m} f_\alpha z^\alpha\rangle\!\rangle = \sum'_{\alpha\in\mathbb{N}^m} p_\alpha^{\mathsf{T}} f_{\alpha+\beta} = \langle\!\langle p, z^\beta \cdot f\rangle\!\rangle$$

by virtue of (8.4).

These two observations lead to the fact that both bilinear forms yield the same duals for $\mathcal{D}$-submodules of $\mathcal{A}^n$ or $\mathcal{D}^n$. Indeed, if $\mathcal{B} \subseteq \mathcal{A}^n$ is a $\mathcal{D}$-submodule, then

$$\begin{aligned} & \{p \in \mathcal{D}^n \mid \langle\langle p, f\rangle\rangle = 0 \; \forall f \in \mathcal{B}\} \\ = \; & \{p \in \mathcal{D}^n \mid \langle\langle p, z^\beta \cdot f\rangle\rangle = 0 \; \forall f \in \mathcal{B} \; \forall \beta \in \mathbb{N}^m\} = \mathcal{B}^\perp. \end{aligned}$$

Similarly, for a $\mathcal{D}$-submodule $\mathcal{C} \subseteq \mathcal{D}^n$ one obtains

$$\begin{aligned} & \{f \in \mathcal{A}^n \mid \langle\langle p, f\rangle\rangle = 0 \; \forall p \in \mathcal{C}\} \\ = \; & \{f \in \mathcal{A}^n \mid \langle\langle z^\beta p, f\rangle\rangle = 0 \; \forall p \in \mathcal{C} \; \forall \beta \in \mathbb{N}^m\} \\ = \; & \{f \in \mathcal{A}^n \mid \langle\langle p, z^\beta \cdot f\rangle\rangle = 0 \; \forall p \in \mathcal{C} \; \forall \beta \in \mathbb{N}^m\} = \mathcal{C}^\perp. \end{aligned}$$

8.4 First-Order Representations for One-Dimensional Codes

In this last section we restrict to the one-dimensional case; thus $\mathcal{D} = \mathbb{F}[z]$ denotes the polynomial ring in one variable over $\mathbb{F}$ and each submodule $\mathcal{C} \in \mathcal{D}^n$ is a convolutional code in the sense of, for example, [6]. Using the duality results from the last section and certain well-studied first-order representations for behaviors, we can derive analogous descriptions for codes along with minimality and uniqueness results.

The main source for this section is the book [4] about behaviors. Although [4] deals with the field $\mathbb{R}$, it can be checked that the results hold true for any field.

We need to introduce the following parameter, called degree, for one-dimensional codes. It is the analogue to the McMillan degree or order of a system. Let $\mathcal{C} = \operatorname{im}_{\mathcal{D}} G$ with $G \in \mathcal{D}^{n \times k}$ and $\operatorname{rank} G = k$, a nonrestrictive assumption. The degree $\delta(\mathcal{C})$ is defined to be the maximum degree of all $k \times k$-minors of G. The degree is sometimes also called the complexity of the code $\mathcal{C}$ (see [6, 2.7]) and it corresponds to the McMillan degree of the associated behavior under the duality studied in the last section; see Theorem 8.1 *(iv)* and [13, p. 276]. Equation (8.1) shows that the degree does not depend on the choice of the encoder G. A code of degree $\delta(\mathcal{C}) = 0$ is in essence a block code.

Theorem 8.5. *Let $\mathcal{C} = \operatorname{im}_{\mathcal{D}} G$ with $G \in \mathcal{D}^{n \times k}$ be a rate k/n code of degree $\delta(\mathcal{C}) = \delta > 0$.*

(a) There exist matrices $(P, Q, R) \in \mathbb{F}^{\delta \times (\delta+k)} \times \mathbb{F}^{\delta \times (\delta+k)} \times \mathbb{F}^{n \times (\delta+k)}$ such that

$$\mathcal{C} = R\big(\ker_{\mathcal{D}}(zP + Q)\big).$$

Moreover,

(i) $\operatorname{rank} P = \delta$,

(ii) rank $\begin{bmatrix} P \\ R \end{bmatrix} = \delta + k$,

(iii) $zP + Q \in \mathcal{D}^{\delta \times (\delta+k)}$ *is left-prime.*

(b) *If* $\mathcal{C} = R(\ker_{\mathcal{D}}(zP+Q)) = \tilde{R}(\ker_{\mathcal{D}}(z\tilde{P}+\tilde{Q}))$ *with matrix triples* (P, Q, R) *and* $(\tilde{P}, \tilde{Q}, \tilde{R})$ *being of the sizes as in (a), then* $(\tilde{P}, \tilde{Q}, \tilde{R}) = (T^{-1}PS, T^{-1}QS, RS)$ *for some* $T \in Gl_\delta(\mathbb{F})$ *and* $S \in Gl_{\delta+k}(\mathbb{F})$.

Proof: *(a)* By Theorem 8.1*(iv)* we have $\mathcal{C}^\perp = \ker_{\mathcal{A}} G^\mathsf{T}$. Without loss of generality we may assume that G is column-reduced; that is, δ is the sum of the column degrees of G. From [4, 5.17] we obtain matrices $(K, L, M) \in \mathbb{F}^{(\delta+k)\times\delta} \times \mathbb{F}^{(\delta+k)\times\delta} \times \mathbb{F}^{(\delta+k)\times n}$ such that $\mathcal{C}^\perp = \{a \in \mathcal{A}^n \mid M \cdot a \in \text{im}_{\mathcal{A}}(zK+L)\}$. Indeed, the parameter ord (Σ) in [4, p. 128] is equal to the degree (cf. [4, 3.11 and 2.22]). Setting $(P, Q, R) = (K^\mathsf{T}, L^\mathsf{T}, M^\mathsf{T})$ and using Theorems 8.2*(b)* and 8.1*(vii)* we obtain the desired representation. Furthermore [4, 5.17] shows that the triple (K, L, M) is minimal with respect to row and column size of the matrix K (or L). Hence, use of [4, 4.32] leads to *(i)* through *(iii)*.

(b) follows from [4, 4.40] and Theorem 8.2*(c)*. □

In fact, the proof shows more. The above given sizes of the matrices (P, Q, R) are minimal among all representations of this type. The minimality is equivalent to the properties *(i)* through *(iii)*. An alternative direct proof, without using duality, is given in the paper [10].

In exactly the same way we can derive so-called (K, L, M)-representations for codes. For this we use [4, 5.10 and 4.3] and dualize these representations using Theorem 8.2*(a)* and *(d)*. This results in [9, Theorems 3.1 and 3.4]:

Theorem 8.6. *Let* $\mathcal{C} = \text{im}_{\mathcal{D}} G$ *with* $G \in \mathcal{D}^{n\times k}$ *be a rate* k/n *code of degree* $\delta(\mathcal{C}) = \delta > 0$.

(a) *There exist matrices* $(K, L, M) \in \mathbb{F}^{(\delta+n-k)\times\delta} \times \mathbb{F}^{(\delta+n-k)\times\delta} \times \mathbb{F}^{(\delta+n-k)\times n}$ *so that*

$$\mathcal{C} = \{p \in \mathcal{D}^n \mid Mp \in \text{im}_{\mathcal{D}}(zK + L)\}.$$

Moreover,

(i) rank $K = \delta$,

(ii) rank $[K, M] = \delta + n - k$,

(iii) $[zK + L \mid M]$ *is left-prime over the polynomial ring* $\mathcal{D}$.

(b) *If* $\mathcal{C} = \{p \in \mathcal{D}^n \mid Mp \in \text{im}_{\mathcal{D}}(zK + L)\} = \{p \in \mathcal{D}^n \mid \tilde{M}p \in \text{im}_{\mathcal{D}}(z\tilde{K} + \tilde{L})\}$ *with matrix triples* (K, L, M) *and* $(\tilde{K}, \tilde{L}, \tilde{M})$ *being of the sizes as in (a), then* $(\tilde{K}, \tilde{L}, \tilde{M}) = (T^{-1}KS, T^{-1}LS, T^{-1}M)$ *for some* $T \in Gl_{\delta+n-k}(\mathbb{F})$, $S \in Gl_\delta(\mathbb{F})$.

Generalized first-order representations as described in the above two theorems are very useful in the design of convolutional codes with large distance that can be encoded in an efficient manner. We refer the interested reader to [9, 8].

8.5 Conclusion

The chapter shows that multidimensional convolutional codes are powerful encoding devices for the transmission of data over a noisy channel. Since these codes are dual objects to multidimensional systems the algebraic theory of linear systems can be profitably applied.

Diederich Hinrichsen, to whom this chapter is dedicated, contributed significantly over the years to algebraic systems theory. As often happens in research, a contribution in one area bears unexpected fruits in another research field. We believe that the recent cross-fertilization between coding theory and systems theory is such an instance.

References

[1] E. Fornasini and M.E. Valcher, Algebraic aspects of 2D convolutional codes, *IEEE Trans. Inf. Theory* IT-40(4):1068–1082, 1994.

[2] E. Fornasini and M.E. Valcher, Multidimensional systems with finite support behaviors: Signal structure, generation, and detection, *SIAM J. Control Optim.* 36(2):760–779, 1998.

[3] G.D. Forney and M.D. Trott, Controllability, observability, and duality in behavioral group systems, *Proc. 34th IEEE Conference on Decision and Control*, New Orleans, LA, 3259–3264, 1995.

[4] M. Kuijper, *First-Order Representations of Linear Systems*, Birkhäuser, Boston, 1994.

[5] U. Oberst, Multidimensional constant linear systems. *Acta Appl. Math.* 20:1–175, 1990.

[6] P. Piret, *Convolutional Codes, an Algebraic Approach*, MIT Press, Cambridge, MA, 1988.

[7] M.P.M. Rocha, Structure and representation of 2D systems, Ph.D. thesis, University of Groningen, 1990.

[8] J. Rosenthal and E.V. York, BCH convolutional codes, *IEEE Trans. Inf. Theory* 45(6):1833–1844, 1999.

[9] J. Rosenthal, J.M. Schumacher, and E.V. York, On behaviors and convolutional codes, *IEEE Trans. Inf. Theory* 42(6):1881–1891, 1996.

[10] R. Smarandache, H. Gluesing-Luerssen, and J. Rosenthal, Generalized first order descriptions and canonical forms for convolutional codes, A. Beghi, L. Finesso, and G. Picci, eds., *Mathematical Theory of Networks and Systems*, 1091–1094, July, 1998, Proc. MTNS-98 Symposium, Padova, Italy.

[11] M.E. Valcher and E. Fornasini, On 2D finite support convolutional codes: an algebraic approach, *Multidim. Syst. Sign. Proc.* 5:231–243, 1994.

[12] P. Weiner, Multidimensional convolutional codes. Ph.D. thesis, University of Notre Dame, 1998.

[13] J.C. Willems. Paradigms and puzzles in the theory of dynamical systems, *IEEE Trans. Auto. Control* AC-36(3):259–294, 1991.

[14] J. Wood and E. Zerz, Notes on the definition of behavioural controllability, *Syst. Control Lett.* 37:31–37, 1999.

[15] J. Wood, E. Rogers, and D.H. Owens, Controllable and autonomous *nD* linear systems, *Multidimens. Syst. Signal Process.* 10(1):33–69, 1999.

Chapter 9

Control of Rate-Bounded Hybrid Systems with Liveness Constraints

Michael Heymann [1]
Feng Lin
George Meyer

ABSTRACT Liveness in hybrid systems is defined as the ability of the system to complete a specified task under all operating conditions and for all possible runs. Liveness is classified in the present chapter into fixed-time, bounded-time, and finite-time liveness. We present an algorithm for synthesis of minimally-interventive controllers that achieve liveness in rate-bounded hybrid systems.

9.1 Introduction

Hybrid systems are dynamical systems in which discrete and continuous behaviors coexist and interact [1] [3] [6]. Such systems frequently arise, for example, from computer aided control of continuous (and discrete) processes in manufacturing, communication networks, flight control systems, traffic control, industrial process control, and the like. Various formalisms have been proposed in the literature for the mathematical description of these systems [6, 3, 1, 12]. Among these, the formalism of hybrid automata [1], which augments the state-machine framework with dynamics to capture timing constraints and continuous dynamics, gained fairly wide acceptance. A formalism related to hybrid automata for modeling hybrid systems, called *hybrid machines*, that differs from the latter in some substantial detail [12], was developed in [10, 11, 12] to capture *open* hybrid systems that interact with their environment both by sharing signals (i.e., by transmission of input-output data), and by event synchronization (through which the

[1]This research is supported in part by the National Science Foundation under grant ECS-9315344 and NASA under grant NAG2- 1043 and in part by the Technion Fund for Promotion of Research. The work by the first author was completed while he was a Senior NRC Research Associate at NASA Ames Research Center, Moffett Field, CA 94035.

system is reconfigured and its structure modified).

Control of hybrid systems can be achieved by employing both interaction mechanisms, to modify and restrict the system's behavior. This flexibility adds significantly to the potential control capabilities of hybrid systems (as compared to either discrete-event or continuous systems), but clearly makes the problem of controller design much more difficult. Indeed, in view of the obvious complexity of hybrid control, even the question of what are tractable and achievable design objectives, is far from easy to resolve. Thus, most attention to date, in control of hybrid systems, has focused either on continuous aspects (i.e., signal-sharing) or discrete aspects (i.e., event-synchronization) but not both.

In the present chapter we examine the control problem for a class of *composite hybrid machines* (CHMs) that consist of the concurrent operation (employing synchronous composition) of *elementary hybrid machines* (EHMs), that allows both signal sharing and event synchronization. A controller can then be coupled with the plant by means of synchronous composition. We confine our attention to controllers that interact with the system only through event synchronization. We further restrict ourselves to a special class of CHMs, where system dynamics is *rate-bounded* and legal guards are conjunctions or disjunctions of atomic formulae in the dynamic variables (of the type $S < C$, $S > C$, $S \leq C$, or $S \geq C$).

The control problem that we focus on in the present chapter is the synthesis of a supervisory controller, where the objective is to guarantee that the system satisfies a set of *legal* specifications. Legal specifications are traditionally partitioned into *safety* specifications that state what the system must be prevented from being able to do, and *liveness* specifications that state what the system is required to do. A typical safety specification is to ensure that the system will never enter a specified set of *illegal* configurations. A typical liveness specification is to ensure that every run of the system will reach a set of *marked* configurations that represent task completion.

The synthesis of legal safety controllers for rate-bounded hybrid machines, was investigated in [10, 11, 12]. Among all legal controllers, we were particularly interested in minimally restrictive (or minimally interventive) ones, that allow the maximal possible set of legal behaviors to survive. Synthesis algorithms for minimally interventive controllers were developed, and the problem of system viability was examined. Synthesis of safety controllers for hybrid systems was also studied in [8, 22].

In this chapter we present an initial investigation of synthesis of liveness controllers for hybrid machines. To this end we define *open* hybrid machines (as opposed to *closed* hybrid machines) as systems that can interact with the environment through event synchronization and can therefore be "driven" to their marked configurations by the user (controller). In view of obvious timing constraints, liveness specifications for hybrid systems must be associated with explicit timing constraints. Thus, we may require that for a specified time limit, every run reach a marked configuration within that time limit. We call such a specification a *fixed-time* liveness specification. Alternatively, a more relaxed specification may be that, for some (unspecified) global time bound, every run of the system reach a marked configuration within that time bound. We call this the *bounded-time* liveness specification. Finally, the least restrictive liveness requirement is that every run reach a marked configuration within a finite time limit (but we do not insist on the existence of a global time bound for all runs). We call this the *finite-time* liveness specification.

It is not hard to see that a fixed-time liveness specification can be readily translated into a safety requirement, by conjoining a global clock to the system, and calling "unsafe" each configuration whose clock value exceeds the (specified) time bound. Therefore, the fixed-time liveness case can be dealt with algorithmically, just as a control problem with safety specifications. In contrast, the bounded-time liveness controller must be handled differently, and this is the focus of the present chapter, where we present a synthesis algorithm for a minimally interventive controller with bounded-time liveness specifications.

9.2 Hybrid Machines

In this section we briefly review the hybrid-machine formalism as described, for example in [12]. An elementary hybrid machine is defined as a tuple

$$EHM = (Q, \Sigma, D, E, I, (q_0, x_0)),$$

whose elements are defined as follows:

- Q is a finite set of vertices.
- Σ is a finite set of event labels. An event is an *input* event, denoted $\underline{\sigma}$ (underline), if it is received by the EHM from its environment; and an *output* event, denoted $\overline{\sigma}$ (overline), if it is generated by the EHM and transmitted to the environment.
- $D = \{d_q = (x_q, y_q, u_q, f_q, h_q) : q \in Q\}$ is the *dynamics* of the EHM, where d_q, the dynamics at the vertex q, is given by:

  $$\dot{x}_q = f_q(x_q, u_q),$$
  $$y_q = h_q(x_q, u_q),$$

  with x_q, u_q, and y_q, respectively, the state, input, and output variables of appropriate dimensions. f_q is a Lipschitz continuous function and h_q a continuous function. (A vertex need not have dynamics associated with it; that is, we permit $d_q = \emptyset$, in which case we say that the vertex is *static*.) Note that the dynamics, and in particular the dimension of x_q, can change from vertex to vertex.

- $E = \{(q, G \wedge \underline{\sigma} \rightarrow \overline{\sigma'}, q', x^0_{q'}) : q, q' \in Q\}$ is a set of *edges* (or *transition-paths*), where q is the vertex exited, q' is the vertex entered, $\underline{\sigma}$ is the input event, and $\overline{\sigma'}$ the output event. G is the guard, formally given as a Boolean combination of inequalities of the form $\Sigma_i a_i s_i \geq C_j$ or $\Sigma_i a_i s_i \leq C_j$, where the s_i are shared (signal) variables, defined shortly, and the a_i and C_j are real constants. Finally, $x^0_{q'}$ is the initialization value for $x_{q'}$ upon entry to q'.

  $(q, G \wedge \underline{\sigma} \rightarrow \overline{\sigma'}, q', x^0_{q'})$ is interpreted as follows. If G is true and the event $\underline{\sigma}$ is received as an input, then the transition to q' takes place at the instant $\underline{\sigma}$ is received,[2] with the assignment of the initial condition $x_{q'}(t_{0,q'}) = x^0_{q'}$ (where $t_{0,q'}$ denotes the time at which the vertex q' is entered and $x^0_{q'}$ is a constant vector[3]). The output event $\overline{\sigma'}$ is transmitted at the same time.

  If $\overline{\sigma'}$ is absent, then no output event is transmitted. If $x^0_{q'}$ is absent (or partially absent), then the initial condition is inherited (or partially inherited) from x_q (assuming x_q and $x_{q'}$ represent the same physical object, and hence are of the same dimension). We often write the transition as $q \overset{G \wedge \underline{\sigma}}{\longrightarrow} q'$ or $(q, G \wedge \underline{\sigma}, q')$ if $\overline{\sigma'}$ and $x^0_{q'}$ are either absent or understood.

  If $\underline{\sigma}$ is absent, then the transition takes place immediately upon G becoming true. Such a transition is called a *dynamic* transition. If G is absent, the guard is always true and the transition will be triggered by the input event $\underline{\sigma}$. Such a transition is called an *event transition.* When both G and $\underline{\sigma}$ are present, the transition is called a *guarded event transition.*

- $I = \{I_q : q \in Q\}$ is a set of invariants. For each $q \in Q$, I_q is defined as $I_q = cl(\neg(G_1 \vee \ldots \vee G_k))$, where $G_1, \ldots, G_k$ are the guards at q, and where $cl(.)$ denotes set closure.[4]

- (q_0, x_0) denote the initialization condition: q_0 is the initial vertex, and $x_{q_0}(t_0) = x_0$.

The invariant I_q of a configuration q expresses the condition under which the EHM is permitted to reside at q; that is, the condition under which none of the guards is true. In particular, from the definition of I_q as $I_q = cl(\neg(G_1 \vee \ldots \vee G_k))$, it follows that each of the vertices of the EHM is *completely guarded.* That is, every invariant violation implies that some

---

[2] If $\underline{\sigma}$ is received as an input while G is false, then no transition is triggered.

[3] More general assignments of the initial conditions such that $x^0_{q'}$ is a function of x_q can also be introduced without much difficulty.

[4] We always insist (especially during computations), that invariants and guards be derived as closed sets by taking their closure.

guard becomes true, triggering a transition out of the current vertex. (It is, in principle, permitted that more than one guard become true at the same instant. In such a case the transition that will actually take place is resolved nondeterministically. It is further permitted that, upon entry into q', one or more of the guards at q' be already true. In such a case, the EHM will immediately exit q' and enter a vertex specified by one of the true guards. Such a transition is considered instantaneous.)

The EHM runs as follows. At a vertex q, the continuous dynamics evolves according to d_q until either a dynamic transition is triggered by a guard becoming true, or an event transition is triggered by the environment through an input event, provided the associated guard is either absent or true.

A *run* of the EHM is a sequence

$$q_0 \xrightarrow{e_1,t_1} q_1 \xrightarrow{e_2,t_2} q_2 \xrightarrow{e_3,t_3} \dots,$$

where e_i is the ith transition and $t_i(\geq t_{i-1})$ is the time when the ith transition takes place. For each run, we define its trajectory, path, and trace as follows.

- The trajectory of the run is the sequence of the vector time functions of the (state) variables:

$$x_{q_0}, x_{q_1}, x_{q_2}, \dots,$$

  where $x_{q_i} = \{x_{q_i}(t) : t \in [t_i, t_{i+1})\}$.
- The path of the run is the sequence of the vertices.
- The input trace of the run is the sequence of the input events.
- The output trace of the run is the sequence of the output events.

To facilitate our ensuing exposition, we standardize EHMs as follows. Recall that our model allows guarded event transitions of the form

$$q \xrightarrow{G \wedge \sigma} q'.$$

However, since for the transition to take place the guard must be true when the event is triggered, a guarded event transition can be decomposed into

$$q_1 \underset{\neg G}{\overset{G}{\rightleftarrows}} q_2 \xrightarrow{\sigma} q',$$

where q has been partitioned into q_1 and q_2, with $I_{q_1} = I_q \wedge \neg G$ and $I_{q_2} = I_q \wedge G$.[5] The dynamics of q_1 and q_2 and the transitions leaving and

[5]Since we use only closed invariants and guards, as described earlier, if I_{q_1}, I_{q_2} or $\neg G$ are not closed, we take their closure.

entering these vertices are the same as for q, except that the transition $(q_1, \underline{\sigma}, q')$ is now impossible. It follows that a guarded event transition can be treated as a combination of a dynamic and an event transition. Thus, in computations, we only need to consider two types of transitions: dynamic transitions, which are labeled by guards only, and event transitions, which are labeled by events only.

A composite hybrid machine consists of several elementary hybrid machines running in parallel:

$$CHM = EHM^1||EHM^2||\dots||EHM^n.$$

Interaction between EHMs is achieved by means of signal transmission (shared variables) and input-output event synchronization (message passing) as described below.

Shared variables consist of output signals from all EHMs as well as signals received from the environment. They are shared by all EHMs in the sense that they are accessible to all EHMs. A shared variable s_i can be the output of at most one EHM. The set of shared variables defines a signal space $S = \{[s_1, s_2, \dots, s_m] \in \mathcal{R}^m\}$.

Transitions are synchronized by an input-output synchronization formalism. That is, if an output event $\overline{\sigma}$ is either generated by one of the EHMs or received from the environment, then all EHMs for which $\underline{\sigma}$ is an active transition label (i.e., $\underline{\sigma}$ is defined at the current vertex with an absent guard or a true guard) will execute $\underline{\sigma}$ (and its associated transition) concurrently with the occurrence of $\overline{\sigma}$. A specific output event can be generated by at most one EHM.

To describe the behavior of

$$CHM = EHM^1||EHM^2||\dots||EHM^n,$$

we define a *configuration* of the CHM to be

$$q =< q^1_{i_1}, q^2_{i_2}, \dots, q^n_{i_n} >\in Q^1 \times Q^2 \times \dots \times Q^n,$$

where Q^j is the set of vertices of EHM^j (components of the EHMs are superscripted).

A transition

$$< q^1_{i_1}, q^2_{i_2}, \dots, q^n_{i_n} > \stackrel{l}{\longrightarrow} < q^1_{i'_1}, q^2_{i'_2}, \dots, q^n_{i'_n} >$$

of a CHM is a triple, where $q =< q^1_{i_1}, q^2_{i_2}, \dots, q^n_{i_n} >$ is the source configuration, $q' =< q^1_{i'_1}, q^2_{i'_2}, \dots, q^n_{i'_n} >$ the target configuration, and l the label that triggers the transition. l can be either an event, or a guard becoming true.[6] Thus, if $l = \underline{\sigma}$ is an event (generated by the environment), then either

[6]This follows from the decomposition of guarded event transitions into dynamic and event transitions as described previously.

$q^j_{i'_j} = q^j_{i_j}$ if $\underline{\sigma}$ is not active at $q^j_{i_j}$, or $q^j_{i'_j}$ is such that $(q^j_{i_j}, \underline{\sigma} \to \overline{\sigma'}, q^j_{i'_j}, x^0_{q^j_{i'_j}})$ is a transition (edge) in E^j. On the other hand, if $l = G$ is a guard, then there must exist a transition $(q^m_{i_m}, G \to \overline{\sigma'}, q^m_{i'_m}, x^0_{q^m_{i'_m}})$ in some EHM^m, and for $j \neq m$, $q^j_{i'_j} = q^j_{i_j}$. The event $\overline{\sigma'}$ (generated as an output event) can trigger a successor event transition if $\underline{\sigma'}$ is active at some vertex $q^k_{i_k}$ $(j \neq m)$; that is, $(q^k_{i_k}, \underline{\sigma'} \to \overline{\sigma''}, q^k_{i'_k}, x^0_{q^k_{i'_k}})$ is a transition in E^k. Note that for simplicity, we do not specify the output events and initial conditions since they are defined in the EHMs.

The transitions are assumed to occur instantaneously, and concurrent vertex changes in parallel components are assumed to occur exactly at the same instant (even when constituting a logically triggered finite chain of transitions). We always assume that only a finite chain of instantaneously triggered transitions can occur in succession.

Based on the above definition, a CHM can be viewed as the same object as an EHM:

$$CHM = (Q, \Sigma, D, E, I, (q_0, x_0)),$$

where

$$\begin{array}{lcl}
Q & = & Q^1 \times Q^2 \times \cdots \times Q^n, \\
\Sigma & = & \Sigma^1 \cup \Sigma^2 \cup \cdots \cup \Sigma^n, \\
D & = & \{(x_q, y_q, u_q, f_q, h_q) : q =< q^1_{i_1}, q^2_{i_2}, \ldots, q^n_{i_n} >\in Q\} \\
 & & \text{combines all the dynamics of } q^j_{i_j}, j = 1, 2, \ldots, n, \\
E & & \text{is defined as above, and} \\
I & = & \{I_{q^1_{i_1}} \wedge I_{q^2_{i_2}} \wedge \ldots \wedge I_{q^n_{i_n}} :< q^1_{i_1}, q^2_{i_2}, \ldots, q^n_{i_n} >\in Q\}, \\
(q_0, x_0) & = & (< q^1_0, q^2_0, \ldots, q^n_0 >, [x^1_0, x^2_0, \ldots, x^n_0]).
\end{array}$$

Therefore, we can define a run of a CHM in the same way as that of an EHM. It can also be easily verified that in view of the fact that the component EHMs are completely guarded, so is the composite CHM.

9.3 Liveness

In our earlier work [10, 12, 11], we developed a synthesis method for designing a safety controller that guarantees the controlled system never to exit a set of specified legal (safety) configurations. Furthermore, the controller was designed to be minimally interventive in the sense that it interferes with the system's operation only when safety violation is otherwise inevitable.

In this chapter, our objective is to synthesize a liveness controller. To define liveness, we first specify a set of marked or *final* configurations $Q_f \subseteq Q$ in the CHM. Liveness is then regarded to be the ability to reach this set of final configurations as discussed below.

To define liveness formally, we must classify hybrid systems into *closed* and *open* systems. A closed system accepts no input events from the environment, and all its transitions are triggered dynamically in its EHMs. Therefore, in the CHM model of a closed system, all transitions are dynamic transitions. On the other hand, an open system accepts input events from its environment, and its CHM model includes event transitions.

Definition 9.1. (Fixed-time liveness for closed systems) *For a fixed-time T, a closed system is* T-live *if every possible run reaches its set of final configurations from the initial configuration within T units of time.*

A weaker version of liveness is the following.

Definition 9.2. (Bounded-time liveness for closed systems) *A closed system is* bounded-time live *if there exists a finite bound T such that every possible run reaches its set of final configurations from the initial configuration within T units of time.*

A still weaker definition of liveness is now given.

Definition 9.3. (Finite-time liveness for closed systems) *A closed system is* finite-time live *if every possible run reaches its set of final configurations from the initial configuration in finite time.*

Clearly, if a closed system is bounded-time live or finite-time live, then it can always reach its set of final configurations from any (reachable) configuration of the system in bounded or finite time, respectively.

The runs of open systems may depend on input events to be triggered by the environment. Thus, we can not insist that open systems reach their final configurations in fixed, bounded, or finite time without considering input events. Therefore, the liveness definitions need to be modified as follows.

Definition 9.4. (Fixed-time liveness for open systems) *For a fixed-time T, an open system is* T-live *if every possible run can be forced (by the environment or user) to reach its set of final configurations from the initial configuration within T units of time.*

Definition 9.5. (Bounded-time liveness for open systems) *An open system is* bounded-time live *if there exists a finite bound T such that every possible run can be forced (by the environment or user) to reach its set of final configurations from the initial configuration within T units of time.*

Definition 9.6. (Finite-time liveness for open systems) *An open system is* finite-time live *if any run of the system can be forced (by the environment or user) to reach its set of final configurations in finite time.*

Clearly, if a system is bounded-time live, then there exists an infimal time bound T_{inf} such that for all $T > T_{inf}$, the system is T-live.

Here we consider only hybrid machines that satisfy the following.

Assumption 9.1. *The dynamics described by f_q and h_q has the following properties: (i) $h_q(x_q, u_q)$ is a linear function; and (ii) $f_q(x_q, u_q)$ is bounded by a lower limit v_q^L and an upper limit v_q^U; that is, the only information given about $f_q(x_q, u_q)$ is that $f_q(x_q, u_q) \in [v_q^L, v_q^U]$.*

Under this assumption, we consider only *rate bounded* hybrid machines in which all the rates are bounded by closed intervals. Such systems are either bounded-time live or not live. Therefore, in the remainder of the chapter, by liveness we simply mean bounded-time liveness. As we stated in the introduction, fixed-time liveness need not be considered further, because it can always be viewed as a special case of safety.

9.4 Control

In this section, we study how to control a hybrid system to achieve bounded-rate liveness. Formally, a *controller* of a CHM is a hybrid machine C that runs in parallel with the CHM. The resultant system

$$CHM||C$$

is called the *controlled* or *closed-loop* system. The objective of control is to force the controlled system to satisfy a prescribed set of behavioral specifications, in this case, to satisfy the liveness constraints. A controller that achieves this objective is then said to be *live*.

In this chapter, we consider only restricted interaction between the controller and the CHM by permitting the controller to interact with the CHM only through input-output event synchronization. Thus, we make the following assumption.

Assumption 9.2. *C can only control the CHM by means of input-output event synchronization. That is, C can only control event transitions in the CHM. Furthermore C can control all the event transitions in the CHM. That is, all the (externally triggered) event transitions are available to the controller.*

The assumption that C can control all the event transitions in the CHM leads to no essential loss of generality because, when some of the events are *uncontrollable*, we can use the methods developed in supervisory control of discrete-event systems [19, 20] to deal with uncontrollable event transitions.

Obviously, there may exist many live controllers with different degrees of restrictiveness. A live controller C is said to be *less interventive* (or restrictive) than another live controller C' if every run permitted by C' is also permitted by C. A live controller is said to be *minimally interventive* if it is less interventive than any live controller. In most cases, we are interested in the minimally interventive live controller.

To synthesize such a controller, we would like to find all the configurations from which the system can be forced to reach the final configurations Q_f in bounded-time (we call these configurations live). We start with all the neighboring configurations of Q_f (that have at least one transition leading to Q_f). For any neighboring configuration q, if it has an event transition leading to Q_f, then clearly q is live. If no such event transitions exist, then we must consider all the dynamic transitions leaving q. If all the dynamic transitions go to Q_f, then q is again live. Otherwise, some of these dynamic transitions go to Q_f. We take the disjunction of their guards and denote it G_g. The remaining dynamic transitions do not go to Q_f. We denote the disjunction of their guards by G_b. Clearly, q will be live if and only if G_g is guaranteed to become true before G_b becomes true. This gives a way to partition the configuration q (or equivalently, its invariant I_q) into live subconfiguration q_1 and unlive subconfiguration q_2.

To describe this partition formally, let us define, for $q \in Q$ and $Q' \subseteq Q$, the set of event transitions from q to Q':

$$ET(q, Q') = \{(q, \underline{\sigma}, q') \in E : q' \in Q'\}.$$

Similarly, the set of dynamic transitions from q to Q' is

$$DT(q, Q') = \{(q, G, q') \in E : q' \in Q'\}.$$

As we said, if $ET(q, Q_f) \neq \emptyset$ or $DT(q, Q-Q_f) = \emptyset$, then q is live. Otherwise, we must partition the invariant I_q into the live part I_{q_1} and the unlive part I_{q_2} as follows.

We first consider the time at which a predicate P will become true. Thus, let $T(P(x(t)))$ $(=T(true(P(x(t)))))$ be the time at which P becomes true for the first time along the trajectory $x(t)$. Since our goal is to guarantee that the liveness specification will not be violated under any condition, we must consider the maximal and minimal values of $T(P(x(t)))$ when evaluated over all possible trajectories of all possible runs. Thus, let us define

$$\begin{aligned} T_{\max}(true(P)) &= \max_{x(t)} T(P(x(t))) \\ T_{\min}(true(P)) &= \min_{x(t)} T(P(x(t))). \end{aligned}$$

These maximum and minimum values can be calculated from the expression of P and the associated dynamics, which is discussed in [10].

Let

$$\begin{aligned} G_g &= \vee_{(q,G,q') \in DT(q,Q_f)} G \\ G_b &= \vee_{(q,G,q') \notin DT(q,Q_f)} G. \end{aligned}$$

We define a precedent condition as

$$pd(q, Q_f) = T_{\max}(true(G_g)) < T_{\min}(true(G_b)).$$

We now split the configuration q into live subconfiguration q_1 and unlive subconfiguration q_2, by partitioning the invariant I_q as

$$\begin{aligned} I_{q_1} &= I_q \wedge pd(q, Q_f) \\ I_{q_2} &= I_q \wedge \neg pd(q, Q_f). \end{aligned}$$

Clearly, the dynamics of q_1 and q_2 and the transitions leaving and entering these configurations are the same as for q, except that all the dynamic transitions in $DT(q_1, Q - Q_f)$ are now impossible (because the dynamic transitions in $DT(q_1, Q_f)$ will take precedence). Also the transition from q_1 to q_2 is dynamic with the guard $\neg pd(q, Q_f)$, and from q_2 to q_1 with guard $pd(q, Q_f)$.

The justification for the above partition is given in the following.

Theorem 9.1. *The precedent condition $pd(q, Q_f)$ is true if and only if for any trajectory $x(t)$ in any run,*

$$T(G_g(x(t))) < T(G_b(x(t))).$$

Proof: If $pd(q, Q_f)$ is true, that is,

$$\max_{x(t)} T(G_g(x(t))) < \min_{x(t)} T(G_b(x(t))),$$

then, clearly,

$$(\forall x(t)) T(G_g(x(t))) < T(G_b(x(t))).$$

On the other hand, if $pd(q, Q_f)$ is not true, then

$$T_{max}(true(G_g)) \geq T_{min}(true(G_b)).$$

Let $x(t)$ be the trajectory achieving $T_{min}(true(G_b))$. Clearly, G_b will become true earlier than G_g along the trajectory $x(t)$; otherwise the system will exit q when G_g becomes true and hence $x(t)$ will not be a valid trajectory. Therefore,

$$(\exists x(t)) T(G_g(x(t))) \geq T(G_b(x(t))),$$

a contradiction. □

The above procedure, of identifying live configurations and calculating live subconfigurations, must be repeated. This is because a configuration from which a live configuration can be forced to be reached in bounded-time is also live. To repeat the procedure, let us consider a transition (q, l, q'). Suppose that q' has been split into its live subconfiguration q'_1 and its unlive subconfiguration q_2'. Then transition from q into q_1' (rather than into q_2') depends on satisfaction, upon entry into q', of the invariant $I_{q_1'}$ (rather than $I_{q_2'}$). Thus, let us define $wp(q, l, q')$ to be the weakest precondition under which the transition (q, l, q') will not violate the invariant $I_{q'}$ upon entry into q'. Since some of the shared variables that appear in $I_{q'}$ are possibly (re-) initialized upon entering q' because $x_{q'}$ is (re-)initialized, the condition $wp(q, l, q')$ can be computed from $I_{q'}$ by substituting into $I_{q'}$ the appropriate initial (entry) values of all the shared variables that are also output variables of q'. That is, if y_j is the jth output variable of q' and $s_i = y_j$ is a shared variable that appears in $I_{q'}$, then the value of s_i must be set to $s_i = h_j(x_{q'}^0, u_{q'})$.

Using this weakest precondition, we can replace each transition (q, l, q') by its equivalence $(q, wp(q, l, q') \wedge l, q')$. That is, a dynamic transition with guard G will be replaced by a dynamic transition with guard $G \wedge wp(q, G, q')$. Similarly, an event transition triggered by $\underline{\sigma}$ will be replaced by a guarded event transition $wp(q, \underline{\sigma}, q') \wedge \underline{\sigma}$, which in turn will be decomposed into event and dynamic transitions.

Therefore, at the beginning of each iteration, we normalize the CHM by performing the following steps.

1. Replace each transition $q \xrightarrow{l} q'$ by $q \overset{wp(q,l,q') \wedge l}{\longrightarrow} q'$,
2. Decompose each guarded event transition $q \overset{G \wedge \underline{\sigma}}{\longrightarrow} q'$ into $q_1 \underset{\overleftarrow{\neg G}}{\overset{G}{\longrightarrow}} q_2 \overset{\underline{\sigma}}{\longrightarrow}$ q',
3. Replace each guard G by its closure $cl(G)$,
4. Replace each invariant I by the closure of the negation of the disjunction of all the guards $cl(\neg(G_1 \vee \ldots \vee G_k))$.

When the iterations terminate (i.e., when there are no more live configurations to be added), the resulting live configurations (and their invariant) have the following property. Either all the possible dynamic transitions are live or there exists at least one event transition to a live configuration that can be forced. Although the forcing can be done at any time when the CHM is in the corresponding configuration, the minimally interventive live controller will not force the event transition, unless a dynamic transition to a blocked configuration (blocked dynamic transition) is about to take place. In other words, the forcing will occur when a blocked guard of a blocked dynamic transition becomes true. We assume that the live guards have precedence over blocked guards. Hence, the forcing will take precedence over the blocked dynamic transition.

In summary, we present our synthesis algorithm as follows.

Algorithm 9.1. *(***Control Synthesis***)*
Input

- The model of the system

$$CHM = (Q, \Sigma, D, E, I, (q_0, x_0)).$$

- The set of final configurations $Q_f \subseteq Q$.

Output

- The controller

$$C = (Q^c, \Sigma^c, D^c, E^c, I^c, (q_0^c, x_0^c)).$$

Initialization

1. Set of live configurations

$$LC := Q_f;$$

2. New set of live configurations

$$NLC := Q_f;$$

3. Set of pending configurations

$$PC := Q - Q_f;$$

Iteration

4. For all $q \in PC$, $e = (q, l, q') \in E$ do

$$E := (E - \{e\}) \cup \{(q, wp(q, l, q') \wedge l, q')\};$$

5. Let

$$Repeat := false;$$

6. For all $q \in PC, e = (q, G \wedge \underline{\sigma}, q') \in E$ do

   begin

$$I_{q_1} := I_q \wedge \neg G;$$
$$I_{q_2} := I_q \wedge G;$$

   If $I_{q_1} = false$, then

$$E := (E - \{e\}) \cup \{(q, \underline{\sigma}, q')\};$$

If $I_{q2} = false$, then

$$E := E - \{e\};$$

Else do

begin

$Repeat := true;$
$PC := (PC - \{q\}) \cup \{q_1, q_2\};$
$E := (E - \{e\}) \cup \{(q_1, G, q_2), (q_2, \neg G, q_1), (q_2, \underline{\sigma}, q')\};$

For all $e' = (q, l, q'') \in E - \{e\}$ do

$$E := (E - \{e'\}) \cup \{(q_1, l, q''), (q_2, l, q'')\};$$

For all $e' = (q'', l, q) \in E$ do

$$E := (E - \{e'\}) \cup \{(q'', l, q_1), (q'', l, q_2)\};$$

end;

end;

7. If $Repeat = true$, go to 4;
8. For all $(q, G, q') \in E$ do

$$G := cl(G);$$

9. For all $q \in PC$ do

$$I_q = cl(\neg \vee_{(q,G,q') \in E} G);$$

10. For all $q \in PC$ do

begin

If $ET(q, Q_f) \neq \emptyset$ or $DT(q, Q - Q_f) = \emptyset$, then

$$NLC := NLC \cup \{q\};$$

Else do

Begin

$G_g := \vee_{(q,G,q') \in DT(q,LC)} G;$
$G_b := \vee_{(q,G,q') \notin DT(q,LC)} G;$
$pd(q, LC) := T_{\max}(true(G_g)) < T_{\min}(true(G_b));;$
$I_{q_1} := I_q \wedge pd(q, LC);$
$I_{q_2} := I_q \wedge \neg pd(q, LC);$

If $I_{q_1} = false$, then

$$skip;$$

If $I_{q_2} = false$, then

$$NLC := NLC \cup \{q\};$$

Else do

begin

$$\begin{aligned}&NLC := NLC \cup \{q_1\};\\&PC := PC \cup \{q_2\};\\&E := E \cup \{(q_1, \neg pd(q, LC), q_2), (q_2, pd(q, LC), q_1)\};\end{aligned}$$

For all $e = (q, l, q') \in E - DT(q, LC)$ do

$$E := (E - \{e\}) \cup \{(q_2, l, q')\};$$

For all $e = (q, l, q') \in DT(q, LC)$ do

$$E := (E - \{e\}) \cup \{(q_1, l, q'), (q_2, l, q')\};$$

For all $e = (q', l, q) \in E$ do

$$E := (E - \{e\}) \cup \{(q', l, q_1), (q', l, q_2)\};$$

11. If $LC \neq NLC$, then

$$LC := NLC;$$

Go to 4;

Construction of C

12. Define vertices, events, dynamics, and invariants:

$$\begin{aligned}&Q^c := LC;\\&\Sigma^c := \Sigma \cup \{\underline{\tilde{\sigma}} : \sigma \in \Sigma\};\\&D^c := \emptyset;\\&I^c := I|_{Q_c};\end{aligned}$$

13. Define transitions:

$$\begin{aligned}&E^c := \{(q, \underline{\tilde{\sigma}} \to \overline{\sigma}, q') : q, q' \in Q^c \wedge (q, \underline{\sigma}, q') \in E\};\\&E^c := E^c \cup \{(q, \vee_{(q,G,q') \in DT(q, Q-Q^c)} G \to \overline{\sigma}, q') : q, q' \in Q^c \wedge (q, \underline{\sigma}, q') \in E\};\end{aligned}$$

14. End.

It is readily seen that the configurations of the controller C consist of the set of all live configurations with their invariants as calculated during the iteration phase of the algorithm. The controller C has no continuous dynamics, so it is "driven" by the dynamics of the CHM. The transitions of C are then triggered when the boundary of some unlive dynamic transitions is reached. The controller thus synthesized is minimally interventive. Its interaction with the system is restricted to the exclusive objective of preventing the system from violating the liveness constraints. The controller is augmented to allow "environment-triggered" transitions labeled by $\underline{\tilde{\sigma}}$, which are allowed to be generated by the environment (possibly by an additional controller) and trigger transitions in C and hence in the CHM whenever such transitions are not disabled or disallowed by C. C will force an (event) transition only if otherwise the live constraint could be violated. We illustrate the algorithm by the following example.

Note that the controlled system $CHM||C$ is also an open system (but with input events $\underline{\sigma}$ replaced by $\underline{\tilde{\sigma}}$). Therefore, we can combine $CHM||C$ with other controller D as follows. First, all the output events $\overline{\sigma}$ in D are replaced by $\overline{\tilde{\sigma}}$ to obtain $\tilde{D}$. Then the composite controlled system is given by

$$CHM||C||\tilde{D}.$$

The following theorem shows the correctness of our algorithm.

Theorem 9.2. *If Algorithm 9.1 terminates in a finite number of steps, then the synthesized controller is a minimally interventive live controller in the following sense.*

(i) $CHM||C$ is live.

(ii) For any live controller D, every run of $CHM||D$ has a corresponding run in $CHM||C||\tilde{D}$.

Proof: Since Algorithm 9.1 terminates in a finite number of steps, by Theorem 9.1, from a live configuration (those that are in Q^c), the system can always be forced to its final configurations in bounded-time. Therefore, to prove Part (i), it is sufficient to show that $CHM||C$ can never go to a configuration outside Q^c. This is obvious because all the event transitions leaving Q^c will be disabled by C and all the dynamic transitions leaving Q^c will be preempted by the forced transitions $(q, \vee_{(q,G,q')\in DT(q,Q-Q^c)} G \rightarrow \overline{\sigma}, q')$.

To prove Part (ii), observe first that Algorithm 9.1 progressively adds live configurations to Q^c until no further addition is possible. Therefore, a controller will be live only if it does not exceed the configurations and invariants of C. Assume that

$$q_0 \overset{e_1,t_1}{\longrightarrow} q_1 \longrightarrow \ldots \longrightarrow q_{n-1} \overset{e_n,t_n}{\longrightarrow} q_n$$

is a possible run of $CHM||D$ and the first $n-1$ transitions are also possible in $CHM||C||\tilde{D}$ but the last transition from q_{n-1} to q_n is impossible in $CHM||C||\tilde{D}$; that is, it is either disabled or preempted by C. Since C only takes action at the boundary of some unlive transitions, the inaction of D at that point implies that for some trajectory associated with some continuation of this run, the invariant of C will be violated, contradicting the hypothesis that D is legal. □

References

[1] R. Alur, C. Courcoubetis, T.A. Henzinger, and P.-H. Ho, Hybrid automata: An algorithmic approach to the specification and verification of hybrid systems, in *Hybrid Systems*, Lecture Notes in Computer Science 736, Springer-Verlag, Berlin, 209–229, 1993.

[2] E. Azarin, O. Maler, and A. Pnueli, Symbolic controller synthesis for discrete and timed systems, in *Hybrid Systems II*, Lecture Notes in Computer Science 999, Springer Verlag, Berlin, 1–20, 1995.

[3] M. S. Branicky, Universal computation and other capabilities of hybrid and continuous dynamical systems. *Theoretical Computer Science* 138:67–100, 1995.

[4] Y. Brave and M. Heymann, On stabilization of discrete-event processes, *Int. J. Control* 51(5):1101–1117, 1990.

[5] Y. Brave and M. Heymann, On optimal attraction in discrete-event processes, *Inf. Sciences* 67:245–267, 1993.

[6] R. W. Brockett, Hybrid models for motion control systems, in H.L. Trentelman and J.C. Willems eds., *Essays in Control: Perspectives in the Theory and its Applications*, Birkhauser, Boston, 29–53, 1993.

[7] J.E.R. Cury, B.H. Krogh, and T. Niinomi, Synthesis of supervisory controllers for hybrid systems based on approximating automata. *IEEE Trans. Auto. Control* 43(4):564–568, 1998.

[8] T. A. Henzinger and P. W. Kopke, Discrete-time control for rectangular hybrid automata. *Theoretical Computer Science* 221:369–392, 1999.

[9] T. Henzinger, P. Kopke, A. Puri, and P. Varaiya, What's decidable about hybrid automata. *Proc. 27th Annual ACM Symposium on the Theory of Computing*, 373–382, 1995.

[10] M. Heymann, F. Lin and G. Meyer, Control synthesis for a class of hybrid systems subject to configuration based safety constraints, NASA Technical Memorandum 112196, 1997.

[11] M. Heymann, F. Lin and G. Meyer, Synthesis of minimally restrictive legal controllers for a class of hybrid systems, in *Hybrid Systems IV*, P. Antsaklis, W. Kohn, A. Nerode, and S. Sastri, Eds. ,Lecture Notes in Computer Science 1273, Springer Verlag, Berlin, 134–159, 1997.

[12] M. Heymann, F. Lin and G. Meyer, Synthesis and viability of minimally interventive legal controllers for hybrid systems. *Discrete Event Dynamic Syst.: Theory Appl.* 8(2):105–135, 1998.

[13] D. Kapur and R. K. Shyamasundar, *Synthesizing Controllers for Hybrid Systems*, Lecture Notes in Computer Science 1201, Springer Verlag, Berlin, 361–375, 1997.

[14] F. Lin and W.M. Wonham, On observability of discrete event systems. *Information Sciences* 44(3):173–198, 1988.

[15] O. Maler, A. Pnueli and J. Sifakis, *On the synthesis of discrete controllers for timed systems*, Lecture Notes in Computer Science 900, Springer-Verlag, Berlin, 229–242, 1995.

[16] D.L. Pepyne and C.G. Cassandras, Modeling, analysis, and optimal control of a class of hybrid systems. *Discrete Event Dynamic Syst.: Theory and Appl.* 8(2):175–202, 1998.

[17] A. Pnueli, *Verifying liveness properties of reactive systems*, Lecture Notes in Computer Science 1201, Springer Verlag, Berlin, 1997.

[18] J. Raisch and S.D. O'Young, Discrete approximation and supervisory control of continuous systems. *IEEE Trans. on Auto. Control* 43(4):569–573, 1998.

[19] R.J. Ramadge and W.M. Wonham, Supervisory control of a class of discrete event processes. *SIAM J. Control Optimization* 25(1):206–230, 1987.

[20] P.J. Ramadge and W.M. Wonham, The control of discrete event systems. *Proc. IEEE* 77(1):81–98, 1989.

[21] A.J. van der Schaft and J.M. Schumacher, Complementarity modeling of hybrid systems. *IEEE Trans. on Auto. Control* 43(4):483–490, 1998.

[22] H. Wong-Toi, The synthesis of controllers for linear hybrid automata. *Proc. 36th IEEE Conference on Decision and Control*, 4607–4613, 1997.

Chapter 10

A General Principle of Marked Extraction

Uli Krause

ABSTRACT For convex sets, in particular convex cones and polyhedra, as well as for commutative monoids, in particular the multiplicative semigroups of integral domains, a method of marked extraction is outlined. By this method, elements can be represented uniquely by distinguished elements such as extreme points and irreducible elements, respectively, if some marking is employed. Geometrically, this yields a particular simplicial subdivision of convex sets and monoids, respectively. Applied to the topic of joint production from economics the famous Nonsubstitution Theorem is generalized to a Substitution Theorem.

10.1 Introduction

During the almost 40 years we have known each other Didi Hinrichsen and I have often talked about economic issues including actual problems of the economy as well as theoretical questions in mathematical economics. When we started some joint work on the intriguing problem of joint production (see Section 10.2) we faced the question of representing points of a possibly unbounded polyhedron in a unique manner by its extreme points and extreme directions. This, however, appeared meaningful only for the special case of a (generalized) simplex — a case the economists were already able to manage by the so-called Nonsubstitution Theorem. To go further, we imposed some marking on the representation process and developed an algorithm that in each step extracts a marked element from a given one. If the algorithm stops it yields a unique representation in terms of marked extreme elements. Thereby, extracting one element from another means to subtract as much as possible of the former from the latter with respect to the convex structure of the polyhedron. (See Section 10.3.) This unique representation by marked extraction geometrically yields a particular subdivision of the polyhedron into (generalized) simplices. Applied to the initial problem of joint production this subdivision results in a Substitution Theorem which describes how optimal production processes have to be substituted for each other if demand for goods changes from the outside (Section 10.2). The Nonsubstitution Theorem then is the limit case where one set of production processes can be chosen once and for all to satisfy all demands.

The principle of marked extraction is, however, much more powerful and works, in addition to polyhedra and joint production, also for quite different settings. One may view, for example, the unique factorization of natural numbers into primes as a kind of Nonsubstitution Theorem and may, therefore, try to find a Substitution Theorem for number rings where unique factorization into primes no longer applies as for certain rings of algebraic numbers. An appropriate setting to work out this idea is given by Krull domains or, more general, Krull monoids. Extraction within Krull domains turns out to be linked with the well–known notion of valuations, a connection which then leads one to consider Krull domains a particular species of the more general extraction domains (Section 10.4).

The present chapter is an extended version of my talk "Didi and Economics" given at the Workshop in Honor of Diederich Hinrichsen on the occasion of his 60th birthday. It is the aim of the chapter to point out the unifying role of the principle of marked extraction in quite different fields ranging from economics over convexity to pure mathematics. For proofs the reader is referred to the references given.

10.2 Joint Production and the Substitution Theorem

Consider the production of n goods by m processes of production that use these goods together with some primary factor (homogeneous labor) as inputs. Economically important and mathematically interesting is the case of *joint production* where each of the processes may produce several different goods and not just one single good. (There is an extended literature on this kind of model which goes back to von Neumann [13] and has been developed further by Morishima [12]. See also Kurz and Salvadori [10].) More precisely, for process i, $1 \leq i \leq m$, let $a_{ij} \geq 0$ denote the input and $b_{ij} \geq 0$ the output of good j for $1 \leq j \leq n$ and let $l_i \geq 0$ be the labor input. $A, B \in \mathbb{R}_+^{m\times n}$ are input and output matrices, respectively, and $l \in \mathbb{R}_+^n$ is the (column) vector of labor inputs. For the net output matrix $C = B - A$ the cone of outputs is given by

$$N = \{x^\top C \mid x \in \mathbb{R}_+^m\}.$$

It is assumed that $m \geq n$, rank $C = n$, and that the production system is *productive*, that is, N contains a (strictly) positive vector. (For two vectors $x \leq y$ means that $x_i \leq y_i$ for all i and $x < y$ means that $x_i < y_i$ for all i; x is positive if $0 < x$. $\mathbb{R}_+^k$ denotes the standard cone in $\mathbb{R}^k$, $\mathbb{R}_+^k = \{x \in \mathbb{R}^k \mid 0 \leq x\}$. Similar notions are employed for matrices.)

The *minimal labor expenditure* to produce a net output $y \in N$ is given by

$$\lambda(y) = \inf\{x^\top l \mid x^\top C = y,\ x \in \mathbb{R}_+^m\}.$$

The main question addressed in this section is how the minimal labor expenditure and the optimal choice of production processes change with demand y from the outside. A famous result obtained by various economists (Arrow, Georgescu–Roegen, Koopmans, and Samuelson) around 1951 (see Kurz and Salvadori [10]) is the following.

Nonsubstitution Theorem: Suppose for the above model that there are *no* joint products, that is, each row of B has exactly one positive entry, and suppose $l_i > 0$ for all $1 \leq i \leq m$. Then there exists *one single productive selection* $I \subset \{1, \dots, m\}$ of n processes such that for all $y \in N,\ y \geq 0$, it holds that

$$\lambda(y) = x_I(y)^\top l_I \text{ with } x_I(y)^\top = yC_I^{-1},$$

where the subscript I means that processes are picked from I. Moreover, $\lambda(y) = y\lambda$ where $\lambda = C_I^{-1} l_I$ is the (column) vector of *individual labor values* λ_j that are *nonnegative.*

An important consequence of the Nonsubstitution Theorem is that it gives way to work, as in the models of Leontief and Sraffa (see Kurz and Salvadori [10]), with a fixed $n \times n$-input matrix A and the identity matrix as output matrix and also that the individual labor values λ_j represent meaningfully the amount of labor needed directly or indirectly to produce one unit of commodity j. (The latter applies similarly to the so–called prices of production if a rate of profit is taken into account.) If, however, joint production is admitted, one single selection of processes is no longer sufficient as stated in the following result.

Theorem 10.1. (**Substitution Theorem,** *Hinrichsen and Krause [3, 4]*) *Suppose for the above model that $l_i > 0$ in the case where there is no $C_i \leq 0$ for the ith row of C for $1 \leq i \leq m$. Then there exists a finite* collection of productive process selections $I \subset \{1, \dots, m\}$ *with $|I| = n$ such that for every $y \in N,\ y \geq 0$, there is some $I = I(y)$ with*

$$\lambda(y) = x_I(y)^\top l_I$$

and such that for all

$$z \in N(I) = \{x_I^\top C_I \mid 0 \leq x_I\}$$

it holds that

$$\lambda(z) = z\lambda(I) \quad with \quad \lambda(I) = C_I^{-1} l_I.$$

Moreover, the mappings $y \longmapsto x(y)$ and $y \longmapsto \lambda(y)$ are continuous and piecewise linear on the pieces $N(I)$ of N.

Before I turn to a method in the next section by which this result can be proven some comments may be in order.

Comments:

1. If there are no joint products the collection of process selections in Theorem 10.1 shrinks to just one member and the Substitution Theorem reduces to the Nonsubstitution Theorem.

2. If there are joint products it may happen that the piecewise labor values $\lambda(I)$ show *negative components.* If $\lambda_j(I)$ is negative, then good j is not separately producible and $-\lambda_j(I)$ is the amount of labor needed directly or indirectly to *dispose of* good j.

3. The framework presented allows for an economics of *disposal processes* as opposed to production processes in the usual sense. In von Neumann [13] the model is cast in terms of inequalities, which implies that the disposal of goods is automatically free. (Thus, in this case, inequalities are a special case of equalities.)

4. The pieces $N(I)$ are simplicial convex cones that together make a simplicial complex exhausting N. (See also the next section.)

10.3 Marked Extraction in Convex Sets

Let K be a convex cone in some real vector space E which is pointed (i.e., $K \cap (-K) = \{0\}$), and lineally closed (i.e., any intersection with a line in E is closed in the ordinary topology of the line).

Definition 10.1. *The function $\lambda(\cdot,\cdot)\colon K \times K \longrightarrow [0,\infty]$ defined by*

$$\lambda(x,y) = \sup\{\lambda \geq 0 \mid y - \lambda x \in K\}$$

is called the order function *on K.*

The mapping $e(\cdot,\cdot)\colon K \times K \longrightarrow K$ defined by

$$e(x,y) = y - \lambda(x,y)x \quad (\textit{with } \infty \cdot 0 = 0)$$

is called the extraction function *on K.*

For $x, y \in K$ the element x is called a component *of y if $\lambda(x,y) > 0$ and $e(x,y)$ is called the* rest *after extracting x from y.*

Remark 10.1. *For $x, y \in K\backslash\{0\}$ an equivalence relation is defined by $x \sim y$ iff $\lambda(x,y) > 0$ and $\lambda(y,x) > 0$, the equivalence classes of which are called the parts of the cone. On every part a metric is given by*

$$p(x,y) = -\log \min\{\lambda(x,y),\ \lambda(y,x)\}$$

and a quasimetric by

$$d(x,y) = -\log[\lambda(x,y)\cdot\lambda(y,x)] \ (d(x,y) = 0 \textit{ iff } y = \lambda x \textit{ for some } \lambda > 0).$$

These two metrics which are called the part metric and Hilbert's projective metric, respectively, play an important role for the stability analysis of self-mappings of the cone (see Krause [9]).

For $x \in K$ let $F(x)$ be the set of all components of x. Equivalently, $F(x)$ is the smallest face of K containing x. Thereby, a *face* F of K is a convex subcone of K such that for $x, y \in K$, $y - x \in K$ and $y \in F$ imply $x \in F$. A face $F(x)$ is also called a *principal face.* The cone K is said to be of *finite type* if there exists no infinite properly descending chain of principal faces $F_1 \supsetneq F_2 \supsetneq \cdots$ of K. A finite dimensional cone is of finite type but not vice versa.

Definition 10.2. *A mapping $\mu\colon K \longrightarrow K$ is called a* marking *on K if it satisfies the following properties for $x, y \in K$:*

(i) $\mu(x)$ is a component of x and $\mu(x) \neq 0$ for $x \neq 0$.

(ii) If x is a component of y and $\mu(y)$ is a component of x, then $\mu(x) = \mu(y)$.

For a marking μ on K the mapping $e\colon K \longrightarrow K$ defined by $e(x) = e(\mu(x), x)$ is called a marked extraction *on K.*

The following example illustrates the concepts introduced.

Example 10.1. Consider finitely many linear functionals $f_i\colon \mathbb{R}^n \longrightarrow \mathbb{R}$, $1 \leq i \leq k$, that are not identically 0 and such that $f_i(x) = 0$ holds for all $1 \leq i \leq k$ iff $x = 0$. Then $K = \{x \in \mathbb{R}^n \mid f_i(x) \geq 0,\ 1 \leq i \leq k\}$ is a convex cone that is pointed and lineally closed. The order function on K is given by $\lambda(x,y) = \min\left\{\frac{(f_i(y))}{(f_i(x))} \mid 1 \leq i \leq k,\ f_i(x) > 0\right\}$. An element x is a component of y whenever $f_i(y) = 0$ implies $f_i(x) = 0$ and the faces of K are given as the joint zero sets of finitely many $f_i, i \in I$, for some subset $\emptyset \subset I \subset \{1, \ldots, k\}$. In particular, all faces are principal and K is of finite type. A marking on K can be constructed as follows. Since the cone K is finitely generated, one can choose some minimal set of generators of K with an arbitrary numbering, say $\{z_1, \ldots, z_r\}$. By $\mu(0) = 0$ and $\mu(x)$ the component z_i of x with smallest index i for $x \neq 0$ a marking μ on K is defined.

A particular example is given by the cone $K = \mathbb{R}^n_+ \cap N$ where N is the net output cone considered in the previous section. Hence, the following results apply to this cone.

Iterating the mapping $e\colon K \longrightarrow K$ of marked extraction defines an *extraction algorithm* that stops after finitely many steps if K is of finite type and that yields for $0 \neq x \in K$ the elements

$$x_i = e^i(x) \quad (e^i \text{ the } i\text{th iterate of } e)$$

$$e_i = \mu(x_i) \text{ and } \lambda_i = \lambda(e_i, x_i) \text{ for } 0 \leq i \leq m(x) \in \mathbb{N}.$$

The e_i are independent in the following sense.

Definition 10.3. *A finite subset $\emptyset \neq M \subset K \backslash \{0\}$ is called* μ–independent *with respect to a marking μ on K if there exists a numbering $M = \{a_0, a_1, \dots, a_m\}$ and a chain of faces $F(y_0) \supsetneqq F(y_1) \supsetneqq \dots \supsetneqq F(y_m)$ such that $a_i = \mu(y_i)$ for all i. A maximal μ–independent set is called a* μ–base.

By pinning down a marking on cone K one obtains the following type of unique representation for convex cones.

Theorem 10.2. *(Hinrichsen und Krause [5]) Let K be a convex cone that is pointed, lineally closed, and of finite type, and let μ be some marking on K. Then every point $0 \neq x \in K$ has exactly one marked representation, that is, a representation*

$$x = \sum_{a \in M} \lambda_a a \text{ with } M \mu\text{–independent and } \lambda_a > 0 \text{ for } a \in M.$$

This representation can be obtained by the extraction algorithm, that is,

$$M = \{e_0, e_1, \dots, e_{m(x)}\},\ e_i = \mu(e^i(x)) \text{ and } \lambda_{e_i} = \lambda(e_i, e^i(x)),$$

where $0 \leq i \leq m(x) \leq \dim F(x) - 1$.

This theorem can be used as follows to obtain a unique representation more generally for convex sets including polytopes. Let C be a nonempty convex set in some real vector space E that is lineally closed and contains no lines. Denote by $D = \text{rec}C$ the *recession cone* of C, that is, the convex cone of directions of C. To distinguish points and directions we identify C with $\{1\} \times C$ and D with $\{0\} \times D$ in the space $\mathbb{R} \times E$. If cone$N$ denotes the convex cone $R_+ N$ generated by a subset $N \subset \mathbb{R} \times E$, then $K = \text{cone}C \cup \text{rec}C$ is a convex cone in $\mathbb{R} \times E$ that is pointed and lineally closed. The set $\mathcal{F}$ consisting of all faces of C and all faces of D is called the *extended face lattice of* C. Define an order relation $\prec$ on $\mathcal{F}$ by $G \prec F$ iff F and G are both faces of C or are both faces of D with $G \subset F$ or G is a face of D and F is a face of C with $G \subset \text{rec}F$. Then $(\mathcal{F}, \prec)$ is a lattice and the mapping that sends a face F of C to cone$F$ $\cup$ rec$F$ and that sends a face of D into itself is a lattice isomorphism of $(\mathcal{F}, \prec)$ onto the face lattice of the convex cone K. Therefore, the cone K is of finite type iff every descending chain of principal faces in $(\mathcal{F}, \prec)$ becomes stationary, in which case C is said to be of *finite type*. The extraction calculus given on cone K by order function $\lambda(\cdot, \cdot)$

and extraction function $e(\cdot,\cdot)$ can be pulled down to $C \cup D$. In particular, x is for C a *component* of y for $x, y \in C \cup D$ if $\lambda(x,y) > 0$, which implies that $x \in C$ is never a component of $y \in D$. A *marking* μ *on* C then is a mapping of $C \cup D$ into itself satisfying properties *(i)* and *(ii)* of Definition 10.2. The *marked extraction on* C is given by the self-mapping e of $C \cup D$ defined by $e(x) = e(\mu(x), x)$. The *extraction algorithm* on cone K also pulls down to C and yields a sequence of points and directions of C, where, however, one has to be careful whether a point or a direction is extracted from a point. Finally, a nonempty finite subset M of $C \cup D$ not containing 0 is μ *independent* if there exists a numbering $M = \{a_0, a_1, \dots, a_m\}$ and a chain $F(y_0) \succ F(y_1) \succ \dots \succ F(y_m)$ of different faces in $\mathfrak{F}$ such that $a_i = \mu(y_i)$ for all i. By using these connections between convex set C and cone K from Theorem 10.2 the following result can be obtained.

Theorem 10.3. *(Hinrichsen and Krause [5]) Let C be a lineally closed convex set of finite type containing no line and let μ be a marking on C. Then every point $x \in C$ has exactly one marked representation, that is, a representation*

$$x = \sum_{a \in M} \lambda_a a$$

with $M \subset C \cup \mathrm{rec}C$ μ–independent, $\lambda_a > 0$ for $a \in M$, and

$$\sum_{a \in M \cap C} \lambda_a = 1.$$

This representation can be obtained by the extraction algorithm and for the length $|M|$ of the representation one has $|M| \le \dim F(x) + 1$.

This theorem covers cones as well as bounded convex sets. The following example illustrates the theorem for convex polytopes, that is, bounded intersections of finitely many half-spaces.

Example 10.2. Let C be a polytope in $\mathbb{R}^n$. Then $\mathrm{rec}C = \{0\}$. The face lattice of C is isomorphic to the face lattice of the cone $K = \{(t, tx) \mid t \ge 0,\ x \in C\}$. For $x, y \in C$ one has $\lambda(x,y) = \sup\{\lambda \in [0,1] \mid y = \lambda x + (1-\lambda)z \text{ for some } z \in C\}$ and $e(x,y) = (1 - \lambda(x,y))^{-1}(y - \lambda(x,y)x)$ for $x \ne y$. Denote by $\{z_1, \dots, z_r\}$ the finite set of extreme points of C. If $\mu(x)$ for $x \in C$ is the extreme point of the smallest index that is a component of x then μ defines a marking on C. Theorem 10.3 yields a unique representation for $x \in C$ as a convex combination of a μ–independent subset of extreme points. If C happens to be a simplex, then this unique representation becomes independent of the marking chosen.

Geometrically, the unique marked representation yields a simplicial subdivision of a convex set that consists of the convex hulls of all μ–bases. In particular, one obtains the following subdivision of polyhedra.

Corollary 10.1. *(Hinrichsen and Krause [5]) Let C be a polyhedron in finite dimensional space containing no lines and let μ be a marking on C that assigns to a point of C an extreme point of C and to a direction of C an extreme direction of C. The finite collection Σ of all convex subsets of C spanned by the μ bases has the following properties.*

(i) Σ consists of generalized simplices having the same dimensions as C.

(ii) C is the union of the simplices in Σ.

(iii) Any two simplices in Σ intersect in a common face.

According to a well-known theorem of Carathéodory every point of a polyhedron containing no lines is contained in a full-dimensional generalized simplex sharing its extreme points and extreme directions with the polyhedron (see Rockafellar [14]). Corollary 10.1 extends this result by showing that the Carathéodory simplices can be chosen in a consistent and minimal way. Moreover, these simplices can be found systematically by the extraction algorithm.

10.4 The Factorial Complex of a Krull Monoid

The method of marked extraction that yields unique representation at least piecewise in cases where global uniqueness is impossible can be developed also in fields different from convexity. It is well-known that for rings of algebraic integers the factorization of nonunits into irreducible elements is not unique in general. A powerful idea to "restore unique factorization" which goes back to Kummer and Dedekind is to embed such a ring faithfully into the smallest possible structure that possesses unique factorization. For Dedekind rings that include the rings of all integers of algebraic number fields such an embedding is possible into the multiplicative semigroup given by the prime ideals of the ring. More generally, for Krull rings such an embedding is possible by constructing a divisor theory. (On Dedekind rings and Krull rings see, e.g., Matsumura [11].) Unfortunately, the structure allowing unique factorization is no longer a ring but a monoid and may be quite big compared to the given ring. The gap between the two is measured by the ideal class group and the divisor class group, respectively. The deviation from unique factorization within the ring is encoded in the class group in a rather intricate manner leading to an array of problems still open today.

Another idea to "restore unique factorization," in a way dual to the embedding idea of Kummer and Dedekind, would be to discover the largest possible regions of unique factorization *within* the given ring. Such an idea is outlined in this section by developing a method of marked extraction for monoids (see [6, 8]). The analysis focuses on commutative and cancellative semigroups with identity element, or *monoids* for short. Such a monoid is mostly written multiplicatively with 1 as the identity element, and for simplicity it is assumed that 1 is the only unit in S and $S \neq \{1\}$.

For $x, y \in S$ the element x *divides* y, $x|y$, if $y = xz$ for some $z \in S$. An element $y \in S \backslash \{1\}$ is *irreducible* if $x|y$ implies that $x = y$ or $x = 1$. Let $\mathrm{Irr}(S)$ denote the (possibly empty) set of irreducible elements of S. A monoid S is *factorial* if every element in $S \backslash \{1\}$ can be uniquely written (up to the ordering of factors) as a product of finitely many irreducible elements. (For monoids see Halter-Koch [2].)

Definition 10.4. *For a monoid S the* order function *is defined for $x, y \in S \backslash \{1\}$ by $\lambda(x, y) = \sup\{\frac{m}{n} \mid x^m \mid y^n;\ m, n \in \mathbb{Z},\ m \geq 0,\ n \geq 1\} \in [0, \infty]$.*

S is called an extraction monoid *if the order function has rational values, that is, for every $x, y \in S \backslash \{1\}$ there exist $m \geq 0$, $n \geq 1$ such that $x^m \mid y^n$ and $\lambda(x, y) = \frac{m}{n}$. For an extraction monoid S the* extraction function *is defined for $x, y \in S \backslash \{1\}$ by $e(x, y) = z$ where z is uniquely given by $y^n = x^m z$ with $\lambda(x, y) = \frac{m}{n}$ and n minimal.*

For $y \in S \backslash \{1\}$ the element x is called a *component* of y if $\lambda(x, y) > 0$ and $e(x, y)$ is called the *rest* after extracting x from y. For $x \in S \backslash \{1\}$ let $F(x)$ be the set of all components of x augmented by 1. One has $y \in F(x)$ iff x is in the radical of the principal ideal generated by y, that is, $x \in \mathrm{rad}(y) = \{z \in S \mid y|z^n \text{ for some } n \geq 1\}$. The monoid S is said to be of *finite type* if there exists no infinite properly ascending chain $\mathrm{rad}(x_1) \subsetneqq \mathrm{rad}(x_2) \subsetneqq \ldots$.

An extremely rich class of extraction monoids is provided by the Krull monoids.

Definition 10.5. *Let S be a monoid, G its quotient group and let F be a family of group homomorphisms from G into the additive group, $\mathbb{Z}$ such that $\{f \in F \mid f(x) \neq 0\}$ is finite for all $x \in G$. The monoid S is called a* Krull monoid *if $S = \{x \in G \mid f(x) \geq 0$ for all $f \in F\}$.*

Let us illustrate this important concept by two quite different classes of examples.

Example 10.3.

1. *Krull rings.* Let R be an integral domain with a field of fractions K. R is called a Krull ring if R is the intersection of a family of discrete valuation rings $A_i \subset K$, $i \in I$, such that for every $0 \neq x \in K$ one has $v_i(x) \neq 0$ only for finitely many $i \in I$ where v_i is a (normalized) discrete additive valuation defining A_i. Thereby, a discrete additive (exponential) valuation v is a mapping $v\colon K \longrightarrow \mathbb{Z} \cup \{\infty\}$ such that $v(xy) = v(x) + v(y)$, $v(x + y) \geq \min\{v(x), v(y)\}$ and $v(x) = \infty$ iff $x = 0$. Obviously, if R is a Krull ring then the multiplicative semigroup $S = R \backslash \{0\}$ (modulo units) of R is a Krull

monoid. It can be shown, conversely, that an integral domain is already a Krull ring if its multiplicative semigroup is a Krull monoid. (See Krause [7], where generalizations of Krull rings and Krull monoids are considered also.) In this sense, being a Krull ring is a purely multiplicative property. This important fact makes the concept of a Krull monoid sensible and this fact does not apply, for example, to Dedekind rings. If R is a Krull ring, then the order function of the multiplicative semigroup (modulo units) S is given by

$$\lambda(x,y) = \min\{\frac{v_i(y)}{v_i(x)} \mid v_i(x) > 0,\ i \in I\}$$

and S is, therefore, an extractive monoid of finite type. In particular, if S is the multiplicative semigroup given by the principal ideals $\neq 0$ of the ring R of all integers of an algebraic number field, then S is an extraction monoid of finite type. S is a Krull monoid defined by the $\mathfrak{p}$-adic valuations where $\mathfrak{p}$ is a prime ideal in R. The $\mathfrak{p}$-adic valuations are given by extraction within the factorial monoid generated by the prime ideals; that is, $v_{\mathfrak{p}}(y) = \lambda(\mathfrak{p},(y))$ for $(y) \in S$. For the ring of rational integers, especially, the p-adic valuations are given by extracting the prime number p. Thus extraction and the order function may be considered as a meaningful extension of valuations. There exist subrings of algebraic numbers, nonprincipal orders such as $R = \mathbb{Z} + 2\mathbb{Z}\sqrt{-5}$, that do not yield Krull monoids but extraction monoids. There also exist subrings such as $\mathbb{Z}+\mathbb{Z}\sqrt{17}$ that do not even yield extraction monoids.

2. As with Krull rings any Krull monoid is an extraction monoid of finite type. In addition, a Krull monoid is generated (as a monoid) by its irreducible elements. There are, however, Krull monoids that can not be realized as the multiplicative semigroup of a Krull ring. An important class of such Krull monoids is given by the monoids S of nonnegative solutions of linear Diophantine equations; that is,

$$S = \{(x_1,\dots,x_n) \in \mathbb{Z}^n \mid \sum_{j=1}^{n} a_{ij}x_j = 0,\ 1 \leq i \leq n,\ x_j \geq 0\}$$

with respect to addition where $a_{ij} \in \mathbb{Z}$. Obviously, S (with addition) is a Krull monoid and, hence, an extraction monoid of finite type. Those monoids may be factorial or not and can show in the latter case a rather difficult behavior with respect to "factorization" into a sum of irreducible (minimal) elements leaving open till today many seemingly simple questions (see Chapman et al. [1]).

Definition 10.6. *For a (multiplicative) monoid S a mapping $\mu\colon S\backslash\{1\} \longrightarrow S\backslash\{1\}$ is called a* marking on S *if it satisfies the following properties for $x, y \in S\backslash\{1\}$.*

(i) $\mu(x)$ is a component of x.

(ii) If x is a component of y and $\mu(y)$ is a component of x then $\mu(x) = \mu(y)$.

For a marking μ on S the mapping $e\colon S\backslash\{1\} \longrightarrow S$ defined by $e(x) = e(\mu(x), x)$ is called a marked extraction *on S.*

Iterating the mapping of a marked extraction defines an *extraction algorithm* which stops after finitely many steps if S is of finite type and which yields for $x \in S\backslash\{1\}$ the elements

$$x_i = e^i(x) \quad (e^i \text{ the } i\text{th iterate of } e)$$

$$e_i = \mu(x_i) \;\text{ and }\; \lambda(e_i, x_i) = \frac{m_i}{n_i} \;\text{ for } 0 \le i \le m(x).$$

The e_i are independent in the following sense.

Definition 10.7. *A finite subset $\emptyset \neq M \subset S\backslash\{1\}$ is called μ-*independent *with respect to a marking μ on S if there exists a numbering $M = \{a_0, a_1, \ldots, a_n\}$ and a chain $\operatorname{rad}(y_0) \subsetneqq \operatorname{rad}(y_1) \subsetneqq \ldots \subsetneqq \operatorname{rad}(y_n)$ such that $a_i = \mu(y_i)$ for all i. An arbitrary subset M of $S\backslash\{1\}$ is called μ-independent if every nonempty finite subset of M is μ-independent. A maximal μ-independent subset is called a μ–*base.

Similarly as in the previous section one obtains for monoids the following type of unique marked representation.

Theorem 10.4. *(Krause [6])*

(i) Let S be an extraction monoid of finite type equipped with some marking μ. Then every $1 \neq x \in S$ has a representation

$$x^n = \prod_{a \in M} a^{n_a}$$

with M μ–independent and natural numbers $n, n_a \geq 1$.
The set M as well as the fractions n_a/n are uniquely determined and the representation can be obtained by the extraction algorithm.

(ii) For a Krull monoid S markings μ can be chosen for which $\mu(x)$ is an irreducible element of S. If μ is such a marking then the family $\mathcal{G}$ of all submonoids $G = G(M)$ of S generated by a μ-base M is a factorial complex *(or Gauß complex) in the following sense.*

(a) Every $G \in \mathcal{G}$ is a maximal factorial submonoid of S.

(b) $\operatorname{Irr}(G) = \operatorname{Irr}(S) \cap G$ for every $G \in \mathcal{G}$.

(c) For $G, G' \in \mathcal{G}$ the monoid $G \cap G'$ is factorial, too, with $\operatorname{Irr}(G \cap G') = \operatorname{Irr}(G) \cap \operatorname{Irr}(G')$. Moreover, the divisibility relations of G, G' and $G \cap G'$ coincide on $G \cap G'$.

(d) *For every element of S some power is contained in the union of all $G \in \mathcal{G}$.*

For the representation given by Theorem 10.4 it is important whether the representation can be done by elements of special interest (e.g., by irreducible elements as in the case of Krull monoids). It would be desirable to have a representation by prime elements, where $p \in S\backslash\{1\}$ is a *prime element* if $p \mid xy$ for $x, y \in S$ implies $p \mid x$ or $p \mid y$.

A prime element is necessarily irreducible whereas the converse does not hold in general. Indeed, it is easily verified that for a monoid S that is generated by its irreducible elements every irreducible element is prime iff S is factorial. Considering nonfactorial monoids, therefore, one has to look for a weaker notion than that of a prime element. Call an element $x \neq 1$ of an extraction monoid S *extractive* if extraction by x is additive, that is,

$$\lambda(x, yz) = \lambda(x, y) + \lambda(x, z) \quad \text{for all } y, z \in S\backslash\{1\}.$$

Obviously, any prime element is extractive whereas the converse does not hold in general. Actually, an element is prime iff it is extractive and $\lambda(x, y) \in \mathbb{N} = \{0, 1, 2, \dots\}$ for all $y \in S\backslash\{1\}$. Weakening the latter property, $x \in S\backslash\{1\}$ is called *discrete* if there exists some $0 \neq n(x) \in \mathbb{N}$ such that $n(x)\lambda(x, y) \in \mathbb{N}$ for all $y \in S\backslash\{1\}$. Thus, dealing with nonfactoriality we employ extractive and discrete elements instead of prime elements. Just to illustrate these notions and their relevance for Theorem 10.4 consider the following simple but famous *Hilbert monoid* $S = \{4n + 1 \mid n \in \mathbb{N}\}$ with the usual multiplication. It can be shown that S is a Krull monoid and, hence, an extraction monoid of finite type. The elements $21, 9$, and 49 are all irreducible in S and because of the equation $21 \cdot 21 = 9 \cdot 49$ the monoid cannot be factorial. In particular, the irreducible elements $21, 9$, and 49 are not prime and 9, and 49 are extractive whereas 21 is not. It is easy to see that $1 \neq x \in S$ is irreducible and extractive (for S) iff x is a prime number (i.e., a prime element of the multiplicative factorial monoid $\mathbb{N}\backslash\{0\}$) or if x is the square of a prime number p with $p \equiv 3 \mod 4$. Since prime elements must be extractive, it follows from the equation $p^2 \cdot 3^2 = (3p) \cdot (3p)$ that the prime elements of S are precisely the prime numbers of the form $4n + 1$. The fact that there are infinitely many prime numbers in the arithmetic progression given by $4n + 1$ implies that S contains infinitely many prime elements. One also has that an irreducible and extractive element (in S) is automatically discrete. Denote by M the set of all irreducible and extractive elements of S. The set M contains the infinitely many prime elements of S but it is much larger and is indeed large enough to represent for every $x \in S\backslash\{1\}$ the element x^2 as a product of finitely many elements from M; that is,

$$x^2 = \prod_{a \in M} a^{n_a}$$

with $n_a \geq 1$ only for finitely many $a \in M$.

Such a representation is neither possible by prime elements, nor can x itself be represented in general by extractive elements. Since $\lambda(a, a') = 0$ for two different elements in M, it follows that $n_a = 2\lambda(a, x)$ and, therefore, the representation by elements of M is unique (up to the ordering of factors). Thus, the monoid $G = G(M)$ generated by M is factorial and $x^2 \in G$ for every $x \in S$ (M is also a μ-base for some marking μ on S.) Furthermore, G is a maximal factorial submonoid of S and $\mathrm{Irr}(G) = \mathrm{Irr}(S) \cap G$. Thus, in light of Theorem 10.4, (ii) it can be said that the factorial complex of the nonfactorial Hilbert monoid consists of just one single monoid. (Although factorization into irreducible elements in S is not unique, it can be shown that for a given element any two such factorizations must have an equal number of factors. Such a monoid is called half-factorial. For criteria of half-factoriality for Krull monoids given by a linear Diophantine equation see Chapman et al. [1]). One may look at the Hilbert monoid also from the outside, in considering the Hilbert monoid as a submonoid of all odd positive integers. Contrary to the Hilbert monoid, the monoid of odd positive integers is factorial and two elements divide each other in the former monoid iff this is the case with respect to the latter. What has been observed for the Hilbert monoid holds true also more generally as shown by theorem 10.5.

A monoid S is *embedded* into a monoid T if there exists a monoid homomorphism $\varphi\colon S \longrightarrow T$ such that for any $x, y \in S$ the element x divides y in S iff $\varphi(x)$ divides $\varphi(y)$ in T. (It follows that φ is injective because 1 is assumed to be the only unit in a monoid.)

Theorem 10.5. *(Krause [8])*

(i) A monoid S admits an embedding φ of a (maximal) factorial monoid S_* into S such that for every element of S some power belongs to $\varphi(S_*)$ iff S is an extraction monoid of finite type for which every nonunit has an extractive component.

(ii) A monoid S admits an embedding φ into a (minimal) factorial monoid S^* such that for every element of S^* some power belongs to $\varphi(S)$ iff S is an extraction monoid of finite type that is integrally closed (i.e., $x^n | y^n$ implies $x|y$) and for which every nonunit has a discrete and extractive component.

This theorem gives a precise meaning to the forementioned dual approaches to "restore unique factorization." Statement *(i)* describes the conditions under which the operation in S can be analyzed by a factorial monoid from the inside. The factorial monoid S_* is unique up to an isomorphism and corresponds to a factorial complex consisting of one single monoid. Statement *(i)* implies that the factorial complex must be larger if S is an extraction monoid of finite type which does not possess enough extractive elements. A simple example is the additive Krull monoid

$S = \{x \in \mathbb{N}^4 \mid x_1 + x_2 = x_3 + x_4\}$. This monoid has four irreducible elements, none of which has an extractive component. The factorial complex of S can be seen to consist of two monoids. (For Krull rings with a factorial complex consisting of two monoids see Wistuba [15].) Statement *(ii)* characterizes the Kummer-Dedekind point of view mentioned in the beginning. In the case where S is the multiplicative semigroup of a Krull ring, the embedding, which is unique up to an isomorphism, is called a divisor theory for S. The property that every element of S^* has a power in $\varphi(S)$ means that the divisor class group $S^*/\varphi(S)$ is a torsion group. From *(ii)* it follows in particular that an integral domain is a Krull ring with torsion divisor class group iff its multiplicative semigroup is an integrally closed extraction monoid of finite type for which every nonunit has a discrete and extractive component.

As is obvious from theorem 10.5, the existence of a divisor theory S^* is more demanding than that of a factorial monoid S_* from the inside. There exist many examples of integral domains, among them nonprincipal orders of algebraic number fields, that do not possess a divisor theory S^* but for which a factorial monoid S_* can be found.

References

[1] S.T. Chapman, U. Krause, and E. Oeljeklaus, Monoids determined by a homogeneous linear diophantine equation and the half-factorial property, *J. Pure Appl. Algebra*, 2000 (to appear).

[2] F. Halter-Koch, *Ideal Systems. An Introduction to Multiplicative Ideal Theory*, M. Dekker Inc., New York, 1998.

[3] D. Hinrichsen and U. Krause, Choice of techniques in joint production models, *Operations Research Verfahren* 34:155–171, 1971.

[4] D. Hinrichsen and U. Krause, A substitution theorem for joint production models with disposal processes, *Operations Research Verfahren* 41:287–291,1981.

[5] D. Hinrichsen and U. Krause, Unique representation in convex sets by extraction of marked components, *Linear Algebra Appl.* 51:73–96, 1983.

[6] U. Krause, Eindeutige Faktorisierung ohne ideale Elemente, *Abh. Braunschw. Wiss. Ges.* (Dedekind Festschrift) 33:169–177, 1982.

[7] U. Krause, On monoids of finite real character, *Proc. Am. Math. Soc.* 105:546–554, 1989.

[8] U. Krause, Semigroups that are factorial from inside or from outside, in J. Almeida et al., eds., *Lattices, Semigroups, and Universal Algebra*, 147–161, Plenum, New York, 1990.

[9] U. Krause, Positive nonlinear systems: Some results and applications, in V. Lakshmikantham, ed., *Proc. 1st World Congress of Nonlinear Analysts*, 1529–1539, de Gruyter, Berlin, 1996.

[10] H.D. Kurz and N. Salvadori, *Theory of Production. A Long–Period Analysis*, Cambridge University Press, Cambridge, 1995.

[11] H. Matsumura, *Commutative Ring Theory*, Cambridge University Press, Cambridge, 1986.

[12] M. Morishima, *Equilibrium, Stability, and Growth. A Multi-Sectoral Analysis*, Oxford University Press, London, 1964.

[13] J. von Neumann, A model of general economic equilibrium, *Rev. Econ. Studies* 13:1–9, 1945 (originally published in German 1937).

[14] R.T. Rockafellar, *Convex Analysis*, Princeton University Press, Princeton, 1970.

[15] H. Wistuba, Faktorielle Komplexe, Ph.D.thesis, University of Bremen, 1991.

Chapter 11

Between Mathematical Programming and Systems Theory: Linear Complementarity Systems

Hans Schumacher

ABSTRACT Complementarity systems arise from the interconnection of an input-output system (of the type well known in mathematical systems theory) with a set of complementarity conditions (of the type well known in mathematical programming). It is shown by means of a list of examples that complementarity systems appear quite naturally in a broad range of applications. A solution concept for linear complementarity systems is provided, and conditions for existence and uniqueness of solutions are given.

11.1 Introduction

Inequalities have played a rather minor role in the powerful development of systems theory that has taken place since about 1960. In contrast to this, they are central to the field of mathematical programming that has likewise seen major advances in the past decades. Of course, systems theory is concerned with differential equations; mixing these with inequalities means giving up the smoothness properties that form the basis of much of the theory of dynamical systems. Technological innovation, however, pushes toward the consideration of systems of a mixed continuous/discrete nature, which are likely to be described by systems of differential equations as well as algebraic equations and inequalities. In fact there are many situations in which there are good reasons to consider dynamics in conjunction with inequalities; think for instance of saturation effects in control systems, unilateral constraints in robotics, piecewise linear dynamics, and so on.

Among the many systems of equations and inequalities that one may imagine, the ones that are in so-called *complementarity* form enjoy particular attention in mathematical programming. More specifically the *linear complementarity problem* (LCP) has been the subject of extensive study because of its wide range of applications; see the book by Cottle et al. [5] for a comprehensive treatment. The LCP may be formulated as follows: given a vector $q \in \mathbb{R}^k$ and a matrix M of size $k \times k$, find vectors y and u

in $\mathbb{R}^k$ that satisfy the affine relation

$$y = q + Mu \tag{11.1}$$

and the complementarity conditions

$$\forall i = 1, \ldots, k: \quad y_i \geq 0, \;\; u_i \geq 0, \quad y_i u_i = 0. \tag{11.2}$$

Under suitable conditions this problem has exactly one solution. This chapter is concerned with the dynamical systems that one obtains when the static relation (11.1) is replaced by a dynamic relation of the form, for instance,

$$\dot{x}(t) = f(x(t), u(t)), \qquad y(t) = h(x(t), u(t)). \tag{11.3}$$

So effectively we have a dynamical input-output system of the type studied in systems and control theory, and we couple it with the complementarity conditions that appear in the LCP. Dynamical systems that are obtained in this way are called *complementarity systems*. Section 11.2 of the paper lists a number of situations in which one finds dynamics of this kind. The fact that the algebraic LCP has unique solutions under suitable circumstances leads to the suggestion that, under certain conditions, complementarity systems may also have unique solutions. Much of this chapter is indeed concerned with finding conditions for existence and uniqueness of solutions of complementarity systems. In particular we consider *linear* complementarity systems in which the dynamic relation (11.3) is specialized to the linear time-homogeneous equations

$$\dot{x}(t) = Ax(t) + Bu(t), \qquad y(t) = Cx(t) + Du(t). \tag{11.4}$$

The idea of applying complementarity conditions to external variables of general dynamical systems originates in [35]. In the specific context of mechanical systems the use of complementarity conditions, which in this case relates to the presence of unilateral constraints, goes back much further and can in fact be traced to work by Fourier and by Farkas for the static (equilibrium) case and papers by Moreau and by Lötstedt for the dynamic case; see [35] for a brief review. The theory of complementarity systems has been further developed and considerably expanded in a number of recent papers, see for instance [36, 21, 20, 29, 7, 16, 18, 19, 17]. It is the purpose of the present chapter to give a survey of results obtained in these papers. Some new material is added in particular with respect to possible applications of complementarity systems.

Parts of this survey have been taken from joint papers with Kanat Çamlıbel, Maurice Heemels, Arjan van der Schaft, and Siep Weiland. It is a pleasure to acknowledge their contributions. Compared to Chapter 4 of [37], we concentrate here on the case of linear complementarity systems.

The chapter is structured as follows. Motivation for the framework of complementarity systems is provided in the next section by means of a number of examples. The issue of existence and uniqueness of solutions is introduced briefly in Section 11.3. We then turn to linear complementarity systems in Section 11.4, where a complete specification of the dynamics (including event rules) is provided. A distributional framework is sketched in Section 11.5, and some results on existence and uniqueness of solutions are given in Section 11.6. Section 11.7 is concerned with an application to relay systems. One of the ways in which linear complementarity systems differ from common nonlinear systems is that there may be discontinuous dependence on initial conditions; a simple example of this is provided in Section 11.8. Conclusions follow in Section 11.9.

For use below we mention here a basic fact about the linear complementarity problem (11.1) and (11.2). The LCP (11.1) and (11.2) has a unique solution (y, u) for all q if and only if all principal minors of the matrix M are positive [34; 5, Theorem 3.3.7]. (Given a matrix M of size $k \times k$ and two nonempty subsets I and J of $\{1, \ldots, k\}$ of equal cardinality, the *(I, J)-minor* of M is the determinant of the square submatrix $M_{IJ} := (m_{ij})_{i \in I, j \in J}$. The *principal minors* are those with $I = J$ [11, p. 2].) A matrix all of whose principal minors are positive is called a P-matrix. For example, all positive definite matrices are P-matrices. This is even true when the term "positive definite" is understood to apply not only to symmetric matrices, but also to nonsymmetric matrices M that are such that $x^T M x > 0$ for all $x \neq 0$.

11.2 Examples

11.2.1 Circuits with ideal diodes

A large amount of electrical network modeling is carried out on the basis of ideal lumped elements: resistors, inductors, capacitors, diodes, and so on. There is not necessarily a one-to-one relation between the elements in a model and the parts of the actual circuit; for instance, a resistor may under some circumstances be better modeled by a parallel connection of an ideal resistor and an ideal capacitor than by an ideal resistor alone. The standard ideal elements should rather be looked at as forming a construction kit from which one can quickly build a variety of models.

Among the standard elements the ideal diode has its own place because of the nonsmoothness of its characteristic. In circuit simulation software that has no ability to cope with mode changes, the ideal diode can not be admitted as a building block and will have to be replaced for instance by a heavily nonlinear resistor; a price will have to be paid in terms of speed of simulation. The alternative is to work with a hybrid system simulator; more specifically, the software will have to be able to deal with complementarity systems.

To write the equations of a network with (say) k ideal diodes in complementarity form, first extract the diodes so that the network appears as a k-port. For each port, we have a choice between denoting voltage by u_i and current by y_i or vice versa (with the appropriate sign conventions). Usually it is possible to make these choices in such a way that the dynamics of the k-port can be written as

$$\dot{x} = f(x,u), \qquad y = h(x,u).$$

For linear networks, one can actually show that it is *always* possible to write the dynamics in this form. To achieve this, it may be necessary to let u_i denote voltage at some ports and current at some other ports; in that case one sometimes speaks of a "hybrid" representation, where of course the term is used in a different sense than in this chapter. Replacing the ports by diodes, we obtain a representation in the semi-explicit complementarity form (11.3).

For electrical networks it would seem reasonable in most cases to assume that there are no jumps in the continuous state variables, so that there is no need to specify event conditions in addition to the flow conditions (11.3). Complementarity systems in general do not always have continuous solutions, so if one wants to prove that electrical networks with ideal diodes do indeed have continuous solutions, one will have to make a connection with certain specific properties of electrical networks. The passivity property is one that immediately comes to mind, and indeed there are certain conclusions that can be drawn from passivity and that are relevant in the study of properties of complementarity systems. To illustrate this, consider the specific case of a linear passive system coupled with a number of ideal diodes. The system is described by equations of the form

$$\begin{aligned} \dot{x} &= Ax + Bu \\ y &= Cx + Du \\ 0 &\leq y \perp u \geq 0, \end{aligned} \tag{11.5}$$

where the last line is a shorthand for the complementarity conditions (11.2). Under the assumption that the system representation is minimal, the passivity property implies (see [39]) that there exists a positive definite matrix Q such that

$$\begin{bmatrix} A^TQ + QA & QB - C^T \\ B^TQ - C & -(D + D^T) \end{bmatrix} \leq 0. \tag{11.6}$$

If now for instance the matrix D is nonsingular, then it follows that D is actually positive definite. Under this condition one can prove that the complementarity system (11.5) has continuous solutions. If on the other hand D is equal to zero, then the passivity condition (11.6) implies that $C = B^TQ$ so that in this case the matrix $CB = B^TQB$ is positive definite (assuming that B has full column rank). One can prove that under this condition the system (11.5) has solutions with continuous state trajectories, if the system is consistently initialized (i. e., the initial condition x_0 satisfies $Cx_0 \geq 0$). The importance of the matrices D and CB derives from the fact that they appear in the power series expansion of the transfer matrix $C(sI - A)^{-1}B + D$ around infinity:

$$C(sI - A)^{-1}B + D \;=\; D + CBs^{-1} + CABs^{-2} + \cdots .$$

We return to this when we discuss linear complementarity systems.

11.2.2 Mechanical systems with unilateral constraints

Mechanical systems with geometric inequality constraints are given by equations of the following form (see [35]), in which $\partial H/\partial p$ and $\partial H/\partial q$ denote column vectors of partial derivatives, and the time arguments of q, p, y, and u have been omitted for brevity.

$$\begin{aligned}
\dot{q} &= \tfrac{\partial H}{\partial p}(q,p) && q \in \mathbb{R}^n,\ p \in \mathbb{R}^n \\
\dot{p} &= -\tfrac{\partial H}{\partial q}(q,p) + \tfrac{\partial C^T}{\partial q}(q)u && u \in \mathbb{R}^k \\
y &= C(q) && y \in \mathbb{R}^k \\
0 &\leq y \perp u \geq 0.
\end{aligned} \tag{11.7}$$

Here, $C(q) \geq 0$ is the column vector of geometric inequality constraints, and $u \geq 0$ is the vector of Lagrange multipliers producing the constraint force vector $(\partial C/\partial q)^T(q)u$. (The expression $(\partial C^T/\partial q)$ denotes an $n \times k$ matrix whose ith column is given by $\partial C_i/\partial q$.) The complementarity conditions in this case express that the i-th component of u_i can be only nonzero if the ith constraint is active; that is, $y_i = C_i(q) = 0$. Furthermore, $u_i \geq 0$ since the constraint forces will be always pushing in the direction of rendering y_i nonnegative. This basic principle of handling geometric inequality constraints can be found, for example, in [32, 26], and dates back to Fourier and Farkas. The Hamiltonian $H(q,p)$ denotes the total energy, generally given as the sum of a kinetic energy $1/2\, p^T M^{-1}(q)p$ (where $M(q)$ denotes the mass matrix, depending on the configuration vector q) and a potential energy $V(q)$. The semiexplicit complementarity system (11.7) is called a Hamiltonian complementarity system, since the dynamics of every mode is Hamiltonian [35]. In particular, every mode is energy-conserving (since the constraint forces are workless); it should be noted though that the model could be easily extended to mechanical systems with dissipation by replacing the second set of equations of (11.7) by

$$\dot{p} = -\frac{\partial H}{\partial q}(q,p) - \frac{\partial R}{\partial \dot{q}}(\dot{q}) + \frac{\partial C^T}{\partial q}(q)u, \tag{11.8}$$

where $R(\dot{q})$ denotes a Rayleigh dissipation function.

11.2.3 Optimal control with state constraints

The purpose of this subsection is to indicate how one may relate optimal control problems with state constraints to complementarity systems. The study of this subject is far from being complete and so we offer some suggestions rather than present a rigorous treatment. Consider the problem of maximizing a functional of the form

$$\int_0^T F(t,x(t),u(t))dt + F_T(x(T)) \tag{11.9}$$

over a collection of trajectories described by

$$\dot{x}(t) = f(t,x(t),u(t)), \qquad x(0) = x_0 \tag{11.10}$$

together with the constraints

$$g(t,x(t),u(t)) \geq 0. \tag{11.11}$$

In the above, g may be a vector-valued function, and then the inequalities are taken componentwise. Under suitable conditions (see [13] for much more detailed information), candidates for optimal solutions can be found by solving a system of equations that is obtained as follows. Let λ be a vector variable of the same length as x, and define the *Hamiltonian* $H(t,x,u,\lambda)$ by

$$H(t,x,u,\lambda) = F(t,x,u) + \lambda^T f(t,x,u). \tag{11.12}$$

Also, let η be a vector of the same length as g, and define the *Lagrangian* $L(t,x,u,\lambda,\eta)$ by

$$L(t,x,u,\lambda,\eta) = H(t,x,u,\lambda) + \eta^T g(t,x,u). \tag{11.13}$$

The system referred to before is now the following.

$$\dot{x}(t) = f(t,x(t),u(t)) \tag{11.14a}$$

$$\dot{\lambda}(t) = -\frac{\partial L}{\partial x}(t,x(t),u(t),\lambda(t),\eta(t)) \tag{11.14b}$$

$$u(t) = \arg \max_{\{u \mid g(t,x(t),u)\geq 0\}} L(t,x(t),u,\lambda(t),\eta(t)) \tag{11.14c}$$

$$0 \leq g(t,x(t),u(t)) \perp \eta(t) \geq 0 \tag{11.14d}$$

with initial conditions

$$x(0) = x_0 \tag{11.15}$$

and final conditions

$$\lambda(T) = \frac{\partial F_T}{\partial x}(x(T)). \tag{11.16}$$

Suppose that $u(t)$ can be solved from (11.14c) so that

$$u(t) = u^*(t, x(t), \lambda(t), \eta(t)), \tag{11.17}$$

where $u^*(t, x, \lambda, \eta)$ is a certain function. Then define $g^*(t, x, \lambda, \eta)$ by

$$g^*(t, x, \lambda, \eta) = g(t, x, u^*(t, x, \lambda, \eta), \lambda, \eta) \tag{11.18}$$

and introduce a variable $y(t)$ by

$$y(t) = g^*(t, x(t), \lambda(t), \eta(t)). \tag{11.19}$$

The system (11.14) can now be rewritten as

$$\begin{aligned}
\dot{x}(t) &= f(t, x(t), u^*(t, x(t), \lambda(t), \eta(t))) \\
\dot{\lambda}(t) &= -\tfrac{\partial L}{\partial x}(t, x(t), u^*(t, x(t), \lambda(t), \eta(t)), \lambda(t), \eta(t)) \\
y(t) &= g^*(t, x(t), \lambda(t), \eta(t)) \\
0 \le y(t) &\perp \eta(t) \ge 0.
\end{aligned} \tag{11.20}$$

Here we have a (time-inhomogeneous) complementarity system with state variables x and λ and complementary variables y and η. The system has mixed boundary conditions (11.15) and (11.16); therefore one will have existence and uniqueness of solutions under conditions that in general will be different from the ones that hold for initial value problems.

A case of special interest is the one in which a quadratic criterion is optimized for a linear time-invariant system, subject to linear inequality constraints on the state. Consider for instance the following problem: minimize

$$\tfrac{1}{2} \int_0^T (x(t)^T Q x(t) + u(t)^T u(t)) dt \tag{11.21}$$

subject to

$$\dot{x}(t) = Ax(t) + Bu(t), \qquad x(0) = x_0 \tag{11.22}$$

$$Cx(t) \ge 0, \tag{11.23}$$

where A, B, and C are matrices of appropriate sizes, and Q is a nonnegative definite matrix. Following the scheme above leads to the system

$$\dot{x} = Ax + Bu, \qquad x(0) = x_0 \tag{11.24a}$$

$$\dot{\lambda} = Qx - A^T\lambda - C^T\eta, \qquad \lambda(T) = 0 \tag{11.24b}$$

$$u = \arg\max\,[-\tfrac{1}{2}u^T u + \lambda^T Bu] \tag{11.24c}$$

$$0 \le Cx \perp \eta \ge 0, \tag{11.24d}$$

where we have omitted the time arguments for brevity. Solving for u from (11.24c) leads to the equations

$$\frac{d}{dt}\begin{bmatrix} x \\ \lambda \end{bmatrix} = \begin{bmatrix} A & BB^T \\ Q & -A^T \end{bmatrix}\begin{bmatrix} x \\ \lambda \end{bmatrix} + \begin{bmatrix} 0 \\ -C^T \end{bmatrix}\eta \tag{11.25a}$$

$$y = [C \quad 0]\begin{bmatrix} x \\ \lambda \end{bmatrix} \tag{11.25b}$$

$$0 \leq y \perp \eta \geq 0. \tag{11.25c}$$

Not surprisingly, this is a linear complementarity system.

The study of optimal control problems subject to state constraints is fraught with difficulties; see Hartl et al. [13] for a discussion. The setting of complementarity systems may be of help at least in categorizing these difficulties.

11.2.4 Variable-structure systems

Consider a nonlinear input-output system of the form

$$\dot{x} = f(x, \bar{u}), \qquad \bar{y} = h(x, \bar{u}) \tag{11.26}$$

in which the input and output variables are adorned with a bar for reasons that become clear in a moment. Suppose that the system is in feedback coupling with a relay element given by the propositional formula

$$\{\{\bar{u} = 1\} \wedge \{\bar{y} \geq 0\}\} \vee \{\{-1 \leq \bar{u} \leq 1\} \wedge \{\bar{y} = 0\}\} \vee \{\{\bar{u} = -1\} \wedge \{\bar{y} \leq 0\}\}. \tag{11.27}$$

Many of the systems considered in the well-known book by Filippov on discontinuous dynamical systems [10] can be rewritten in this form. At first sight, relay systems do not seem to fit in the complementarity framework. However, let us introduce new variables y_1, y_2, u_1, and u_2, together with the following new equations.

$$\begin{aligned} u_1 &= \tfrac{1}{2}(1 - \bar{u}) \\ u_2 &= \tfrac{1}{2}(1 + \bar{u}) \\ \bar{y} &= y_1 - y_2. \end{aligned} \tag{11.28}$$

Instead of considering (11.26) together with (11.27), we can also consider (11.26) together with the standard complementarity conditions for the vectors $y = \mathrm{col}(y_1, y_2)$ and $u = \mathrm{col}(u_1, u_2)$:

$$\begin{aligned} &\{\{\{y_1 = 0\} \wedge \{u_1 \geq 0\}\} \vee \{\{y_1 \geq 0\} \wedge \{u_1 = 0\}\}\} \wedge \\ &\qquad \{\{\{y_2 = 0\} \wedge \{u_2 \geq 0\}\} \vee \{\{y_2 \geq 0\} \wedge \{u_2 = 0\}\}\}. \end{aligned} \tag{11.29}$$

It can be easily verified that the trajectories of (11.26), (11.28), and (11.29) are the same as those of (11.26) and (11.27). Note in particular that, although (11.29) in principle allows four modes, the conditions (11.28) imply that $u_1 + u_2 = 1$ so that the mode in which both u_1 and u_2 vanish is excluded, and the actual number of modes is three.

So it turns out that we can rewrite a relay system as a complementarity system, at least if we are willing to accept that some algebraic equations appear in the system description. It is possible to eliminate the variables $\bar{y}$ and $\bar{u}$ and obtain equations in the form

$$\begin{aligned} \dot{x} &= f(x, u_2 - u_1) \\ y_1 - y_2 &= h(x, u_2 - u_1) \\ u_1 + u_2 &= 1 \end{aligned} \tag{11.30}$$

together with the complementarity conditions (11.29), but (11.30) is not in standard input-state-output form but rather in a DAE type of form

$$F(\dot{x}, x, y, u) = 0. \tag{11.31}$$

If the relay is a part of a model whose equations are built up from submodels, then it is likely anyway that the system description will already be in terms of both differential and algebraic equations, and then it may not be much of a problem to have a few algebraic equations added (depending on how the "index" of the system is affected). Alternatively however one may replace the equations (11.28) by

$$\begin{aligned} u_1 &= \tfrac{1}{2}(1 - \bar{u}) \\ y_2 &= \tfrac{1}{2}(1 + \bar{u}) \\ \bar{y} &= y_1 - u_2 \end{aligned} \tag{11.32}$$

which are the same as (11.28) except that y_2 and u_2 have traded places. The equations (11.30) can now be rewritten as

$$\begin{aligned} \dot{x} &= f(x, 1 - 2u_1) \\ y_1 &= h(x, 1 - 2u_1) + u_2 \\ y_2 &= 1 - u_1 \end{aligned} \tag{11.33}$$

and this system does appear in standard input-output form. The only concession one has to make here is that (11.33) will have a feedthrough term (i. e., the output y depends directly on the input u) even when this is not the case in the original system (11.26).

11.2.5 Piecewise linear systems

Suppose that a linear system is coupled with a control device that switches among several linear low-level controllers depending on the state of the controlled system, as is the case for instance in many gain scheduling controllers; then the closed-loop system may be described as a piecewise linear system. Another way in which piecewise linear systems may arise is as approximations to nonlinear systems. Modeling by means of piecewise linear systems is attractive because it combines the relative tractability of linear dynamics with a flexibility that is often needed for a precise description of dynamics through a range of operating conditions.

There exist definitions of piecewise linear systems at various levels of generality. Here we limit ourselves to systems of the following form (time arguments omitted for brevity).

$$\dot{x} = Ax + Bu \tag{11.34a}$$

$$y = Cx + Du \tag{11.34b}$$

$$(y_i, u_i) \in \operatorname{graph}(f_i) \qquad (i = 1, \ldots, k), \tag{11.34c}$$

where, for each i, f_i is a piecewise linear function from $\mathbb{R}$ to $\mathbb{R}^2$. As is common usage, we use the term "piecewise linear" to refer to functions that would in fact be more accurately described as being piecewise affine. We consider functions f_i that are continuous, although from some points of view it would be natural to include also discontinuous functions; for instance systems in which the dynamics is described by means of piecewise constant functions have attracted attention in hybrid systems theory.

The model (11.34) is natural for instance as a description of electrical networks with a number of piecewise linear resistors. Descriptions of this form are quite common in circuit theory (cf. [28]). Linear relay systems are also covered by (11.34); note that the "sliding mode" corresponding to the vertical part of the relay characteristic is automatically included. Piecewise linear friction models are often used in mechanics (for instance Coulomb friction), which again leads to models like (11.34).

One needs to define a solution concept for (11.34); in particular, one has to say in what function space one will be looking for solutions. With an eye on the intended applications, it seems reasonable to require that the trajectories of the variable x should be continuous and piecewise differentiable. As for the variable u, some applications suggest that it may be too much to require continuity for this variable as well. For an example of this, take a mass point that is connected by a linear spring to a fixed wall, and that can move in one direction subject to Coulomb friction. In a model for this situation the variable u would play the role of the friction force which, according to the Coulomb model, has constant magnitude as long as the mass point is moving, and has sign opposite to the direction of motion. If the mass point is given a sufficiently large initial velocity away from the fixed wall, it will come to a standstill after some time and then immediately be pulled back toward the wall, so that in this case the friction force jumps instantaneously from one end of its interval of possible values to the other. Even allowing jumps in the variable u, we can still define a solution of (11.34) to be a triple (x, u, y) such that (11.34b) and (11.34c) hold for almost all t, and (11.34a) is satisfied in the sense of Carathéodory; that is to say,

$$x(t) = x(0) + \int_0^t [Ax(\tau) + Bu(\tau)]d\tau \tag{11.35}$$

for all t.

The first question that should be answered in connection with the system (11.34) is whether solutions exist and are unique. For this, one should first of all find conditions under which, for a given initial condition $x(0) = x_0$, there exists a unique continuation in one of the possible "modes" of the systems (corresponding to all possible combinations of the different branches of the piecewise linear characteristics of the system). This can be a highly nontrivial problem; for instance in a mechanical system with many friction points, it may not be so easy to say at which points sliding will take place and at which points stick will occur. It turns out to be possible to address the problem on the basis of the theory of the linear complementarity problem and extensions of it. For the case of Coulomb friction, also in combination with nonlinear dynamics, this is worked out in [33]. The general case can be developed on the basis of a theorem by Kaneko and Pang [25], which states that any piecewise linear characteristic can be described by means of the so-called *extended horizontal linear complementarity problem*. On this basis, the piecewise linear system (11.34) may also be described as an extended horizontal linear complementarity system. Results on the solvability of the EHLCP have been given by Sznajder and Gowda [38]. Using these results, one can obtain sufficient conditions for the existence of unique solution starting at a given initial state; see [4] for details.

11.2.6 Projected dynamical systems

The concept of equilibrium is central to mathematical economics. For instance, one may consider an oligopolistic market in which several competitors determine their production levels so as to maximize their profits; it is of interest to study the equilibria that may exist in such a situation. On a wider scale, one may discuss general economic equilibrium involving production, consumption, and prices of commodities. In fact in all kinds of competitive systems the notion of equilibrium is important.

The term "equilibrium" can actually be understood in several ways. For instance, the celebrated Nash equilibrium concept of game theory is defined as a situation in which no player can gain by unilaterally changing his position. Similar notions in mathematical economics lead to concepts of equilibria that can be characterized in terms of systems of *algebraic* equations and inequalities. On the other hand, we have the classical notion of equilibrium in the theory of dynamical systems, where the concept is defined in terms of a given set of *differential* equations. It is natural to expect, though, that certain relations can be found between the static and dynamic equilibrium concepts.

In 1993, Dupuis and Nagurney [9] have proposed a general strategy to embed a given static equilibrium problem into a dynamic system. Dupuis and Nagurney assume that the static equilibrium problem can be formulated in terms of a *variational equality*; that is to say, the problem is specified by giving a closed convex subset K of $\mathbb{R}^k$ and a function F from K to $\mathbb{R}^k$, and $\bar{x} \in K$ is an equilibrium if

$$\langle F(\bar{x}), x - \bar{x} \rangle \geq 0 \tag{11.36}$$

for all $x \in K$. The formulation in such terms is standard within mathematical programming. With the variational problem they associate a discontinuous dynamical system that is defined by $\dot{x} = -F(x)$ on the interior of K but which is defined differently on the boundary of K in such a way as to make sure that solutions will not leave the convex set K. They then prove that the stationary points of the so-defined dynamical system coincide with the solutions of the variational equality.

In some more detail, the construction proposed by Dupuis and Nagurney can be described as follows. The space $\mathbb{R}^k$ in which state vectors take their values is taken as a Euclidean space with the usual inner product. Let P denote the mapping that assigns to a given point x in $\mathbb{R}^k$ the (uniquely defined) point in K that is closest to x; that is to say,

$$P(x) = \arg \min_{z \in K} ||x - z||. \tag{11.37}$$

For $x \in K$ and a velocity vector $v \in \mathbb{R}^k$, let

$$\pi(x, v) = \lim_{\delta \to 0} \frac{P(x + \delta v) - x}{\delta}. \tag{11.38}$$

If x is in the interior of K, then clearly $\pi(x, v) = v$; however if x is on the boundary of K and v points outwards then $\pi(x, v)$ is a modification of v. The dynamical system considered by Dupuis and Nagurney is now defined by

$$\dot{x} = \pi(x, -F(x)) \tag{11.39}$$

with initial condition x_0 in K. The right-hand side of this equation is in general subject to a discontinuous change when the state vector reaches the boundary of K. The state may then follow the boundary along a $(k-1)$-dimensional surface or a part of the boundary characterized by more than one constraint, and after some time it may reenter the interior of K after which it may again reach the boundary, and so on.

In addition to the expression (11.38) Dupuis and Nagurney also employ a different formulation which has been used in [8]. For this, first introduce the set of inward normals that is defined, for a boundary point x of K, by

$$n(x) = \{\gamma \mid ||\gamma|| = 1, \text{ and } \langle \gamma, x - y \rangle \leq 0, \quad \forall y \in K\}. \tag{11.40}$$

If K is a convex polyhedron then we may write

$$\pi(x, v) = v + \langle v, -\gamma^* \rangle \gamma^*, \tag{11.41}$$

where γ^* is defined by

$$\gamma^* := \arg \max_{\gamma \in n(x)} \langle v, -\gamma \rangle. \tag{11.42}$$

A further reformulation is possible by introducing the "cone of admissible velocities." To formulate this concept, first recall that a *curve* in $\mathbb{R}^k$ is a smooth mapping from an interval, say $(-1, 1)$, to $\mathbb{R}^k$. An *admissible velocity* at a point x with respect to the closed convex set $K \subset \mathbb{R}^k$ is any vector that appears as a directional derivative at 0 of a C^∞ function $f(t)$ that satisfies $f(0) = x$ and $f(t) \in K$ for $t \geq 0$. One can show that the set of admissible velocities is a closed convex cone for any x in the boundary of K; of course, the set of admissible velocities is empty when $x \notin K$ and coincides with $\mathbb{R}^k$ if x belongs to the interior of K. One can furthermore show (see [22]) that the mapping defined in (11.41) for given x is in fact just the projection to the cone of admissible velocities. In this way we get an alternative definition of projected dynamical systems. The new formulation is more "intrinsic" in a differential-geometric sense than the original one which is based on the standard coordinatization of k-dimensional Euclidean space. Indeed it would be possible in this way to formulate projected dynamics for systems defined on Riemannian manifolds; the inner product on tangent spaces that is provided by the Riemannian structure makes it possible to define the required projection. One possible application would be the use of projected gradient flows to find minima subject to constraints (cf. for instance [23] for the unconstrained case).

Assume now that the set K is given as an intersection of convex sets of the form $\{x \mid h_i(x) \geq 0\}$ where the h_is are smooth functions. This is actually the situation that one typically finds in applications. It is then possible to reformulate the projected dynamical system as a complementarity system. The construction is described in [22] and we summarize it briefly here. Let $H(x)$ denote the gradient matrix defined by the functions $h_i(x)$; that is to say, the (i, j)th element of $H(x)$ is

$$(H(x))_{ij} = \frac{\partial h_i}{\partial x_j}(x).$$

For $x \in K$, let $I(x)$ be the set of "active" indices; that is,

$$I(x) = \{i \mid h_i(x) = 0\}.$$

We denote by $H_{I(x)\bullet}$ the matrix formed by the rows of $H(x)$ whose indices are active; it will be assumed that this matrix has full row rank for all x in the boundary of K ("independent constraints"). One can then show that for each $x \in K$ the cone of admissible velocities is given by $\{v \mid H_{I(x)\bullet}v \geq 0\}$. Moreover, the set of inward normals as defined in (11.40) is given by $\{\gamma \mid ||\gamma|| = 1 \text{ and } \gamma = H^T_{I(x)\bullet}u \text{ for some } u\}$. Consequently, the projection of an arbitrary vector v_0 on the cone of admissible velocities is obtained by solving the minimization problem

$$\min_v \{||v_0 - v|| \mid H_{I(x)\bullet}v \geq 0\}.$$

By standard methods, one finds that the minimizer is given by $H^T_{I(x)\bullet}u$ where u is the (unique) solution of the complementarity problem

$$0 \leq H_{I(x)\bullet}v_0 + H_{I(x)\bullet}H^T_{I(x)\bullet}u \perp u \geq 0. \tag{11.43}$$

Now, compare the projected dynamical system (11.39) to the complementarity system defined by

$$\dot{x} = -F(x) + H^T(x)u \tag{11.44a}$$
$$y = h(x) \tag{11.44b}$$
$$0 \leq y \perp u \geq 0, \tag{11.44c}$$

where $h(x)$ is a vector defined in the obvious manner by $(h(x))_i = h_i(x)$, and where the trajectories of all variables are required to be continuous. Suppose that the system is initialized at $t = 0$ at a point x_0 in K. For indices i such that $h_i(x_0) > 0$, the complementarity conditions imply that we must have $u_i(0) = 0$. For indices that are active at x_0 we have $y_i(0) = 0$; to satisfy the inequality constraints also for positive t we need $\dot{y}_i(0) \geq 0$. Moreover, it follows from the complementarity conditions and the continuity conditions that we must have $u_i(0) = 0$ for indices i such that $\dot{y}_i(0) = 0$, and, vice versa, $\dot{y}_i(0) = 0$ for indices i such that $u_i(0) > 0$. Since

$$\begin{aligned}\dot{y}_{I(x_0)}(0) &= H_{I(x_0)\bullet}(-F(x_0) + H^T u(0)) \\ &= H_{I(x_0)\bullet}(-F(x_0) + H^T_{I(x_0)\bullet}u_{I(x_0)}(0))\end{aligned}$$

the vector $u_{I(x_0)}(0)$ must be a solution of the complementarity problem (11.43). It follows that $H^T u(0)$ is of the form appearing in (11.41). The reverse conclusion follows as well, and moreover one can show that "local" equality of solutions as just shown implies "global" equality [22].

11.2.7 *Diffusion with a free boundary*

In this subsection we consider a situation in which a complementarity system arises as an approximation to a partial differential equation with a free boundary. We take a specific example that arises in the theory of option pricing. For this we first need to introduce some terminology. A *European put option* is a contract that gives the holder the right, but not the obligation, to sell a certain asset to the counterparty in the contract for a specified price (the "exercise price") at a specified time in the future ("time of maturity"). The underlying asset can for instance be a certain amount of stocks, or a certain amount of foreign currency. For a concrete example, consider an investor who has stocks that are worth 110 now and who would like to turn these stocks into cash in one year's time. Of course it is hard to predict what the value of the stocks will be at that time; to make sure that the proceeds will be at least 100, the investor may buy a put option with exercise price 100 that matures in one year. In this way the investor is sure that she can sell the stocks for at least 100.

Of course one has to pay a price to buy such protection, and it is the purpose of option theory to determine "reasonable" option prices. The modern theory of option pricing started in the early 1970s with the seminal work by Black, Scholes, and Merton. This theory is not based on the law of large numbers, but rather on the observation that the risk that goes with conferring an option contract is not as big as it would seem to be at first sight. By following an active trading strategy in the underlying asset, the seller ("writer") of the option will be able to reduce the risk. Under suitable model assumptions the risk can even be completely eliminated; that is to say, the cost of providing protection becomes independent of the evolution of the value of the underlying asset and hence can be predicted in advance. The "no-arbitrage" argument then states that this fixed cost must, by the force of competition, be the market price of the option. The model assumptions under which one can show that the risk of writing an option can be eliminated are too strong to be completely realistic; nevertheless, they provide a good guideline for devising strategies that at least are able to reduce risk substantially.

One of the assumptions made in the original work of Black and Scholes [3] is that the price paths of the underlying asset may be described by a stochastic differential equation of the form

$$dS(t) \;=\; \mu S(t)dt + \sigma S(t)dw(t), \tag{11.45}$$

where $w(t)$ denotes a standard Wiener process. (See any textbook on SDEs, for instance [31], for the meaning of the above.) Under a number of additional assumptions (for instance: the underlying asset can be traded continuously and without transaction costs, and there is one fixed interest rate r that holds both for borrowing and for lending), Black and Scholes derived a partial differential equation that describes the price of the option at any time before maturity as a function of two variables, namely, time t and the price S of the underlying asset. The Black–Scholes equation for the option price $C(S,t)$ is (with omission of the arguments)

$$\frac{\partial C}{\partial t} + \tfrac{1}{2}\sigma^2 S^2 \frac{\partial^2 C}{\partial S^2} + rS\frac{\partial C}{\partial S} - rC = 0 \tag{11.46}$$

with end condition for time of maturity T and exercise price K

$$C(S,T) = \max(K-S,0) \tag{11.47}$$

and boundary conditions

$$C(0,t) = e^{-r(T-t)}K, \qquad \lim_{S\to\infty} C(S,t) = 0. \tag{11.48}$$

It turns out that the "drift" parameter μ in equation (11.45) is immaterial, whereas the "volatility" parameter σ is very important since it determines the diffusion coefficient in the PDE (11.46).

So far we have been discussing a *European* put option. An *American* put option is the same except that the option may be exercised at any time until the maturity date, rather than only at the time of maturity. (The terms "European" and "American" just serve as a way of distinction; both types of options are traded on both continents.) The possibility of early exercise brings a discrete element into the discussion since at any time the option may be in two states: "alive" or "exercised." For American options, the Black–Scholes equation (11.46) is replaced by an inequality

$$\frac{\partial C}{\partial t} + \tfrac{1}{2}\sigma^2 S^2 \frac{\partial^2 C}{\partial S^2} + rS\frac{\partial C}{\partial S} - rC \le 0 \tag{11.49}$$

in which equality holds if the option is not exercised, that is, if its value exceeds the revenues of exercise. For the put option, this is expressed by the inequality

$$C(S,t) > \max(K-S,0). \tag{11.50}$$

The Black–Scholes equation (11.46) is a nonlinear partial differential equation. A simple substitution, however, will transform it to a linear PDE. To this end, express the price of the underlying asset S in terms of a new (dimensionless) independent variable x by

$$S = Ke^x.$$

To make the option price dimensionless as well, introduce $v := C/K$. We also change the final value problem (11.46) and (11.47) to an initial value problem by setting set $\tau = T - t$. After some computation, we find that the equation (11.46) is replaced by

$$-\frac{\partial v}{\partial \tau} + \tfrac{1}{2}\sigma^2 \frac{\partial^2 v}{\partial x^2} + (r - \tfrac{1}{2}\sigma^2)\frac{\partial v}{\partial x} - rv = 0 \tag{11.51}$$

with the initial and boundary conditions (for the European put option)

$$v(x,0) = \max(1-e^x,0), \quad \lim_{x\to-\infty} v(x,\tau) = e^{-r\tau}, \quad \lim_{x\to\infty} v(x,\tau) = 0. \tag{11.52}$$

For the American put option, we get the set of inequalities

$$\frac{\partial v}{\partial \tau} - \tfrac{1}{2}\sigma^2 \frac{\partial^2 v}{\partial x^2} - (r - \tfrac{1}{2}\sigma^2)\frac{\partial v}{\partial x} + rv \geq 0 \tag{11.53}$$

$$v \geq \max(1-e^x,0) \tag{11.54}$$

with the boundary conditions

$$\lim_{x\to\infty} v(x,\tau) = 0 \tag{11.55}$$

and

$$v(x,\tau) = 1 - e^x \quad \text{for } x \leq x_f(\tau), \tag{11.56}$$

where $x_f(\tau)$ is the location at time τ of the free boundary that should be determined as part of the problem on the basis of the so-called "smooth pasting" or "high contact" conditions which require that v and $\partial v/\partial x$ should be continuous as functions of x across the free boundary.

Define the function $g(x)$ by

$$g(x) = \max(1-e^x,0). \tag{11.57}$$

The partial differential inequality (11.53) and its associated boundary conditions may then be written in the following form which is implicit with respect to the free boundary.

$$\left(\tfrac{\partial v}{\partial \tau} - \tfrac{1}{2}\sigma^2 \tfrac{\partial^2 v}{\partial x^2} - (r - \tfrac{1}{2}\sigma^2)\tfrac{\partial v}{\partial x} + rv\right)(v-g) = 0, \tag{11.58a}$$

$$\tfrac{\partial v}{\partial \tau} - \tfrac{1}{2}\sigma^2 \tfrac{\partial^2 v}{\partial x^2} - (r - \tfrac{1}{2}\sigma^2)\tfrac{\partial v}{\partial x} + rv \geq 0, \quad v - g \geq 0. \tag{11.58b}$$

This already suggests a complementarity formulation. Indeed, the above might be considered as an example of an infinite-dimensional complementarity system, both in the sense that the state space dimension is infinite and in the sense that there are infinitely many complementary variables. A complementary system of the type studied here is obtained by approximating the infinite-dimensional system by a finite-dimensional system. For the current application it seems reasonable to carry out the approximation by what is known in numerical analysis as the "method of lines." In this approach the space variable is discretized but the time variable is not. Specifically, take a grid of, say, N points in the space variable (in our case this is the dimensionless price variable x), and associate with each grid point x_i a variable $x_i(t)$ which is intended to be an approximation to $v(x_i,t)$. The action of the linear differential operator

$$v \mapsto \tfrac{1}{2}\sigma^2 \frac{\partial^2 v}{\partial x^2} + (r - \tfrac{1}{2}\sigma^2)\frac{\partial v}{\partial x} - rv \tag{11.59}$$

can be approximated by a linear mapping acting on the space spanned by the variables $x_1, \ldots, x_N$. For instance, on an evenly spaced grid an approximation to the first-order differential operator $\partial/\partial x$ is given by

$$A_1 = \frac{1}{2h}\begin{bmatrix} -2 & 2 & 0 & \cdots & \cdots & 0 \\ -1 & 0 & 1 & 0 & & \vdots \\ 0 & -1 & 0 & 1 & \ddots & \vdots \\ \vdots & \ddots & \ddots & \ddots & \ddots & 0 \\ \vdots & & 0 & -1 & 0 & 1 \\ 0 & \cdots & \cdots & 0 & -2 & 2 \end{bmatrix}, \tag{11.60}$$

where h is the mesh size, and the second-order differential operator $\partial^2/\partial x^2$ is approximated by

$$A_2 = \frac{1}{h^2}\begin{bmatrix} -2 & 1 & 0 & \cdots & \cdots & 0 \\ 1 & -2 & 1 & 0 & & \vdots \\ 0 & 1 & -2 & 1 & \ddots & \vdots \\ \vdots & \ddots & \ddots & \ddots & \ddots & 0 \\ \vdots & & 0 & 1 & -2 & 1 \\ 0 & \cdots & \cdots & 0 & 1 & -2 \end{bmatrix}. \tag{11.61}$$

The mapping (11.59) is then approximated by the matrix

$$A = \tfrac{1}{2}\sigma^2 A_2 + (r - \tfrac{1}{2}\sigma^2)A_1 - rI. \tag{11.62}$$

The function $g(x)$ is represented by the vector g with entries $g_i = g(x_i)$. Consider now the linear complementary system with $N+1$ state variables and N pairs of complementary variables given by

$$\dot{x} = \begin{bmatrix} A & 0 \\ 0 & 0 \end{bmatrix} x + \begin{bmatrix} I \\ 0 \end{bmatrix} u \tag{11.63a}$$

$$y = [I \;\; -g]\, x \tag{11.63b}$$

$$0 \leq y \perp u \geq 0. \tag{11.63c}$$

The system is initialized at

$$x_i(0) = g_i \qquad (i = 1, \ldots, N), \qquad x_{N+1}(0) = 1. \tag{11.64}$$

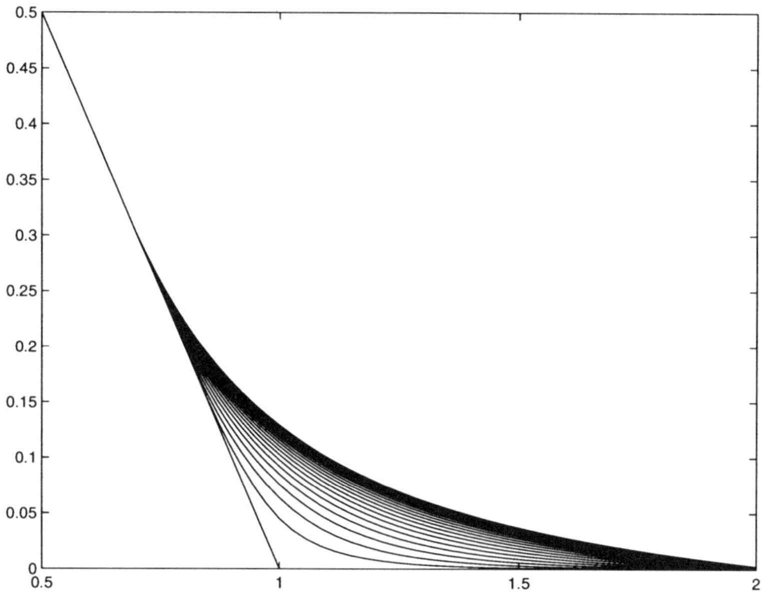

FIGURE 11.1. Value of an American put option at different times before expiry.

The complementarity system defined in this way is a natural candidate for providing an approximate solution to the diffusion equation with a free boundary that corresponds to the Black–Scholes equation for an American put option.

A natural idea for generating approximate solutions of complementarity systems is to use an implicit Euler method. For linear complementarity systems of the form (11.4) and (11.2), the method may be written as follows.

$$\frac{x((k+1)\Delta t) - x(k\Delta t)}{\Delta t} = Ax((k+1)\Delta t) + Bu((k+1)\Delta t) \tag{11.65a}$$

$$y((k+1)\Delta t) = Cx((k+1)\Delta t) + Du((k+1)\Delta t) \tag{11.65b}$$

$$0 \leq y((k+1)\Delta t) \perp u((k+1)\Delta t) \geq 0. \tag{11.65c}$$

At each step this gives rise to a complementarity problem which under suitable assumptions has a unique solution. The results of applying the above method to the equations for an American put option are shown in Figure 11.1, which presents solutions for various times before expiry as a function of the value of the underlying asset (the variable S).

A more standard numerical method for dealing with the American-style Black-Scholes equation is the finite difference method in which both the time and space variables are discretized; see, for instance, [40]. In general, an advantage of a "semidiscretization" approach over an approach in which all variables are discretized simultaneously is that one may make use of the highly developed theory of step-size control for numerical solutions of ODEs, rather than using a uniform grid. We are of course working here with a complementarity system rather than an ODE, and it must be said that the theory of step-size control for complementarity systems is at an early stage of development. Moreover, the theory of approximation of free boundary problems by complementarity systems has been presented here only for a special case and on the basis of plausibility rather than formal proof, and much further work on this topic is needed.

11.2.8 Max-plus systems

From the fact that the relation

$$z = \max(x, y) \tag{11.66}$$

may also be written in the form

$$z = x + a = y + b, \qquad 0 \leq a \perp b \geq 0 \tag{11.67}$$

it follows that any system that can be written in terms of linear operations and the "max" operation can also be written as a complementarity system. In particular it follows that the so-called max-plus systems (see [1]), which are closely related to timed Petri nets, can be written as complementarity systems. The resulting equations appear in discrete time, as opposed to the other examples in this section which are all in continuous time; note however that the "time" parameter in a max-plus system is in the standard applications a cycle counter rather than actual time. For further discussion of the relation between the max algebra and the complementarity problem see [6].

11.3 Existence and uniqueness of solutions

Complementarity systems can be looked at as a class of *hybrid* systems; this term is understood here as referring to systems that are described in terms of both continuous (real-valued) and discrete (finite-valued) variables. Hybrid systems provide a rather wide modeling context, so that there are no easily verifiable necessary and sufficient conditions for well-posedness of general hybrid dynamical systems. It is already of interest to give sufficient conditions for well-posedness of particular classes of hybrid systems, such as complementarity systems. The advantage of considering special classes is that one can hope for conditions that are relatively easy to verify. In a number of special cases, such as mechanical systems or electrical network models, there are moreover natural candidates for such sufficient conditions.

Uniqueness of solutions is always understood below in the sense of what is sometimes called *right uniqueness*, that is, uniqueness of solutions defined on an interval $[t_0, t_1)$ given an initial state at t_0. It can easily happen in general hybrid systems and even in complementarity systems that uniqueness holds in one direction of time but not in the other; this is one of the points in which discontinuous dynamical systems differ from smooth systems. We also allow for the possibility of an initial jump.

We have to distinguish between *local* and *global* existence and uniqueness. Local existence and uniqueness, for solutions starting at t_0, holds if there exists an $\varepsilon > 0$ such that on $[t_0, t_0+\varepsilon)$ there is a unique solution starting at the given initial condition. For global existence and uniqueness, we require that for any given initial condition there is a unique solution on $[t_0, \infty)$. If local uniqueness holds for all initial conditions and existence holds globally, then uniqueness must also hold globally since there is no point at which solutions can split. However local existence does not imply global existence. This phenomenon is already well known in the theory of smooth dynamical systems; for instance the differential equation $\dot{x}(t) = x^2(t)$ with $x(0) = x_0$ has the unique solution $x(t) = x_0(1-x_0 t)^{-1}$ which for positive x_0 is defined only on the interval $[0, x_0^{-1})$. Some growth conditions have to be imposed to prevent this "escape to infinity." In hybrid systems, there may be additional complications; in particular we may have an accumulation of mode switches such as in the following example that has been adapted from Filippov [10, p. 116].

Example 11.1. Consider the relay system given by

$$\begin{aligned} \dot{x}_1 &= -\operatorname{sgn} x_1 + 2 \operatorname{sgn} x_2 \\ \dot{x}_2 &= -2 \operatorname{sgn} x_1 - \operatorname{sgn} x_2. \end{aligned} \tag{11.68}$$

This is a *piecewise constant* system; in each quadrant of the (x_1, x_2)-plane the right-hand side is a constant vector. From each initial point there exists locally a unique solution. The solutions are spiraling towards the origin, which is an equilibrium point. It can be verified that $d/dt\,(|x_1(t)| + |x_2(t)|) = -2$ which means that solutions starting at $\operatorname{col}(x_{10}, x_{20})$ can not stay away from the origin for longer than $(|x_{10}| + |x_{20}|)/2$ units of time. However, solutions can not arrive at the origin without going through an infinite number of mode switches; since these mode switches have to occur in a finite time interval, we do not have a global solution in the space of piecewise differentiable functions (if only finitely many pieces are allowed). Some solutions of the system (11.68) are plotted in Figure 11.2.

Although the occurrence of an accumulation of mode switches would seem to be exceptional, no general conditions are known at present that exclude this phenomenon. Below we use the term *well-posedness* to refer to local existence and uniqueness of solutions for all feasible initial conditions (i. e., initial conditions for which none of the inequality constraints are

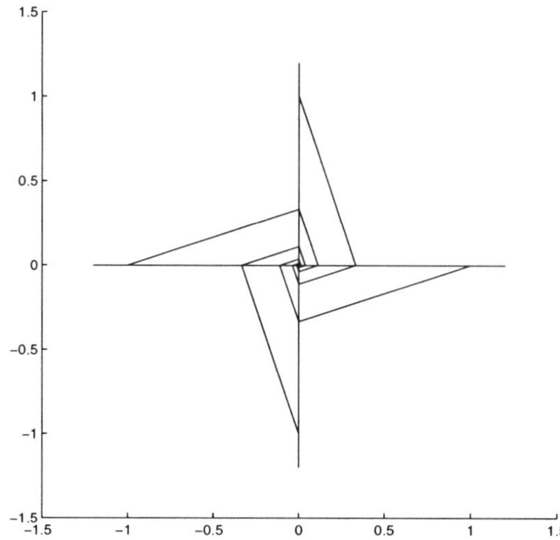

FIGURE 11.2. Solutions of Filippov's example (11.68).

violated).

11.4 Linear complementarity systems

Consider the following system of linear differential and algebraic equations and inequalities

$$\dot{x}(t) = Ax(t) + Bu(t) \tag{11.69a}$$

$$y(t) = Cx(t) + Du(t) \tag{11.69b}$$

$$0 \le y(t) \perp u(t) \ge 0. \tag{11.69c}$$

The equations (11.69a) and (11.69b) constitute a linear system in state space form; the number of inputs is taken equal to the number of outputs. The relations (11.69c) are the usual *complementarity conditions*. The set of indices for which $y_i(t) = 0$ (we call this the *active index set*) need not be constant in time, so that the system may switch from one "operating mode" to another. To define the dynamics of (11.69) completely, we have to specify when these mode switches occur, what their effect will be on the state variables, and how a new mode will be selected. A proposal for answering these questions (cf. [21]) is explained below. The specification of the complete dynamics of (11.69) defines a class of dynamical systems called *linear complementarity systems*.

Let n denote the length of the vector $x(t)$ in the equations (11.69a) and (11.69b) and let k denote the number of inputs and outputs. There are then 2^k possible choices for the active index set. The equations of motion when the active index set is I are given by

$$\begin{aligned} \dot{x}(t) &= Ax(t) + Bu(t) \\ y(t) &= Cx(t) + Du(t) \\ y_i(t) &= 0, \quad i \in I \\ u_i(t) &= 0, \quad i \in I^c, \end{aligned} \tag{11.70}$$

where I^c denotes the index set that is complementary to I; that is, $I^c = \{i \in \{1,\dots,k\} \mid i \notin I\}$. We say that the above equations represent the system in *mode* I. An equivalent and somewhat more explicit form is given by the (generalized) state equations

$$\begin{aligned} \dot{x}(t) &= Ax(t) + B_{\bullet I}u_I(t) \\ 0 &= C_{I\bullet}x(t) + D_{II}u_I(t) \end{aligned} \tag{11.71}$$

together with the output equations

$$\begin{aligned} y_{I^c}(t) &= C_{I^c\bullet}x(t) + D_{I^cI}u_I(t) \\ u_{I^c}(t) &= 0. \end{aligned} \tag{11.72}$$

Here and below, the notation $M_{\bullet I}$, where M is a matrix of size $m \times k$ and I is a subset of $\{1,\dots,k\}$, denotes the submatrix of M formed by taking the columns of M whose indices are in I. The notation $M_{I\bullet}$ denotes the submatrix obtained by taking the rows with indices in the index set I.

The system (11.71) in general does not have solutions in a classical sense for all possible initial conditions. The initial values of the variable x for which there does exist a continuously differentiable solution are called *consistent states*. Under conditions that are specified below, each consistent initial state gives rise to a unique solution of (11.71). The system (11.69) follows the path of such a solution (it "stays in mode I") as long as the variables $u_I(t)$ defined implicitly by (11.71) and the variables $y_{I^c}(t)$ defined by (11.72) are all nonnegative. As soon as continuation in mode I would lead to a violation of these inequality constraints, a switch to a different mode has to occur. If moreover the value of the variable $x(t)$ at which violation of the constraints has become imminent is not a consistent state for the new mode, then a state jump is necessary. So both concerning the dynamics in a given mode and concerning the transitions between different modes there are a number of questions to be answered. For this we rely on the geometric theory of linear systems (see [41, 2, 4] for the general background).

Denote by V_I the *consistent subspace* of mode I; that is to say, V_I is the set of initial conditions x_0 for which there exist smooth functions $x(\cdot)$ and $u_I(\cdot)$, with $x(0) = x_0$, such that (11.71) is satisfied. The space V_I can be computed as the limit of the sequence defined by

$$V_I^0 = \mathbb{R}^n$$

$$V_I^{i+1} = \{x \in V_I^i \mid \exists u \in \mathbb{R}^{|I|} \text{ s.t. } Ax + B_{\bullet I}u \in V_I^i,\ C_{I\bullet}x + D_{II}u = 0\}. \tag{11.73}$$

There exists a linear mapping F_I such that (11.71) will be satisfied for $x_0 \in V_I$ by taking $u_I(t) = F_I x(t)$. The mapping F_I is uniquely determined, and more generally the function $u_I(\cdot)$ that satisfies (11.71) for given $x_0 \in V_I$ is uniquely determined, if the full column rank condition

$$\ker \begin{bmatrix} B_{\bullet I} \\ D_{II} \end{bmatrix} = \{0\} \tag{11.74}$$

holds and moreover we have

$$V_I \cap T_I = \{0\}, \tag{11.75}$$

where T_I is the subspace that can be computed as the limit of the following sequence.

$$\begin{aligned} T_I^0 &= \{0\} \\ T_I^{i+1} &= \{x \in \mathbb{R}^n \mid \exists \tilde{x} \in T_I^i,\ \tilde{u} \in \mathbb{R}^{|I|} \text{ s.t.} \\ &\qquad\qquad x = A\tilde{x} + B_{\bullet I}\tilde{u},\ C_{I\bullet}\tilde{x} + D_{II}\tilde{u} = 0\}. \end{aligned} \tag{11.76}$$

As indicated below, the subspace T_I is best thought of as the *jump space* associated with mode I, that is, as the space along which fast motions will occur that take an inconsistent initial state instantaneously to a point in the consistent space V_I; note that under the condition (11.75) this projection is uniquely determined. The projection can be used to define a *jump rule*. However, there are 2^k possible projections, corresponding to all possible subsets of $\{1,\dots,k\}$; which one of these to choose should be determined by a *mode selection rule*.

For the formulation of a mode selection rule we have to relate index sets in some way to continuous states. Such a relation can be established on the basis of the so-called *rational complementarity problem* (RCP). The RCP is defined as follows. Let a rational vector $q(s)$ of length k and a rational matrix $M(s)$ of size $k \times k$ be given. The rational complementarity problem is to find a pair of rational vectors $y(s)$ and $u(s)$ (both of length k) such that

$$y(s) = q(s) + M(s)u(s) \tag{11.77}$$

and moreover for all indices $1 \le i \le k$ we have either $y_i(s) = 0$ and $u_i(s) > 0$ for all sufficiently large s, or $u_i(s) = 0$ and $y_i(s) > 0$ for all sufficiently large s. The vector $q(s)$ and the matrix $M(s)$ are called the *data* of the RCP, and we write RCP($q(s)$,$M(s)$). We also consider an RCP whose data are a quadruple of constant matrices (A, B, C, D) (such as could be used to define (11.69a) and (11.69b)) and a constant vector x_0, namely, by setting

$$q(s) = C(sI - A)^{-1}x_0 \quad \text{and} \quad M(s) = C(sI - A)^{-1}B + D.$$

We say that an index set $I \subset \{1, \ldots, k\}$ *solves* the RCP (11.77) if there exists a solution $(y(s), u(s))$ with $y_i(s) = 0$ for $i \in I$ and $u_i(s) = 0$ for $i \notin I$. The collection of index sets I that solve RCP$(A, B, C, D; x_0)$ is denoted $\mathcal{S}(A, B, C, D; x_0)$ or simply $\mathcal{S}(x_0)$ if the quadruple (A, B, C, D) is given by the context.

After these preparations, we can now proceed to a specification of the complete dynamics of linear complementarity systems. We assume that a quadruple (A, B, C, D) is given whose transfer matrix $G(s) = C(sI - A)^{-1}B + D$ is *totally invertible*; that is, for each index set I the $k \times k$ matrix $G_{II}(s)$ is nonsingular. Under this condition (see Theorem 11.1), the two subspaces V_I and T_I as defined above form for all I a direct sum decomposition of the state space $\mathbb{R}^n$, so that the projection along T_I onto V_I is well defined. We denote this projection P_I. The interpretation that we give to the equations (11.69) is the following.

$$\begin{array}{llll} \dot{x} = Ax + Bu, & y = Cx + Du & & \\ u_I \geq 0, & y_I = 0, & u_{I^c} = 0, & y_{I^c} \geq 0 \\ I^\sharp \in \mathcal{S}(x), & x^\sharp = P_{I^\sharp} x, & & \end{array} \tag{11.78}$$

where the symbol $\sharp$ denotes "next value." In the expression above, we use a continuous state x and the continuous complementary variables y and u, as well as a discrete state I. The discrete state must switch when this is necessary to prevent violation of the inequality constraints; in general we allow the continuous state to jump, possibly even several times at the same instant (in this case we have an "event with multiplicity"). Below we always consider the system (11.69) in the interpretation (11.78).

11.5 A distributional interpretation

The interpretation of T_I as a jump space can be made precise by introducing the class of *impulsive-smooth distributions* that was studied by Hautus [14] (see also [15, 12]). The general form of an impulsive-smooth distribution ϕ is

$$\phi \;=\; p(d/dt)\delta \;+\; f \tag{11.79}$$

where $p(\cdot)$ is a polynomial, d/dt denotes the distributional derivative, δ is the delta distribution with support at zero, and f is a distribution that can be identified with the restriction to $(0, \infty)$ of some function in $C^\infty(\mathbb{R})$. The class of such distributions is denoted C_{imp}. For an element of C_{imp} of the form (11.79), we write $\phi(0^+)$ for the limit value $\lim_{t \downarrow 0} f(t)$. Having introduced the class C_{imp}, we can replace the system of equations (11.71) by its distributional version

$$\begin{aligned} \tfrac{d}{dt}x &= Ax + B_{\bullet I} u_I + x_0 \delta \\ 0 &= C_{I\bullet} x + D_{II} u_I \end{aligned} \tag{11.80}$$

in which the initial condition x_0 appears explicitly, and we can look for a solution of (11.80) in the class of vector-valued impulsive-smooth distributions. It was shown in [15] that if the conditions (11.74) and (11.75) are satisfied, then there exists a unique solution $(x, u_I) \in C_{\text{imp}}^{n+|I|}$ to (11.80) for each $x_0 \in V_I + T_I$; moreover, the solution is such that $x(0^+)$ is equal to $P_I x_0$, the projection of x_0 onto V_I along T_I. The solution is most easily written down in terms of its Laplace transform:

$$\hat{x}(s) = (sI - A)^{-1}x_0 + (sI - A)^{-1}B_{\bullet I}\hat{u}_I(s) \tag{11.81}$$

$$\hat{u}_I(s) = -G_{II}^{-1}(s)C_{I\bullet}(sI - A)^{-1}x_0, \tag{11.82}$$

where

$$G_{II}(s) := C_{I\bullet}(sI - A)^{-1}B_{\bullet I} + D_{II}. \tag{11.83}$$

Note that the notation is consistent in the sense that $G_{II}(s)$ can also be viewed as the (I, I)-submatrix of the transfer matrix $G(s) := C(sI - A)^{-1}B + D$. It is shown in [15] (see also [30]) that the transfer matrix $G_{II}(s)$ associated with the system parameters in (11.71) is left invertible when (11.74) and (11.75) are satisfied. Since the transfer matrices $G_{II}(s)$ that we consider are square, left invertibility is enough to imply invertibility, and so (by duality) we also have $V_I + T_I = \mathbb{R}^n$. Summarizing, we can list the following equivalent conditions.

Theorem 11.1. *Consider a time-invariant linear system with k inputs and k outputs, given by standard state space parameters (A, B, C, D). The following conditions are equivalent.*

(i) For each index set $I \subset \bar{k}$, the associated system (11.71) admits for each $x_0 \in V_I$ a unique smooth solution (x, u) such that $x(0) = x_0$.

(ii) For each index set $I \subset \bar{k}$, the associated distributional system (11.80) admits for each initial condition x_0 a unique impulsive-smooth solution (x, u).

(iii) The conditions (11.74) and (11.75) are satisfied for all $I \subset \bar{k}$.

(iv) The transfer matrix $G(s) = C(sI - A)^{-1}B + D$ is totally invertible (as a matrix over the field of rational functions).

In connection with the system (11.69) it makes sense to introduce the following definitions.

Definition 11.1. *An impulsive-smooth distribution $\phi = p(d/dt)\delta + f$ as in (11.79) is called* initially nonnegative *if the leading coefficient of the polynomial $p(\cdot)$ is positive, or, in the case $p = 0$, the smooth function f is nonnegative on an interval of the form $(0, \varepsilon)$ with $\varepsilon > 0$. A vector-valued impulsive-smooth* *distribution is called* initially nonnegative *if each of its components is initially nonnegative in the above sense.*

Definition 11.2. *A triple of vector-valued impulsive-smooth distributions* (u, x, y) *is called an* initial solution *to* (11.69) *with* initial state x_0 *and* solution mode I *if*

(i) the triple (u, x, y) *satisfies the distributional equations*

$$\begin{aligned}\frac{d}{dt}x &= Ax + Bu + x_0\delta\\ y &= Cx + Du;\end{aligned}$$

(ii) both u *and* y *are initially nonnegative;*

(iii) $y_i = 0$ *for all* $i \in I$ *and* $u_i = 0$ *for all* $i \notin I$.

For an impulsive-smooth distribution w that has a rational Laplace transform $\hat{w}(s)$ (such as in (11.81) and (11.82)), we have that w is initially nonnegative if and only if $\hat{w}(s)$ is nonnegative for all sufficiently large real values of s. From this it follows that the collection of index sets I for which there exists an initial solution to (11.69) with initial state x_0 and solution mode I is exactly $\mathcal{S}(x_0)$ as we defined this set before in terms of the rational complementarity problem.

11.6 Well-posedness

Given the interpretation (11.78) of the dynamics (11.69), of course the first question to ask is under which conditions we have solutions for all initial conditions, and under which conditions these solutions are unique. Well-posedness is understood here in the sense of existence of a unique solution on some interval of nonzero length for all initial conditions. Uniqueness is understood in the sense of piecewise differentiable solutions whose points of nondifferentiability do not accumulate (sometimes called "non-Zeno" solutions). Actually it is in some sense natural to allow accumulations "forward in time" and to exclude accumulations "backward in time," but we do not enter this discussion here; see [37] for more information.

We start with a result from [35], which gives necessary and sufficient conditions for well-posedness be it only for a rather limited class of linear complementarity systems. The result is concerned with *bimodal* linear complementarity systems, that is, systems with only two modes ($k = 1$). Such a system of the form (11.69) has a transfer function $g(s) = C(sI - A)^{-1}B + D$ which is a rational function. In this case the conditions of Theorem 11.1 apply if $g(s)$ is nonzero. The system is said to have *no feedthrough term* if the matrix D vanishes. The system is called *degenerate* if the transfer matrix $g(s)$ is of the form $g(s) = 1/q(s)$ where $q(s)$ is a polynomial; in this case the consistent subspace in the constrained mode is just the origin. The *Markov parameters* of the system are the coefficients of the expansion of $g(s)$ around infinity,

$$g(s) = g_0 + g_1 s^{-1} + g_2 s^{-2} + \cdots .$$

The *leading Markov parameter* is the first parameter in the sequence g_0, $g_1, \ldots$ that is nonzero. Having introduced this terminology, we can now formulate the following result [35, Theorem 4.8].

Theorem 11.2. *A nondegenerate bimodal linear complementarity system without feedthrough term and with nonzero transfer function is well-posed if and only if its leading Markov parameter is positive.*

It is typical to find that well-posedness of complementarity systems is linked to a positivity condition. If the number of pairs of complementary variables is larger than one, an appropriate matrix version of the positivity condition has to be used. As might be expected, the type of positivity that we need is the "P-matrix" property from the theory of the LCP. Recall (see the end of Section 11.1) that a square real matrix is said to be a P-matrix if all its principal minors are positive.

To state a result on well-posedness for multivariable linear complementarity systems, we again need some concepts from linear systems theory. Recall (see, for instance, [24, p. 384] or [27, p. 24]) that a square rational matrix $G(s)$ is said to be *row proper* if it can be written in the form

$$G(s) = \Delta(s)B(s), \tag{11.84}$$

where $\Delta(s)$ is a diagonal matrix whose diagonal entries are of the form s^k for some integer k that may be different for different entries, and $B(s)$ is a proper rational matrix that has a proper rational inverse (i.e., $B(s)$ is *bicausal*). A proper rational matrix $B(s) = B_0 + B_1 s^{-1} + \cdots$ has a proper rational inverse if and only if the constant matrix B_0 is invertible. This constant matrix is uniquely determined in a factorization of the above form; it is called the *leading row coefficient matrix* of $G(s)$. In a completely similar way one defines the notions of column properness and of the leading column coefficient matrix. We can now state the following result [21, Theorem 6.3].

Theorem 11.3. *The linear complementarity system* (11.69) *is well-posed if the associated transfer matrix $G(s) = C(sI - A)^{-1}B + D$ is both row and column proper, and if both the leading row coefficient matrix and the leading column coefficient matrix are P-matrices. Moreover, in this case the multiplicity of events is at most one; that is, at most one reinitialization takes place at times when a mode change occurs.*

An alternative sufficient condition for well-posedness can be based on the rational complementarity problem (RCP) that was already used above (Section 11.4). For a given set of linear system parameters (A, B, C, D), we denote by RCP(x_0) the rational complementarity problem RCP$(q(s), M(s))$ with data $q(s) = C(sI - A)^{-1}x_0$ and $M(s) = C(sI - A)^{-1}B + D$. For the purposes of simplicity, the following result is stated under somewhat stronger hypotheses than were used in the original paper [20, Theorems 5.10, 5.16].

Theorem 11.4. *Consider the linear complementarity system* (11.69), *and assume that the associated transfer matrix is totally invertible. The system* (11.69) *is well-posed if the problem RCP(x_0) has a unique solution for all x_0.*

A connection between the rational complementarity problem and the standard linear complementarity problem can be established in the following way [20, Theorem 4.1, Corollary 4.10].

Theorem 11.5. *For given $q(s) \in \mathbb{R}^k(s)$ and $M(s) \in \mathbb{R}^{k\times k}(s)$, the problem RCP($q(s), M(s)$) is uniquely solvable if and only if there exists $\mu \in \mathbb{R}$ such that for all $\lambda > \mu$ the problem LCP($q(\lambda), M(\lambda)$) is uniquely solvable.*

The above theorem provides a convenient way of proving well-posedness for several classes of linear complementarity systems. The following example is taken from [20].

Example 11.2. A linear mechanical system may be described by equations of the form

$$M\ddot{q} + D\dot{q} + Kq = 0, \tag{11.85}$$

where q is the vector of generalized coordinates, M is the generalized mass matrix, D is the damping matrix, and K is the elasticity matrix. The mass matrix M is positive definite. Suppose now that we subject the above system to unilateral constraints of the form

$$Fq \geq 0, \tag{11.86}$$

where F is a given matrix. Under the assumption of inelastic collisions, the dynamics of the resulting system may be described by

$$M\ddot{q} + D\dot{q} + Kq = F^T u, \qquad y = Fq \tag{11.87}$$

together with complementarity conditions between y and u. The associated RCP is the following.

$$y(s) = F(s^2M + sD + K)^{-1}[(sM + D)q_0 + M\dot{q}_0] + F(s^2M + sD + K)^{-1}F^T u(s). \tag{11.88}$$

If F has full row rank, then the matrix $F(s^2M+sD+K)^{-1}F^T$ is positive definite (although not necessarily symmetric) for all sufficiently large s because the term with s^2 becomes dominant. By combining the standard result on solvability of LCPs with Theorem 11.5, it follows that RCP is solvable and we can use this to prove the well-posedness of the constrained mechanical system; this provides some confirmation for the validity of the model that has been used, since physical intuition certainly suggests that a unique solution should exist.

In the above example, one can easily imagine cases in which the matrix F does not have full row rank so that the fulfillment of some constraints already implies that some other constraints will also be satisfied; think, for instance, of a chair having four legs on the ground. In such cases the basic result on solvability of LCPs does not provide enough information, but there are alternatives available that make use of the special structure that is present in equations like (11.88). On the basis of this, one can still prove well-posedness; in particular the trajectories of the coordinate vector $q(t)$ are uniquely determined, even though the trajectories of the constraint force $u(t)$ are not.

11.7 Relay systems

For piecewise linear relay systems of the form

$$\dot{x} = Ax + Bu, \qquad y = Cx + Du, \qquad u_i = -\operatorname{sgn} y_i \;\; (i = 1, \dots, k) \tag{11.89}$$

Theorem 11.5 can be applied, but the application is not straightforward for the following reason. As noted above, it is possible to rewrite a relay system as a complementarity system (in several ways actually). Using the method (11.28), one arrives at a relation between the new inputs $\operatorname{col}(u_1, u_2)$ and the new outputs $\operatorname{col}(y_1, y_2)$ that may be written in the frequency domain as follows (ι denotes the vector all of whose entries are 1, and $G(s)$ denotes the transfer matrix $C(sI - A)^{-1}B + D$).

$$\begin{bmatrix} u_1(s) \\ u_2(s) \end{bmatrix} = \begin{bmatrix} -G^{-1}(s)C(sI - A)^{-1}x_0 + s^{-1}\iota \\ G^{-1}(s)C(sI - A)^{-1}x_0 + s^{-1}\iota \end{bmatrix} + \\ + \begin{bmatrix} G^{-1}(s) & -G^{-1}(s) \\ -G^{-1}(s) & G^{-1}(s) \end{bmatrix} \begin{bmatrix} y_1(s) \\ y_2(s) \end{bmatrix}. \tag{11.90}$$

The matrix that appears on the right-hand side is singular for all s and so the corresponding LCP does not always have a unique solution. However the vector that we find at the right-hand side is of a special form and we only need to ensure existence of a unique solution for vectors of this particular form. On the basis of this observation, the following result is obtained.

Theorem 11.6. [29, 20] *The piecewise linear relay system* (11.89) *is well-posed if the transfer matrix* $G(s)$ *is a P-matrix for all sufficiently large* s.

This result gives a criterion that is straightforward to verify (compute the determinants of all principal minors of $G(s)$, and check the signs of the leading Markov parameters), but that is restricted to piecewise linear systems. Filippov [10, §2.10] gives a criterion for well-posedness that works for general nonlinear systems, but needs to be verified on a point-by-point basis.

11.8 Discontinuous dependence on initial conditions

Finally, let us briefly consider a property that for smooth dynamical systems is often taken into the definition of well-posedness, namely, the continuous dependence of solutions on initial conditions. In a hybrid system context this property would obviously have to be restricted to the trajectories of the continuous variables. Even then, the continuous dependence on initial conditions can easily be violated, even within such a limited class as linear Hamiltonian complementarity systems. This is shown in the following example from [21].

Example 11.3. Consider the equations

$$\ddot{x}_1 = -2x_1 + x_2 + u_1 + u_2 \tag{11.91}$$

$$\ddot{x}_2 = x_1 - x_2 - u_2 \tag{11.92}$$

$$y_1 = x_1 \tag{11.93}$$

$$y_2 = x_1 - x_2 \tag{11.94}$$

together with the standard complementarity conditions $0 \leq y \perp u \geq 0$. These equations arise from a system with two carts connected by springs to each other and to a fixed wall. The variables x_1 and x_2 indicate the positions of the first and the second cart, respectively. There are inelastic stops in the system that prevent x_1 and $x_1 - x_2$ from becoming negative. There are arbitrarily close initial conditions for which the two constraints become active in a different order, which results in quite different trajectories.

The phenomenon of discontinuous dependence on initial conditions should be viewed as a result of an idealization, reflecting a very strong dependence on initial conditions in a corresponding smooth model. Certainly the fact that such discontinuities appear is a problem in numerical simulation, but numerical problems would also occur when, for instance, the strict unilateral constraints in the example above would be replaced by very stiff springs. So the hybrid model in itself cannot be held responsible for the (near-)discontinuity problem; one should rather say that it clearly exposes this problem.

11.9 Conclusions

In this chapter we have discussed the class of nonsmooth dynamical systems that arises from connecting a set of complementarity conditions to an input-output dynamical system. It has been shown by a number of examples that systems of this type arise naturally in a number of applications, and that in some other cases a connection can be made through suitable rewriting operations. While a solution concept for complementarity systems with general nonlinear dynamics is not available at present due to the difficulties associated with state jumps in a nonlinear context, it is possible to formulate a complete specification of the dynamics when we restrict to linear complementarity systems under suitable hypotheses. Concepts of linear systems theory play an important role in the development of the theory of linear complementarity systems; for instance, the field of rational functions is used extensively just as in the linear systems case, with an extra element that comes from the imposition of an order structure on this field.

For a proper understanding of linear complementarity systems it should be emphasized that they are in fact nonlinear systems and may well exhibit, for instance, chaotic behavior. Moreover, they are discontinuous systems whose trajectories are nonsmooth; state jumps may occur as well as accumulations of event times. On the other hand, linear complementarity systems are close to linear systems and can be analyzed using methods that have been developed in linear systems theory over the past decades. It is a pleasure to make this observation in a volume dedicated to one of the foremost proponents of this field of study. So, here's to Didi.

References

[1] F. Baccelli, G. Cohen, G.J. Olsder, and J.P. Quadrat, *Synchronization and Linearity*, Wiley, New York, 1992.

[2] G. Basile and G. Marro, *Controlled and Conditioned Invariants in Linear System Theory*, Prentice Hall, Englewood Cliffs, NJ, 1992.

[3] F. Black and M. Scholes, The pricing of options and corporate liabilities, *Journal of Political Economy* 81:637–659, 1973.

[4] M.K. Çamlıbel and J.M. Schumacher, Well-posedness of a class of piecewise linear systems, In *Proc. European Control Conference '99*, Karlsruhe, Aug. 31 – Sept. 3, 1999. (CD-ROM)

[5] R.W. Cottle, J.-S. Pang, and R.E. Stone, *The Linear Complementarity Problem*, Academic Press, Boston, 1992.

[6] B. De Schutter and B. De Moor, Minimal realization in the max algebra is an extended linear complementarity problem, *Syst. Control Letters* 25:103–111, 1995.

[7] B. De Schutter and B. De Moor, The linear dynamic complementarity problem is a special case of the extended linear complementarity problem, *Syst. Control Letters* 34:63–75, 1998.

[8] P. Dupuis, Large deviations analysis of reflected diffusions and constrained stochastic approximation algorithms in convex sets, *Stochastics* 21:63–96, 1987.

[9] P. Dupuis and A. Nagurney, Dynamical systems and variational inequalities, *Annals Operations Research* 44:9–42, 1993.

[10] A.F. Filippov, *Differential Equations with Discontinuous Righthand Sides*, Kluwer, Dordrecht, 1988.

[11] F.R. Gantmacher, *The Theory of Matrices, Vol. I*, Chelsea, New York, 1959.

[12] A.H.W. Geerts and J.M. Schumacher, Impulsive-smooth behavior in multimode systems. Part I: State-space and polynomial representations, *Automatica* 32:747–758, 1996.

[13] R.F. Hartl, S.P. Sethi, and R.G. Vickson, A survey of the maximum principles for optimal control problems with state constraints, *SIAM Review* 37:181–218, 1995.

[14] M.L.J. Hautus, The formal Laplace transform for smooth linear systems, in *Mathematical Systems Theory*, G. Marchesini and S.K. Mitter, eds., Lect. Notes Econ. Math. Syst. 131, Springer, New York, 29–47, 1976.

[15] M.L.J. Hautus and L.M. Silverman, System structure and singular control, *Lin. Alg. Appl.* 50:369–402, 1983.

[16] W.P.M.H. Heemels, J.M. Schumacher, and S. Weiland, Complete description of dynamics in the linear complementary-slackness class of hybrid systems, *Proc. IEEE Conf. Dec. Contr.*, San Diego, 1997.

[17] W.P.M.H. Heemels, J.M. Schumacher, and S. Weiland, Complementarity problems in linear complementarity systems, *Proc. American Control Conf. 1998*, Philadelphia, 706–710, 1998.

[18] W.P.M.H. Heemels, J.M. Schumacher, and S. Weiland, Dissipative systems and complementarity conditions, *Proc. IEEE Conf. Dec. Contr. 1998*, Tampa, FL, 4127–4132, 1998.

[19] W.P.M.H. Heemels, J.M. Schumacher, and S. Weiland, Mode selection in linear complementarity systems, *Proc. IFAC Conf. System Structure and Control 1998*, Nantes, 153–158, 1998.

[20] W.P.M.H. Heemels, J.M. Schumacher, and S. Weiland, The rational complementarity problem, *Linear Algebra Appl.* 294:93–135, 1999.

[21] W.P.M.H. Heemels, J.M. Schumacher, and S. Weiland, Linear complementarity systems, *SIAM J. Appl. Math.* 60:1234–1269, 2000.

[22] W.P.M.H. Heemels, J.M. Schumacher, and S. Weiland, Projected dynamical systems in a complementarity formalism, *Operations Research Letters* 2000, to appear.

[23] U. Helmke and J.B. Moore, *Optimization and Dynamical Systems*, Springer, London, 1994.

[24] T. Kailath, *Linear Systems*, Prentice-Hall, Englewood Cliffs, N.J., 1980.

[25] I. Kaneko and J.S. Pang, Some n by dn linear complementarity problems, *Linear Algebra Appl.* 34:297–319, 1980.

[26] C. W. Kilmister and J.E. Reeve, *Rational Mechanics*, Longmans, London, 1966.

[27] M. Kuijper, *First-Order Representations of Linear Systems*, Birkhäuser, Boston, 1994.

[28] D.M.W. Leenaerts and W.M.G. van Bokhoven, *Piecewise Linear Modeling and Analysis*, Kluwer, Dordrecht, 1998.

[29] Y.J. Lootsma, A.J. van der Schaft, and M.K. Çamlıbel, Uniqueness of solutions of relay systems, *Automatica* 35:467–478, 1999.

[30] A.S. Morse and W.M. Wonham, Status of noninteracting control, *IEEE Trans. Automat. Contr.* AC-16:568–581, 1971.

[31] B. Øksendal, *Stochastic Differential Equations. An Introduction with Applications*, 5th ed., Springer, Berlin, 1998.

[32] J. Pérès, *Mécanique Générale*, Masson & Cie., Paris, 1953.

[33] F. Pfeiffer and C. Glocker, *Multibody Dynamics with Unilateral Contacts*, Wiley, Chichester, 1996.

[34] H. Samelson, R.M. Thrall, and O. Wesler, A partition theorem for Euclidean n-space, *Proc. Amer. Math. Soc.* 9:805–807, 1958.

[35] A.J. van der Schaft and J.M. Schumacher, The complementary-slackness class of hybrid systems, *Math. Control Signals Syst.* 9:266–301, 1996.

[36] A.J. van der Schaft and J.M. Schumacher, Complementarity modeling of hybrid systems, *IEEE Trans. Auto. Control* AC-43:483–490, 1998.

[37] A.J. van der Schaft and J.M. Schumacher, *An Introduction to Hybrid Dynamical Systems*, Springer, London, 2000.

[38] R. Sznajder and M.S. Gowda, Generalizations of p_0- and p-properties; extended vertical and horizontal linear complementarity problems, *Linear Algebra Appl.* 223/224:695–715, 1995.

[39] J.C. Willems, Dissipative dynamical systems. Part II: Linear systems with quadratic supply rates, *Arch. Rational Mech. Anal.* 45:352–393, 1972.

[40] P. Wilmott, S.D. Howison, and J.N. Dewynne, *The Mathematics of Financial Derivatives: A Student Introduction*, Cambridge University Press, Cambridge, 1995.

[41] W.M. Wonham, *Linear Multivariable Control: A Geometric Approach*, 3rd ed., Springer Verlag, New York, 1985.

Chapter 12

Exact Controllability of C_0-groups with One-Dimensional Input Operators

Birgit Jacob
Hans Zwart

ABSTRACT *Assume that A generates a C_0-group $T(t)$ on the complex, separable Hilbert space H, and that $b \in D(A^*)'$ is an admissible control operator for the semigroup $T(t)$. It is known that exact controllability of the system defined by A and b implies that every element of the spectrum of A is an eigenvalue. We develop equivalent conditions for exact controllability of the system defined by A and b. These conditions are given in terms of the eigenvalues and eigenvectors of A and the control operator b. Also necessary conditions are included. A necessary condition is that one over the eigenvalues is a sequence in $\ell_{1+\varepsilon}$, $\varepsilon > 0$. If additionally, A is a diagonal operator, then we prove that the conjecture of Russell and Weiss [12] holds.*

12.1 Introduction

This chapter is concerned with exact controllability of infinite-dimensional systems of the following kind,

$$\dot{x}(t) = Ax(t) + bu(t), \tag{12.1}$$

where $A : D(A) \to H$ is the generator of a C_0-group $T(t)$ on the complex, separable Hilbert space H and b is an admissible one-dimensional control operator. The input function u is assumed to be in $\boldsymbol{L}_2^{loc}(0,\infty;\mathbb{C})$. Moreover, we assume that $-A$ generates an exponentially stable C_0-semigroup. We later show that the assumption that the C_0-semigroup generated by $-A$ is exponentially stable is not restrictive.

In general, the class of *admissible control operators for $T(t)$* consists of those elements b for which

$$\int_0^{t_1} T(t_1-\rho)bu(\rho)\,d\rho$$

defines a bounded linear operator from $\boldsymbol{L}_2(0,t_1;\mathbb{C})$ to H. If b is such an admissible control operator for $T(t)$, then for every x_0 and for every $u \in \boldsymbol{L}_2^{loc}(0,\infty;\mathbb{C})$ (12.1) has a unique solution $x(\cdot)$ which is given by

$$x(t) = T(t)x_0 + \int_0^t T(t-\rho)bu(\rho)\,d\rho, \quad t \geq 0.$$

Weiss [14] showed that $x(\cdot)$ lies in H and is continuous.

We say system (12.1) is *exactly controllable* if there is a time $t_0 > 0$ such that for every $x \in H$ there exists an input function $u \in \boldsymbol{L}_2(0,\infty)$ such that

$$x = \int_0^{t_0} T(t_0-\rho)bu(\rho)\,d\rho.$$

Note that system (12.1) cannot be exactly controllable if b is an element of H (see Curtain and Zwart [2, Theorem 4.1.5]); that is, system (12.1) can only be exactly controllable if b is an unbounded operator.

In the literature a variety of necessary and sufficient conditions have been derived that ensure exact controllability of the system (12.1), see, for example, Avdonin and Ivanov [1], Grabowski [5], Grabowski and Callier [6], Jacob and Zwart [9], Rebarber and Weiss [11], and Russell and Weiss [12].

The starting point for our consideration is the following result which partly can be found in Jacob and Zwart [9].

Proposition 12.1. *Assume that the system (12.1) is exactly controllable. Then*

(i) the spectrum of A consists only of countable many points; that is, $\sigma(A) = \{\lambda_n\}_n$;

(ii) every λ_n is an isolated point of the spectrum of A;

(iii) every λ_n is an eigenvalue of A with finite algebraic multiplicity and geometric multiplicity one;

(iv) $\sigma(A^*) = \{\bar{\lambda}_n\}_n$ and every $\bar{\lambda}_n$ is an isolated eigenvalue of A^* with finite algebraic multiplicity and geometric multiplicity one;

(v) $H = \overline{\operatorname{span}_n P^A_{\lambda_n} H} = \overline{\operatorname{span}_n P^{A^*}_{\bar{\lambda}_n} H}$, where $P^A_{\lambda_n} H$ denotes the spectral subspace corresponding to the eigenvalue λ_n of A;

(vi) if the algebraic multiplicity of every eigenvalue is one then

$$H = \overline{\operatorname{span}_n \ker(A-\lambda_n I)} = \overline{\operatorname{span}_n \ker(A^*-\bar{\lambda}_n I)},$$

where ker *denotes the kernel.*

Thus the spectrum of A is of a very special form if the system (12.1) is exactly controllable. We introduce the following assumption.

Assumption 12.1. *A satisfies Parts (i) through (v) of Proposition* 12.1 *and the algebraic multiplicity of every eigenvalue of A is one.*

If the system (12.1) is exactly controllable, then A always satisfies Parts *(i)* through *(v)* of Proposition 12.1. Moreover, if the system (12.1) is exactly controllable or if A satisfies Assumption 12.1, we denote by $\{\lambda_n\}_n$ the spectrum of A, by ϕ_n a normalized eigenvector of A corresponding to the eigenvalue λ_n, and by ψ_n a normalized eigenvector of A^*. One of the main results of this chapter is the following theorem.

Theorem 12.1. *Let A satisfy Assumption* 12.1. *Then the following statements are equivalent.*

(i) System (12.1) *is exactly controllable.*

(ii) For all $x_0 \in H$ there exists a function $\omega \in \boldsymbol{H}_2(\mathbb{C}^+)$ such that

$$\omega(\lambda_n) = -\frac{\langle x_0, \psi_n\rangle}{\langle b, \psi_n\rangle}, \quad n \in \mathbb{N}.$$

(iii) For every $x \in H$ there exists a constant $M_x > 0$ such that

$$\sum_{i,j=1}^{n} c_i \bar{c}_j \left[\frac{M_x^2}{\lambda_i + \overline{\lambda}_j} - \frac{\langle x, \psi_i\rangle \overline{\langle x, \psi_j\rangle}}{\langle b, \psi_i\rangle \overline{\langle b, \psi_j\rangle}}\right] \geq 0,$$

for every $n \in \mathbb{N}$ and all $c_1, \ldots, c_n \in \mathbb{C}$.

Theorem 12.1 shows that the system is exactly controllable if and only if a certain interpolation problem in $\boldsymbol{H}_2(\mathbb{C}^+)$ is solvable. The equivalence of Parts *(ii)* and *(iii)* has already been proved by Duren and Williams [3]. However, for a given system it is not easy to check if Part *(iii)* holds. If we additionally assume that the eigenvalues of A are *properly spaced* (i.e., $\inf_{n\neq m} |\lambda_n - \lambda_m| > 0$) then we derive simple checkable equivalent conditions for exact controllability of the system $\Sigma(A, b)$; see Theorem 12.2. In Section 12.5, we give an example which shows that in general the eigenvalues of an exactly controllable system are not properly spaced. A sequence $\{\psi_n\}_n \subset H$ is called a *Bessel sequence*, if

$$\sum_{n\in\mathbb{N}} |\langle x, \psi_n\rangle|^2 \leq M\|x\|^2, \quad x \in H, \tag{12.2}$$

for some constant $M > 0$ and $\{\psi_n\}_n \subset H$ is called a *Riesz basis* of H, if $\overline{\operatorname{span}_{n\in\mathbb{N}}\psi_n} = H$ and

$$m\sum_{n\in\mathbb{N}} |\langle x, \psi_n\rangle|^2 \leq \|x\|^2 \leq M\sum_{n\in\mathbb{N}} |\langle x, \psi_n\rangle|^2, \quad x \in H, \tag{12.3}$$

for some constants $m, M > 0$.

Theorem 12.2. *Assuming that A satisfies Assumption* 12.1 *and that the eigenvalues of A are properly spaced, the following statements are equivalent.*

(i) System (12.1) *is exactly controllable.*

(ii) $\sum_{n\in\mathbb{N}} \left|\frac{\langle x, \psi_n\rangle}{\langle b, \psi_n\rangle}\right|^2 < \infty$ *for every* $x \in H$.

(iii) $0 < \inf_{n\in\mathbb{N}} |\langle b, \psi_n\rangle| \le \sup_{n\in\mathbb{N}} |\langle b, \psi_n\rangle| < \infty$ *and* $\{\psi_n\}_n$ *forms a Bessel sequence in* H.

(iv) $0 < \inf_{n\in\mathbb{N}} |\langle b, \psi_n\rangle| \le \sup_{n\in\mathbb{N}} |\langle b, \psi_n\rangle| < \infty$ *and* $\{\psi_n\}_n$ *forms a Riesz basis of* H.

If the eigenvectors of A^* form a Riesz basis, then we are able to prove further equivalent conditions for exact controllability. Especially, we show that the conjecture of Russell and Weiss [12] holds and that the result of Avdonin and Ivanov [1, p. 164] can be extended to this situation. This result is formulated in the following theorem.

Theorem 12.3. *Assume that A satisfies Assumption* 12.1 *and that* $\{\psi_n\}_n$ *is a Riesz basis. Then the following statements are equivalent.*

(i) $\Sigma(A, b)$ *is exactly controllable.*

(ii) There is an $M > 0$ *such that for any* $s \in \mathbb{C}^+$ *and any* $x \in D(A^*)$

$$\frac{1}{|\mathrm{Re}\, s|^2}\|(sI - A^*)x\|^2 + \frac{1}{|\mathrm{Re}\, s|}|\langle b, x\rangle|^2 \ge M\|x\|^2. \tag{12.4}$$

(iii) $\inf_{n\neq m} |\lambda_n - \lambda_m| > 0$ *and* $0 < \inf_{n\in\mathbb{N}} |\langle b, \psi_n\rangle| \le \sup_{n\in\mathbb{N}} |\langle b, \psi_n\rangle| < \infty$.

(iv) $\{\langle b, \psi_n\rangle e^{-\overline{\lambda}_n \cdot}\}_{n\in\mathbb{N}}$ *is a Riesz basis of the closure of its span in* $\boldsymbol{L}_2(0, t_0)$ *for some* $t_0 > 0$.

The equivalence of Parts *(i)* and *(iv)* have already been proved by Avdonin and Ivanov [1, p. 164] if the eigenvectors of A form an orthonormal basis. Moreover, Russell and Weiss [12] proved the equivalence of Parts *(i)* and *(ii)* if A is a bounded operator or if the eigenvectors form an orthonormal basis, $\mathrm{Re}\,\lambda_n = \delta$, and $\inf_{n\neq m} |\lambda_n - \lambda_m| > 0$ holds. Note, that Part *(iii)* is the most easily verifiable condition. We conclude this chapter with the following easily checkable necessary condition. This condition even holds without the assumption that the algebraic multiplicity of each eigenvalue λ_n is one.

Theorem 12.4. *Assuming that the system* (12.1) *is exactly controllable, we have*

$$\sum_{n\in\mathbb{N}} \frac{k_n}{|\lambda_n|^{1+\varepsilon}} < \infty, \quad \varepsilon > 0.$$

Here k_n denote the algebraic multiplicity of the eigenvalue λ_n.

This theorem has already been proved by Rebarber and Weiss [11] for a different class of systems. In Rebarber and Weiss [11] the input space is only required to be finite-dimensional and $T(t)$ is allowed to be an exponentially stable C_0-semigroup, but the eigenvectors of A have to form a Riesz basis of H.

The organization of the chapter is as follows. In the next two sections we introduce the necessary background and notation. The proof of Proposition 12.1 can be found in Section 12.3, while the proofs of Theorems 12.1 through 12.4 are given in Section 12.4. Finally, in Section 12.5 we present an example which shows that there is an exactly controllable system that satisfies Assumption 12.1 but the eigenvalues are not properly spaced.

12.2 System Description

In this section, we describe the general class of systems discussed in this paper. We first need to introduce some notation.

$\mathbb{C}^+$	$\{z \in \mathbb{C} : \mathrm{Re}\,(z) > 0\}$,
H	separable, complex Hilbert space,
$D(A)$	domain of A,
$\sigma(A)$	spectrum of A,
$\rho(A)$	resolvent set of A,
$\ker(A)$	$\{x \in D(A) \mid Ax = 0\}$,
$\mathrm{Im}(A)$	$\{x \in H \mid \exists y \in D(A) : Ay = x\}$,
$\boldsymbol{H}_\infty(\mathbb{C}^+; H)$	set of holomorphic and bounded functions from $\mathbb{C}^+$ to H,
$\boldsymbol{H}_2(\mathbb{C}^+; H)$	set of holomorphic functions $f : \mathbb{C}^+ \to H$ with $\sup_{x>0} \left(\int_{-\infty}^{\infty} \|f(x+iy)\|^2 dy \right)^{1/2} < \infty.$

We deal with infinite-dimensional, time-invariant systems of the following kind

$$\dot{x}(t) = Ax(t) + bu(t). \tag{12.5}$$

Here $A : D(A) \to H$ is the generator of a C_0-group $T(t)$ on H and the input function u is assumed to be in $\boldsymbol{L}_2^{loc}(0, \infty; \mathbb{C})$.

In order to define our class of control operators b we have to introduce some notation. We define the space H_{-1} to be the completion of H with respect to the norm

$$\|x\|_{-1} := \|(\beta I - A)^{-1}x\|$$

and the space H_1 to be $D(A)$ with the norm

$$\|x\|_1 := \|(\beta I - A)x\|,$$

where $\beta \in \rho(A)$. It is easy to verify that H_{-1} and H_1 do not depend on $\beta \in \rho(A)$. Moreover, $\|\cdot\|_1$ is equivalent to the graph norm on $D(A)$, so H_1 is complete. In Weiss [14, Remark 3.4] it is shown that $T(t)$ has a restriction to a C_0-semigroup on H_1 whose generator is the restriction of A to $D(A)$, and $T(t)$ can be extended to a C_0-semigroup on H_{-1} whose generator is an extension of A with domain H. Therefore we get

$$A \in \mathcal{L}(H_1, H) \quad \text{and} \quad A \in \mathcal{L}(H, H_{-1}).$$

H_{-1} equals $D(A^*)'$ the dual of $D(A^*)$, where we have equipped $D(A^*)$ with the graph norm (see Weiss [14]). Following Weiss [14] we introduce admissible control operators for $T(t)$.

Definition 12.1. *For $b \in H_{-1}$ and $t \geq 0$ we define $\mathcal{B}_t : \boldsymbol{L}_2(0,\infty;\mathbb{C}) \to H_{-1}$ by*

$$\mathcal{B}_t u := \int_0^t T(t-\rho) b u(\rho)\, d\rho.$$

Then b is called an admissible control operator for $T(t)$, *if for some (and hence any) $t > 0$, $\mathcal{B}_t \in \mathcal{L}(\boldsymbol{L}_2(0,\infty;\mathbb{C}), H)$.*

By a solution of (12.1) with initial condition $x(0) = x_0 \in H$ we mean the function defined by the variation of parameters formula

$$x(t) = T(t)x_0 + \int_0^t T(t-\rho) b u(\rho)\, d\rho, \quad t \geq 0. \tag{12.6}$$

Note that the admissibility of b guarantees $x(t) \in H$ for $t \geq 0$ and in Weiss [14] it is shown that x is continuous.

If A generates a C_0-group $T(t)$ on H, $-A$ generates an exponentially stable C_0-semigroup, and b is an admissible control operator for $T(t)$, we denote system (12.1) by $\Sigma(A, b)$.

It is easy to see that b is an admissible control operator for the C_0-semigroup $T_\tau(t)$, $\tau \in \mathbb{R}$, given by

$$T_\tau(t) := e^{\tau t} T(t).$$

We end this section with the following property of admissible control operators for $T(t)$.

Proposition 12.2. *Every system* $\Sigma(A,b)$ *satisfies*

(i) $(\cdot I - A)^{-1}x \in \boldsymbol{L}_2(i\mathbb{R};H)$ *for every* $x \in H$*;*

(ii) $(\cdot I - A)^{-1}b \in \boldsymbol{L}_\infty(i\mathbb{R};H)$.

Proof:

(i) We choose $x \in H$ arbitrarily. The exponential stability of $T(-t)$ implies $T(-\cdot)x \in \boldsymbol{L}_2(0,\infty;H)$. Since the Laplace transform of $T(-\cdot)x$ is given by $(\cdot I + A)^{-1}x$, the Paley–Wiener Theorem (Curtain and Zwart [2, Theorem A.6.21]) shows

$$(\cdot I + A)^{-1}x \in \boldsymbol{H}_2(\mathbb{C}^+;H).$$

Thus $(\cdot I + A)^{-1}x \in \boldsymbol{L}_2(i\mathbb{R};H)$ or equivalently $(\cdot I - A)^{-1}x \in \boldsymbol{L}_2(i\mathbb{R};H)$.

(ii) It is easy to see that b also is an admissible control operator for the semigroup $T(-t)$. Weiss [15] has proved that

$$(\cdot I + A)^{-1}b \in \boldsymbol{H}_\infty(\mathbb{C}^+;H).$$

Thus $(\cdot I + A)^{-1}b \in \boldsymbol{L}_\infty(i\mathbb{R};H)$ or equivalently we get $(\cdot I - A)^{-1}b \in \boldsymbol{L}_\infty(i\mathbb{R};H)$.

□

12.3 Exact Controllability

In this section we recall the notion of exact controllability and we present basic properties of exactly controllable systems. In particular, the proof of Proposition 12.1 is given in this section.

Definition 12.2. *We call the system* $\Sigma(A,b)$ exactly controllable in time $t_0 > 0$ *if for every* $x \in H$ *there exists an input function* $u \in \boldsymbol{L}_2(0,\infty)$ *such that*

$$x = \int_0^{t_0} T(t_0 - \rho)bu(\rho)\,d\rho.$$

We call the system $\Sigma(A,b)$ exactly controllable (in finite time) *if there is a time* $t_0 > 0$ *such that the system* $\Sigma(A,b)$ *is exactly controllable in time* $t_0 > 0$.

Since $\Sigma(A,b)$ is exactly controllable in time t_0 if and only if $\Sigma(A+\delta I,b)$ is exactly controllable in time t_0 for all real δ, the assumption that $-A$

generates an exponentially stable C_0-semigroup is not restrictive. We start with the proof of Proposition 12.1.

Proposition 12.3. *If $\Sigma(A,b)$ is exactly controllable, there exists an operator $K \in \mathcal{L}(D(A^*)', D(A)')$ such that $\Sigma(A^*, Kb)$ is exactly controllable.*

Proof: The proof of this theorem is based on the following result, which was first proved by Grabowski [5]. He showed that b is an admissible control operator for $T(t)$ and $\Sigma(A,b)$ is exactly controllable in finite time if and only if there exists an operator $L \in \mathcal{L}(H)$ such that L is self–adjoint, coercive, and a solution of the Lyapunov equation with terms in $\mathcal{L}(D(A^*), D(A^*)')$,

$$AL + LA^* = -bb^*. \tag{12.7}$$

Thus by the assumptions of the theorem there exists such an operator L. We now define the operator $K \in \mathcal{L}(D(A^*)', D(A)')$ by

$$K = A^* L^{-1} A^{-1}.$$

Multiplying equation (12.7) from the left with K and from the right with K^* we get the following Lyapunov equation with terms in $\mathcal{L}(D(A), D(A)')$,

$$A^* L^{-1} + L^{-1} A = -(Kb)(Kb)^*.$$

Again using Grabowski's result, the theorem is proved. □

We are now in the position to prove Proposition 12.1.

Proof: Parts *(i)* through *(iv)* and $H = \overline{\text{span}_{\lambda_n} P^A_{\lambda_n} H}$ have already been proved by Jacob and Zwart [9]. Moreover, Part *(vi)* follows directly from Part *(v)*. Thus it remains to prove $H = \overline{\text{span}_n P^{A^*}_{\bar{\lambda}_n} H}$. By Lemma 12.3 there exists an operator $K \in \mathcal{L}(D(A^*)', D(A)')$ such that $\Sigma(A^*, Kb)$ is exactly controllable. Now Jacob and Zwart [9] show $H = \overline{\text{span}_n P^{A^*}_{\bar{\lambda}_n} H}$, which completes the proof. □

Next, we present a series of propositions which are useful for the proof of our main results. However, we stated these results separately since they are also of independent interest.

Proposition 12.4. *If the system $\Sigma(A,b)$ is exactly controllable in time $t_0 > 0$, then the system $\Sigma(A,b)$ is exactly controllable in time t_1, $t_1 > t_0$.*

Proof: Let $t_1 > t_0$ and $x \in H$. By assumption there exists an input function $\tilde{u} \in \boldsymbol{L}_2(0, t_0)$ such that

$$T(t_1 - t_0)^{-1} x = \int_0^{t_0} T(t_0 - \rho) b \tilde{u}(\rho)\, d\rho.$$

Defining $u \in \boldsymbol{L}_2(0, t_1)$ by $u(t) = \tilde{u}(t)$ if $t \in [0, t_0)$ and $u(t) = 0$; otherwise, we get

$$x = \int_0^{t_1} T(t_1 - \rho) b u(\rho)\, d\rho.$$

□

In Jacob and Zwart [9] the following equivalent conditions for exact controllability are proved.

Proposition 12.5. *The following statements are equivalent.*

(i) The system $\Sigma(A, b)$ is exactly controllable.

(ii) For every $x_0 \in H$ there exists an input function $u \in \boldsymbol{L}_2(0, \infty)$ such that the solution x as given by (12.6) is in $\boldsymbol{L}_2(0, \infty; H)$.

(iii) For every $x_0 \in H$ there exist $\xi \in \boldsymbol{H}_2(\mathbb{C}^+; H)$ and $\omega \in \boldsymbol{H}_2(\mathbb{C}^+; \mathbb{C})$ such that

$$x_0 = (sI - A)\xi(s) - b\omega(s), \quad s \in \mathbb{C}^+. \tag{12.8}$$

Proposition 12.6. *Assume that the system $\Sigma(A, b)$ is exactly controllable. Then we have*

(i) $0 < \inf_{n \in \mathbb{N}} |\langle b, \psi_n \rangle| \le \sup_{n \in \mathbb{N}} |\langle b, \psi_n \rangle| < \infty$.

(ii) $\langle \phi_n, \psi_m \rangle = 0$ *if* $n \neq m$.

(iii) Let $x_0 \in H$ and let $\xi_{x_0} \in \boldsymbol{H}_2(\mathbb{C}^+; H)$ and $\omega_{x_0} \in \boldsymbol{H}_2(\mathbb{C}^+; \mathbb{C})$ satisfy (12.8). Then we have

$$\omega_{x_0}(\lambda_n) = -\frac{\langle x_0, \psi_n \rangle}{\langle b, \psi_n \rangle}, \quad n \in \mathbb{N}.$$

Proof:

(i) This follows from Jacob and Zwart [9].

(ii) The calculation

$$(\lambda_m - \lambda_n)\langle \phi_n, \psi_m \rangle = \langle \phi_n, A^* \psi_m \rangle - \langle A\phi_n, \psi_m \rangle = 0$$

shows $\langle \phi_n, \psi_m \rangle = 0$ if $n \neq m$.

(iii) By Part *(i)* we have $\langle b, \psi_n \rangle \neq 0$ for every $n \in \mathbb{N}$. We choose $n \in \mathbb{N}$ arbitrarily. Choosing $s := \lambda_n$ and taking the inner product of (12.8) with ψ_n we get

$$\begin{aligned} \langle x_0, \psi_n \rangle &= \langle \xi_{x_0}(\lambda_n), (\overline{\lambda}_n I - A^*)\psi_n \rangle - \langle b, \psi_n \rangle \omega_{x_0}(\lambda_n) \\ &= -\langle b, \psi_n \rangle \omega_{x_0}(\lambda_n), \end{aligned}$$

and thus $\omega_{x_0}(\lambda_n) = -\langle x_0, \psi_n \rangle / \langle b, \psi_n \rangle$.

□

Proposition 12.7. *Assume that the system $\Sigma(A,b)$ is exactly controllable. Let $\xi_{\phi_n} \in \boldsymbol{H}_2(\mathbb{C}^+;H)$ and $\omega_{\phi_n} \in \boldsymbol{H}_2(\mathbb{C}^+)$ satisfy*

$$\phi_n = (\lambda I - A)\xi_{\phi_n}(\lambda) - b\omega_{\phi_n}(\lambda), \quad \lambda \in \mathbb{C}^+. \tag{12.9}$$

Then we have $\omega_{\phi_n}^{(l)}(\lambda_m) = 0$ if $n \neq m$, $l \in \{0, \dots, k_m - 1\}$, and $\omega_{\phi_n} \not\equiv 0$. Here k_m denotes the algebraic multiplicity of the eigenvalue λ_m.

Proof: Let $n, m \in \mathbb{N}$ with $n \neq m$. By the definition of the algebraic multiplicity of λ_m, there exists an element $\tilde{\psi}_m \in \ker((\overline{\lambda}_m I - A^*)^{k_m}) \setminus \ker((\overline{\lambda}_m I - A^*)^{k_m - 1})$. Moreover, since $(\overline{\lambda}_n I - A^*)^{k_m-1}\tilde{\psi}_m \in (\ker(\overline{\lambda}_n I - A^*))\backslash\{0\}$ and $\ker(\overline{\lambda}_n I - A^*) = \{\alpha\psi_m \mid \alpha \in \mathbb{C}\}$, we can choose $\tilde{\psi}_m$ in such a way that $(\overline{\lambda}_n I - A^*)^{k_m-1}\tilde{\psi}_m = \psi_m$.

We now prove by induction over $l \in \{0, \dots, k_m - 1\}$ that

$$\langle \phi_n, (\overline{\lambda}_m I - A^*)^l \tilde{\psi}_m \rangle = 0 \tag{12.10}$$

holds. For $l := k_m - 1$, equation (12.10) is proved by Proposition 12.6. We now assume that the statement is true for $l+1 \in \{1, \dots, k_m - 1\}$. Then the calculation

$$\begin{aligned} &\lambda_n \langle \phi_n, (\overline{\lambda}_m I - A^*)^l \tilde{\psi}_m \rangle \\ &= \langle A\phi_n, (\overline{\lambda}_m I - A^*)^l \tilde{\psi}_m \rangle \\ &= -\langle \phi_n, (\overline{\lambda}_m I - A^*)^{l+1} \tilde{\psi}_m \rangle + \lambda_m \langle \phi_n, (\overline{\lambda}_m I - A^*)^l \tilde{\psi}_m \rangle \\ &= \lambda_m \langle \phi_n, (\overline{\lambda}_m I - A^*)^l \tilde{\psi}_m \rangle \end{aligned}$$

shows that the statement holds for l.

Next, we prove by induction over $l \in \{0, \dots, k_m - 1\}$ that

$$\begin{aligned} &\omega_{\phi_n}^{(l)}(\lambda_m) = 0, \\ &\qquad \phi_n = (-1)^l \frac{1}{l!}(\lambda_m I - A)^{l+1} \xi_{\phi_n}^{(l)}(\lambda_m) \end{aligned}$$

hold. First, we consider $l := 0$. Proposition 12.6 *(iii)* implies $\omega_{\phi_n}(\lambda_m) = 0$, and thus equation (12.9) shows

$$\phi_n = (\lambda_m I - A)\xi_{\phi_n}(\lambda_m).$$

Therefore the statement holds for $l = 0$. Assume that the statement is true for $l - 1 \in \{0, \dots, k_m - 2\}$. Differentiating equation (12.9) l times, we obtain

$$0 = (\lambda I - A)\xi_{\phi_n}^{(l)}(\lambda) + l\xi_{\phi_n}^{(l-1)}(\lambda) - b\omega_{\phi_n}^{(l)}(\lambda).$$

Choosing $\lambda := \lambda_m$, multiplying both sides with $(\lambda_m I - A)^l$ and using that the statement holds for $l - 1$, we get

$$l!(-1)^l \phi_n = (\lambda_m I - A)^{l+1} \xi_{\phi_n}^{(l)}(\lambda_m) - (\lambda_m I - A)^l b\omega_{\phi_n}^{(l)}(\lambda_m). \tag{12.11}$$

Taking the inner product of (12.11) with $(\overline{\lambda}_m I - A^*)^{k_m-l-1}\tilde{\psi}_m$ we obtain with (12.10) that

$$0 = \langle (\lambda_m I - A)\xi_{\phi_n}^{(l)}(\lambda_m), \psi_m\rangle - \langle b, \psi_m\rangle \omega_{\phi_n}^{(l)}(\lambda_m) = -\langle b, \psi_m\rangle \omega_{\phi_n}^{(l)}(\lambda_m),$$

and therefore $\omega_{\phi_n}^{(l)}(\lambda_m) = 0$. Equation (12.11) now implies that the statement holds for l.

Using (12.9) the assumption $\omega_{\phi_n} \equiv 0$ would imply $(\lambda I - A)^{-1}\phi_n \in \boldsymbol{H}_2(\mathbb{C}^+; H)$. However, since ϕ_n is an eigenvector we have that $(\lambda I - A)^{-1}\phi_n = (1/(\lambda - \lambda_n))\phi_n$. Since λ_n has real part larger than zero, this function is not in $\boldsymbol{H}_2(\mathbb{C}^+; H)$ and so $\omega_{\phi_n} \not\equiv 0$. □

12.4 Proofs of the Main Results

We start with the proof of Theorem 12.1.

Proof:

(i)⇒(ii) This follows from Propositions 12.6 and 12.5.

(ii)⇒(i) We choose $x \in H$. By Part *(ii)* there exists a function $\omega_x \in \boldsymbol{H}_2(\mathbb{C}^+; \mathbb{C})$ such that

$$\omega_x(\lambda_n) = -\frac{\langle x, \psi_n\rangle}{\langle b, \psi_n\rangle}, \quad n \in \mathbb{N}.$$

We define the function $\xi_x : \mathbb{C}^+ \backslash \{\lambda_n\}_{n\in\mathbb{N}} \to H$ by

$$\xi_x(s) := (sI - A)^{-1}x + (sI - A)^{-1}b\omega_x(s).$$

For $n \in \mathbb{N}$ we have

$$\langle \xi_x(s), \psi_n\rangle = (s - \lambda_n)^{-1}\left[\langle x, \psi_n\rangle + \omega_x(s)\langle b, \psi_n\rangle\right].$$

Since $\omega_x \in \boldsymbol{H}_2(\mathbb{C}^+; \mathbb{C})$ and $\omega_x(\lambda_n) = -\langle x, \psi_n\rangle/\langle b, \psi_n\rangle$, it is easy to see that

$$\langle \xi_x(\cdot), \psi_n\rangle \in \boldsymbol{H}_2(\mathbb{C}^+; \mathbb{C}).$$

Furthermore, the property $\omega_x|_{i\mathbb{R}} \in \boldsymbol{L}_2(i\mathbb{R}; \mathbb{C})$ together with Proposition 12.2 proves $\xi_x \in \boldsymbol{L}_2(i\mathbb{R}; H)$. Thus it is the Fourier transform of a function $x \in \boldsymbol{L}_2(\mathbb{R}; H)$. For every $n \in \mathbb{N}$, we have that the Fourier transform of $\langle x(\cdot), \psi_n\rangle$ is $\langle \xi_x(\cdot), \psi_n\rangle$. Since the latter is in $\boldsymbol{H}_2(\mathbb{C}^+; \mathbb{C})$, we have that $\langle x(\cdot), \psi_n\rangle$ is zero almost everywhere on $(-\infty, 0)$. From $\overline{\operatorname{span}_{n\in\mathbb{N}}\psi_n} = H$ and the countability of $\{\psi_n\}_n$ we conclude that $x(\cdot)$ must be zero almost everywhere on $(-\infty, 0)$, which implies that $\xi_x \in \boldsymbol{H}_2(\mathbb{C}^+; \mathbb{C})$. By the definition of ξ_x we have

$$x = (sI - A)\xi_x(s) - b\omega_x(s), \quad s \in \mathbb{C}^+.$$

Thus Proposition 12.5 shows that the system $\Sigma(A, b)$ is exactly controllable.

(ii)⇔(iii) We define $\{\rho_n\}_n \subset D$, $D := \{z \in \mathbb{C} \mid |z| < 1\}$ by

$$\rho_n := \frac{\lambda_n - 1}{\lambda_n + 1}, \quad n \in \mathbb{N}.$$

By Hoffman [7, p. 130] Part *(ii)* is equivalent to: for every $x \in H$ there exists a function $g \in \boldsymbol{H}_2(D;\mathbb{C})$ ($\boldsymbol{H}_2(D;\mathbb{C})$ denotes the Hardy space on the disk), with

$$g(\rho_n) = -\frac{\langle x, \psi_n\rangle}{\langle b, \psi_n\rangle} \frac{1}{1-\rho_n}, \quad n \in \mathbb{N}.$$

Now Duren and Williams [3] show that this is equivalent to the following statement: For every $x \in H$ there exists a constant $M_x > 0$ such that

$$\sum_{i,j=1}^{n} c_i \overline{c}_j \left[\frac{M_x^2}{1-\rho_i\overline{\rho}_j} - \frac{\langle x, \psi_i\rangle \overline{\langle x, \psi_j\rangle}}{\langle b, \psi_i\rangle \overline{\langle b, \psi_j\rangle}} \frac{1}{(1-\rho_i)(1-\overline{\rho}_j)} \right] \geq 0 \tag{12.12}$$

for every $n \in \mathbb{N}$ and all $c_1, \ldots, c_n \in \mathbb{C}$. Inserting the definition of ρ_n in (12.12) we get that this statement is equivalent to Part *(iii)*. □

The following lemmas are used in the proof of Theorem 12.2.

Lemma 12.1. *Let $\{\lambda_n\}_n$ be a sequence of points in $\mathbb{C}^+$ that satisfies $0 < \inf_{n\in\mathbb{N}} \operatorname{Re}(\lambda_n) \leq \sup_{n\in\mathbb{N}} \operatorname{Re}(\lambda_n) < \infty$. Then the following statements are equivalent.*

(i) $\inf_{n\neq m} |\lambda_n - \lambda_m| > 0$.

(ii) $\inf_{n\neq m} \left|(\lambda_n - \lambda_m)/(\lambda_n + \overline{\lambda}_m)\right| > 0$.

(iii) *$\{\lambda_n\}_n$ is an $\boldsymbol{H}_\infty$-interpolation sequence; that is, the operator T defined by $T(\omega) = \{\omega(\lambda_n)\}_n$ is a bounded surjective mapping from $\boldsymbol{H}_\infty(\mathbb{C}^+)$ to ℓ_∞.*

(iv) *The operator T_2 defined by $T_2(\omega) = \{\omega(\lambda_n)\}_n$ is a bounded surjective mapping from $\boldsymbol{H}_2(\mathbb{C}^+)$ to ℓ_2.*

Proof: In Garnett [4, p. 287] it is shown that Part *(iii)* is equivalent to

$$\inf_{m\in\mathbb{N}} \prod_{n\neq m} \left| \frac{\lambda_n - \lambda_m}{\lambda_n + \overline{\lambda}_m} \right| > 0, \tag{12.13}$$

and in Avdonin and Ivanov [1, p. 53] it is shown that Part *(i)*, Part *(ii)* and (12.13) are equivalent. Thus Parts *(i)* through *(iii)* are equivalent. Finally, the equivalence of Parts *(iii)* and *(iv)* has been proved in Garnett [4, p. 316]. □

If U, H are complex, separable Hilbert spaces, we denote by $\mathcal{GL}(U,H)$ the space of bounded invertible operators from U to H. We define by $\mathbb{L} \in$

$\mathcal{GL}(\boldsymbol{L}_2(0,\infty), \boldsymbol{H}_2(\mathbb{C}^+;\mathbb{C}))$ the Laplace transform. Furthermore, we define $\hat{u} := \mathbb{L}u$, $u \in \boldsymbol{L}_2(0,\infty)$, and $\check{\omega} := \mathbb{L}^{-1}\omega$, $\omega \in \boldsymbol{H}_2(\mathbb{C}^+,\mathbb{C})$. If $u \in \boldsymbol{L}_2(0,t_0)$, then $\hat{u}$ is the Laplace transform of u extended with zero after t_0.

Lemma 12.2. *Assume that the system $\Sigma(A,b)$ is exactly controllable in time t_0 and that A satisfies Assumption 12.1. We define $\mathbb{B}_{t_0} \in \mathcal{L}(\boldsymbol{L}_2(0,t_0), H)$ by*

$$\mathbb{B}_{t_0}u := -\int_0^{t_0} T(-\rho)bu(\rho)\,d\rho, \quad u \in \boldsymbol{L}_2(0,t_0).$$

Then for $U \subset \mathbb{L}_2(0,t_0)$ defined by $U = (\ker \mathbb{B}_{t_0})^\perp$, we have that $\mathbb{B}_{t_0} \in \mathcal{GL}(U,H)$ and

$$\hat{u}(\lambda_n) = -\frac{\langle \mathbb{B}_{t_0}u, \psi_n\rangle}{\langle b, \psi_n\rangle}, \quad n \in \mathbb{N}, u \in U.$$

Proof: We define $W(x) \subseteq \boldsymbol{L}_2(0,t_0)$, $x \in H$, by

$$W(x) := \{u \in \boldsymbol{L}_2(0,t_0) \mid x = \mathbb{B}_{t_0}u\}, \quad x \in H.$$

The exact controllability in time t_0 of system $\Sigma(A,b)$ implies $W(x) \neq \emptyset$ for every $x \in H$. Moreover, $W(x)$ is affine linear for every $x \in H$. Therefore, for every $x \in H$ there exists a unique $u_x \in W(x)$ with minimal norm. Moreover, $\{u_x\} = W(0)^\perp \cap W(x)$. Defining $U := W(0)^\perp = (\ker \mathbb{B}_{t_0})^\perp$ we have $\mathbb{B}_{t_0} \in \mathcal{GL}(U,H)$. We choose $u \in U$ arbitrarily and let u be extended to $\boldsymbol{L}_2(0,\infty)$ by defining $u(\rho) = 0$ if $\rho > t_0$. Then the state corresponding to the initial condition $x_0 := \mathbb{B}_{t_0}u$ and the input function u is given by

$$x(t) = T(t)x_0 + \int_0^t T(t-\rho)bu(\rho)\,d\rho, \quad t \geq 0. \tag{12.14}$$

$\mathbb{B}_{t_0}u = x_0$ now implies $x(t) = 0$ for $t \geq t_0$. Thus $x \in \boldsymbol{L}_2(0,\infty;H)$ and $\hat{x} \in \boldsymbol{H}_2(\mathbb{C}^+;H)$. Taking the Laplace transform of equation (12.14) we get

$$\hat{x}(\lambda) = (\lambda I - A)^{-1}x_0 + (\lambda I - A)^{-1}b\hat{u}(\lambda), \quad \text{for } \operatorname{Re}\lambda \geq \omega, \tag{12.15}$$

for some $\omega > 0$. Multiplying equation (12.15) by $(\lambda I - A)$ we get

$$x_0 = (\lambda I - A)\hat{x}(\lambda) - b\hat{u}(\lambda), \quad \lambda \in \mathbb{C}^+.$$

Thus Proposition 12.6 shows

$$\hat{u}(\lambda_n) = -\frac{\langle x_0, \psi_n\rangle}{\langle b, \psi_n\rangle}, \quad n \in \mathbb{N}.$$

□

Lemma 12.3. *Assume that the system $\Sigma(A,b)$ is exactly controllable in time t_0, that A satisfies Assumption 12.1, and $\inf_{n\neq m}|\lambda_n - \lambda_m| > 0$. Let $U \subset \boldsymbol{L}_2(0,t_0)$ be the subspace given by Lemma 12.2. We define $\hat{U} \subset \boldsymbol{H}_2(\mathbb{C}^+;\mathbb{C})$ by*

$$\hat{U} := \{\omega \in \boldsymbol{H}_2(\mathbb{C}^+;\mathbb{C}) \mid \check{\omega}|_{[0,t_0]} \in U \text{ and } \check{\omega}|_{(t_0,\infty)} \equiv 0\}.$$

Then the mapping T_2, defined by

$$T_2(\omega) = \{\omega(\lambda_n)\}_n, \quad \omega \in \hat{U},$$

satisfies $T_2 \in \mathcal{GL}\left(\hat{U}, \ell_{2,H}\right)$, *where* $\ell_{2,H} = \left\{\{-\langle x, \psi_n\rangle / \langle b, \psi_n\rangle\}_n \mid x \in H\right\}$ *is equipped with the* ℓ_2*-norm.*

Proof: Since $\mathbb{L}$ is invertible in $\mathcal{L}(\boldsymbol{L}_2(0,\infty), \boldsymbol{H}_2(\mathbb{C}^+;\mathbb{C}))$ and U is a closed subspace of $\boldsymbol{L}_2(0,t_0)$, we get that $\hat{U}$ is a closed subspace of $\boldsymbol{H}_2(\mathbb{C}^+;\mathbb{C})$. By Lemma 12.1 we get $T_2 \in \mathcal{L}(\hat{U}, \ell_2)$. We choose $\omega \in \hat{U}$ arbitrarily. Defining $x \in H$ as $x := \mathbb{B}_{t_0}\check{\omega}$, then by Lemma 12.2 we have that

$$\omega(\lambda_n) = -\frac{\langle x, \psi_n\rangle}{\langle b, \psi_n\rangle}, \quad n \in \mathbb{N}.$$

Thus T_2 is a linear and bounded operator from $\hat{U}$ to $\ell_{2,H}$ and $\ell_{2,H}$ is a subspace of ℓ_2. In order to prove that T_2 is injective, we assume that there exist $\omega_1, \omega_2 \in \hat{U}$, $\omega_1 \neq \omega_2$, such that $\omega_1(\lambda_n) = \omega_2(\lambda_n)$ for every $n \in \mathbb{N}$. Defining $x_1, x_2 \in H$ as $x_i = \mathbb{B}_{t_0}\check{\omega}_i$, $i = 1, 2$, by Lemma 12.2 we have that

$$\frac{\langle x_1, \psi_n\rangle}{\langle b, \psi_n\rangle} = \frac{\langle x_2, \psi_n\rangle}{\langle b, \psi_n\rangle}, \quad n \in \mathbb{N}.$$

This would imply $\langle x_1 - x_2, \psi_n\rangle = 0$ for every $n \in \mathbb{N}$ and thus $\overline{\operatorname{span}_{n\in\mathbb{N}}\psi_n} = H$ proves $x_1 = x_2$. Since the Laplace transform is bijective from U to $\hat{U}$ and $\mathbb{B}_{t_0}$ is bijective from U to H we get $\omega_1 = \omega_2$.

In order to prove that T_2 is surjective, we choose $x \in H$ arbitrarily. Since the system is exactly controllable, by Lemma 12.2 there exists a $u \in U$ such that $x = \mathbb{B}_{t_0}u$ and

$$\hat{u}(\lambda_n) = -\frac{\langle x, \psi_n\rangle}{\langle b, \psi_n\rangle}, \quad n \in \mathbb{N}.$$

Therefore, $T_2 : \hat{U} \longrightarrow \ell_{2,H}$ is linear, continuous, and bijective.

Now it remains to prove that $\ell_{2,H}$ is a closed subspace of ℓ_2. Then the open mapping theorem implies that

$$T_2 \in \mathcal{GL}(\hat{U}, \ell_{2,H})$$

holds.

Lemma 12.1 implies that the operator $T_2 \in \mathcal{L}(H_2(\mathbb{C}^+;\mathbb{C}), \ell_2)$ is surjective. Thus the operator $\tilde{T}_2 \in \mathcal{L}(\boldsymbol{H}_2(\mathbb{C}^+;\mathbb{C})/\ker T_2, \ell_2)$, defined by

$$\tilde{T}_2([\omega]) := T_2(\omega), \quad \omega \in H_2(\mathbb{C}^+, \mathbb{C}),$$

is bijective and so the open mapping theorem implies that $\tilde{T}_2^{-1}$ also is a bounded operator. Since T_2 is a linear, bounded, and bijective operator from $\hat{U}$ to $\ell_{2,H}$, we get

$$\hat{U} = \hat{U}/\ker T_2 \cong \ell_{2,H}$$

and so the boundedness of $\tilde{T}_2^{-1}$ shows that $\ell_{2,H}$ is a closed subspace of ℓ_2. □

Now we are in the position to prove Theorem 12.2.

Proof: It is easy to see that Part *(iv)* implies Part *(iii)* and that Part *(iii)* implies Part *(ii)*.

(i)⇒(iv) Proposition 12.6 shows that $0 < \inf_{n\in\mathbb{N}} |\langle b, \psi_n\rangle| \leq \sup_{n\in\mathbb{N}} |\langle b, \psi_n\rangle| < \infty$ holds. Thus it remains to prove that $\{\psi_n\}_n$ forms a Riesz basis of H. Let $U \subset \boldsymbol{L}_2(0, t_0)$ be the subspace given by Lemma 12.2 and we define $\hat{U} \subset \boldsymbol{H}_2(\mathbb{C}^+; \mathbb{C})$ as in Lemma 12.3. Thus $\mathbb{L} \in \mathcal{GL}(U, \hat{U})$. Moreover, by Lemma 12.2 we have $\mathbb{B}_{t_0} \in \mathcal{GL}(U, H)$ and Lemma 12.3 shows $T_2 \in \mathcal{GL}\left(\hat{U}, \ell_{2,H}\right)$. Therefore, the map $S : H \to \ell_{2,H}$, given by

$$Sx := T_2 \mathbb{L} \mathbb{B}_{t_0}^{-1} x = \left\{ -\frac{\langle x, \psi_n\rangle}{\langle b, \psi_n\rangle} \right\}_n,$$

satisfies $S \in \mathcal{GL}(H, \ell_{2,H})$. So there exist constants $m, M > 0$ such that

$$m \sum_{n\in\mathbb{N}} \left| \frac{\langle x, \psi_n\rangle}{\langle b, \psi_n\rangle} \right|^2 \leq \|x\|^2 \leq M \sum_{n\in\mathbb{N}} \left| \frac{\langle x, \psi_n\rangle}{\langle b, \psi_n\rangle} \right|^2, \qquad x \in H.$$

Taking into account that $0 < \inf_{n\in\mathbb{N}} |\langle b, \psi_n\rangle| \leq \sup_{n\in\mathbb{N}} |\langle b, \psi_n\rangle| < \infty$ holds, we get that $\{\psi_n\}_n$ forms a Riesz basis of H.

(ii)⇒(i) We choose $x \in H$ arbitrarily. Since $\{-\langle x, \psi_n\rangle / \langle b, \psi_n\rangle\}_n \in \ell_2$, Lemma 12.1 shows that there exists a function $\omega \in \boldsymbol{H}_2(\mathbb{C}^+)$ such that

$$\omega(\lambda_n) = -\frac{\langle x, \psi_n\rangle}{\langle b, \psi_n\rangle}, \qquad n \in \mathbb{N}.$$

Thus by Theorem 12.1, Part *(i)* holds. □

As a corollary of Theorem 12.2 we get the following result.

Corollary 12.1. *Assume that the system $\Sigma(A, b)$ is exactly controllable and that A satisfies Assumption 12.1. Then the following statements are equivalent.*

(i) $\inf_{n\neq m} |\lambda_n - \lambda_m| > 0$.

(ii) $\inf_{n\in\mathbb{N}} |\langle \phi_n, \psi_n\rangle| > 0$.

(iii) *$\{\phi_n\}_n$ forms a Riesz basis of H.*

(iv) *$\{\psi_n\}_n$ forms a Riesz basis of H.*

(v) *$\{\phi_n\}_n$ as well as $\{\psi_n\}_n$ forms a Riesz basis of H.*

Proof:

(i)⇒(iv) This follows from Theorem 12.2.

(iv)⇒(v) By Proposition 12.6 we get $\langle \phi_n, \psi_m \rangle = 0$ if $n \neq m$. Thus the Riesz basis property of $\{\psi_n\}_{n\in\mathbb{N}}$ implies $\langle \phi_n, \psi_n \rangle \neq 0$ for every $n \in \mathbb{N}$. Therefore we are able to choose constants $\{\alpha_n\}_{n\in\mathbb{N}} \subset \mathbb{C}$ such that

$$\langle \alpha_n \phi_n, \psi_n \rangle = 1, \quad n \in \mathbb{N}.$$

Hence $\{\alpha_n \phi_n\}_n$ and $\{\psi_n\}_n$ are biorthogonal and thus by Avdonin and Ivanov [1, p. 27] $\{\alpha_n \phi_n\}_n$ is also a Riesz basis in H. The Riesz basis property of $\{\psi_n\}_{n\in\mathbb{N}}$ further implies the existence of $m, M > 0$ such that (12.3) holds. Thus

$$m \leq \|\alpha_n \phi_n\|^2 = |\alpha_n|^2 \leq M, \quad n \in \mathbb{N}.$$

Now the boundedness of $|\alpha_n|$ and $|\alpha_n^{-1}|$ together with the fact that the sequence $\{\alpha_n \phi_n\}_n$ is a Riesz basis in H shows that $\{\phi_n\}_n$ forms a Riesz basis of H.

(v)⇒(iii) Trivial.

(iii)⇒(ii) By Proposition 12.6 we get $\langle \phi_n, \psi_m \rangle = 0$ if $n \neq m$. Thus the Riesz basis property of $\{\phi_n\}_{n\in\mathbb{N}}$ implies $\langle \phi_n, \psi_n \rangle \neq 0$ for every $n \in \mathbb{N}$. Therefore we are able to choose constants $\{\alpha_n\}_{n\in\mathbb{N}} \subset \mathbb{C}$ such that

$$\langle \phi_n, \alpha_n \psi_n \rangle = 1, \quad n \in \mathbb{N}.$$

Furthermore, there exist constants $m, M > 0$ such that (12.3) holds with ϕ_n instead of ψ_n. Thus

$$m \leq \|\alpha_n \psi_n\|^2 = |\alpha_n|^2 \leq M, \quad n \in \mathbb{N},$$

and therefore we get

$$|\langle \phi_n, \psi_n \rangle| = \frac{1}{|\alpha_n|} |\langle \phi_n, \alpha_n \psi_n \rangle| = \frac{1}{|\alpha_n|} \geq \frac{1}{\sqrt{M}}, \quad n \in \mathbb{N}.$$

(ii)⇒(i) Jacob and Zwart [9, Propositions 3.6 and 3.9] show that the exact controllability implies the existence of $\xi_{\phi_n} \in \boldsymbol{H}_\infty(\mathbb{C}^+; H)$ and $\omega_{\phi_n} \in \boldsymbol{H}_\infty(\mathbb{C}^+)$ satisfying equation (12.8) and $\|\xi_{\phi_n}\|_{\boldsymbol{H}_\infty}, \|\omega_{\phi_n}\|_{\boldsymbol{H}_\infty} \leq M$ for some constant $M > 0$ independent of $n \in \mathbb{N}$. For $n \neq m$, Proposition 12.6 implies $\omega_{\phi_n}(\lambda_m) = 0$ and thus by (12.8) we obtain $\phi_n = (\lambda_m I - A)\xi_{\phi_n}(\lambda_m)$, $n \neq m$. This shows

$$\langle \phi_n, \psi_n \rangle = \langle (\lambda_m I - A)\xi_{\phi_n}(\lambda_m), \psi_n \rangle = (\lambda_m - \lambda_n)\langle \xi_{\phi_n}(\lambda_m), \psi_n \rangle$$

and therefore

$$|\lambda_m - \lambda_n| = \frac{|\langle \phi_n, \psi_n \rangle|}{|\langle \xi_{\phi_n}(\lambda_m), \psi_n \rangle|} \geq \frac{\delta}{M}, \quad n, m \in \mathbb{N}, n \neq m,$$

for some $\delta > 0$ independent of $n, m \in \mathbb{N}$.

□

The following lemma is useful.

Lemma 12.4.

(i) *$\{\psi_n\}_n$ is a Riesz basis of H if and only if there exist constants $m, M > 0$ such that*

$$m \sum_{i=1}^{N} \|c_i\|^2 \le \left\| \sum_{i=1}^{N} c_i \psi_i \right\|^2 \le M \sum_{i=1}^{N} \|c_i\|^2$$

for every $N \in \mathbb{N}$ and for every $c_1, \dots, c_N \in \mathbb{C}$.

(ii) *Let H_1, H_2 be two complex, separable Hilbert spaces, let $\{\psi_n\}_n$ be a Riesz basis of H_1, and $T \in \mathcal{GL}(H_1, H_2)$. Then $\{T\psi_n\}_n$ is a Riesz basis of H_2.*

The proof of Part *(i)* can be found in Avdonin and Ivanov [1, p. 27] and Part *(ii)* follows immediately from Part *(i)*.

Proposition 12.8. *Let $\{\lambda_n\}_{n\in\mathbb{N}} \subset \mathbb{C}^+$ be a sequence with $0 < \inf_{n\in\mathbb{N}} \operatorname{Re}\lambda_n \le \sup_{n\in\mathbb{N}} \operatorname{Re}\lambda_n < \infty$. Then the following statements are equivalent.*

(i) $\inf |\lambda_n - \lambda_m| > 0$.

(ii) *There exists a $t_0 > 0$ such that $\{e^{-\overline{\lambda}_n \cdot}\}_{n\in\mathbb{N}}$ is a Riesz basis of the closure of $\operatorname{span}_{n\in\mathbb{N}} e^{-\overline{\lambda}_n \cdot}$ in $\boldsymbol{L}_2(0, t_1)$ for every $t_1 \ge t_0$.*

(iii) *There exists a $t_0 > 0$ such that $\{e^{-\overline{\lambda}_n \cdot}\}_{n\in\mathbb{N}}$ is a Riesz basis of the closure of $\operatorname{span}_{n\in\mathbb{N}} e^{-\overline{\lambda}_n \cdot}$ in $\boldsymbol{L}_2(0, t_0)$.*

Proof:

(ii)⇒(iii) Trivial.

(i)⇒(ii) We choose H as a separable Hilbert space with orthonormal basis $\{\phi_n\}_{n\in\mathbb{N}}$ and define $T(t) = \operatorname{diag}(e^{\overline{\lambda}_n t})$ on H. Moreover, we choose $b := \sum_{n\in\mathbb{N}} \phi_n$. Weiss [13] has proved that b is an admissible control operator for $T(t)$ and Theorem 12.2 shows that the system $\Sigma(A, b)$ is exactly controllable in finite time. Moreover, there exists a time $t_0 > 0$ such that the system is exactly controllable in time t_1 for every $t_1 \ge t_0$ (use Proposition 12.4).

We choose $t_1 \ge t_0$ arbitrarily. By Lemma 12.2 we have that $\mathbb{B}_{t_1} \in \mathcal{GL}(U, H)$, with $U = (\ker \mathbb{B}_{t_1})^\perp$. Thus the previous lemma implies that $\{u_n\}_n$, $u_n := \mathbb{B}_{t_1}^{-1}\phi_n$, is a Riesz basis of U. Moreover, Lemma 12.2 together with Proposition 12.6 shows

$$\hat{u}_n(\lambda_m) = \delta_{nm}, \quad n, m \in \mathbb{N}.$$

Defining $f_n(t) := e^{-\overline{\lambda}_n t}$ on $\boldsymbol{L}_2(0,t_1)$ we obtain $\langle u_n, f_m\rangle = \hat{u}_n(\lambda_m) = \delta_{nm}$, $n,m \in \mathbb{N}$, and so if we can show that f_n is an element of U, then this equality shows that $\{f_n\}_n$ is the biorthogonal sequence of $\{u_n\}_n$. Next we can conclude from Avdonin and Ivanov [1, p. 27] that $\{f_n\}_n$ is a Riesz basis of U. Since $U = (\ker \mathbb{B}_{t_1})^\perp$, we have that $f_n \in U$ if and only if $\langle v, f_n\rangle = 0$ for all $v \in \ker \mathbb{B}_{t_1}$. For $n \in \mathbb{N}$ and $v \in \mathbb{B}_{t_1}$ we get

$$\begin{aligned}\langle v, f_n\rangle &= \int_0^{t_1} e^{-\lambda_n \rho} v(\rho)\, d\rho = \int_0^{t_1} \langle T(-\rho)bv(\rho), \phi_n\rangle\, d\rho \\ &= \langle \int_0^{t_1} T(-\rho)bv(\rho)\, d\rho, \phi_n\rangle = 0,\end{aligned}$$

where we have used that $v \in \ker \mathbb{B}_{t_1}$. Thus $f_n \in U$.

(iii)⇒(i) Since $\{e^{-\overline{\lambda}_n \cdot}\}_{n\in\mathbb{N}}$ is a Riesz basis in the closure of its span in $\boldsymbol{L}_2(0,t_0)$, we know that there exists a $c > 0$ such that

$$\|\alpha e^{-\overline{\lambda}_n \cdot} + \beta e^{-\overline{\lambda}_m \cdot}\|^2 = \int_0^{t_0} |\alpha e^{-\overline{\lambda}_n t} + \beta e^{-\overline{\lambda}_m t}|^2 dt \geq c[|\alpha|^2 + |\beta|^2].$$

Now we choose $\alpha = \sqrt{\operatorname{Re}\lambda_n}$ and $\beta = \sqrt{\operatorname{Re}\lambda_m}$; then we see that

$$\begin{aligned}&\int_0^{\infty} |\sqrt{\operatorname{Re}\lambda_n} e^{-\overline{\lambda}_n t} + \sqrt{\operatorname{Re}\lambda_m} e^{-\overline{\lambda}_m t}|^2 dt \\ &\quad \geq \int_0^{t_0} |\sqrt{\operatorname{Re}\lambda_n} e^{-\overline{\lambda}_n t} + \sqrt{\operatorname{Re}\lambda_m} e^{-\overline{\lambda}_m t}|^2 dt \geq c[\operatorname{Re}\lambda_n + \operatorname{Re}\lambda_m].\end{aligned}$$

This last expression is by assumption uniformly bounded away from zero. Now Lemma II.1.20 in Avdonin and Ivanov [1] shows that *(i)* holds. □

Next we prove Theorem 12.3.

Proof:

(iii)⇒(i) This follows from Theorem 12.2.

(i)⇒(ii) See Russell and Weiss [12].

(ii)⇒(iii) First of all we prove $\inf_{n\neq m} |\lambda_n - \lambda_m| > 0$. Let λ_n, λ_m, $n \neq m$, be two eigenvalues of A. Then there exist two numbers $\alpha, \beta \in \mathbb{C}$ such that

$$|\alpha|^2 + |\beta|^2 = 1 \quad \text{and} \quad \overline{\alpha}\langle b, \psi_m\rangle + \overline{\beta}\langle b, \psi_n\rangle = 0.$$

We define

$$x := \alpha\psi_m + \beta\psi_n.$$

Since $\{\psi_k\}$ is a Riesz basis in H, there exists a constant $c > 0$ independent of n and m such that $\|x\|^2 \geq c$. Then for every $s \in \mathbb{C}^+$ we have

$$\begin{aligned} cM &\leq M\|x\|^2 \\ &\leq \frac{1}{|\operatorname{Re} s|^2}\|(sI - A^*)x\|^2 + \frac{1}{|\operatorname{Re} s|}\|\langle b, x\rangle\|^2 \\ &\leq \frac{1}{|\operatorname{Re} s|^2}\|(s-\overline{\lambda}_m)\alpha\psi_m + (s-\overline{\lambda}_n)\beta\psi_n\|^2 + \\ &\qquad \frac{1}{|\operatorname{Re} s|}\|\langle b, \alpha\psi_m + \beta\psi_n\rangle\|^2 \\ &= \frac{1}{|\operatorname{Re} s|^2}\|(s-\overline{\lambda}_m)\alpha\psi_m + (s-\overline{\lambda}_n)\beta\psi_n\|^2. \end{aligned}$$

Choosing $s := \overline{\lambda}_m$ we get

$$C|\lambda_m - \lambda_n|^2 \geq \frac{1}{|\operatorname{Re}\lambda_m|^2}\|(\overline{\lambda}_m - \overline{\lambda}_n)\beta\psi_n\|^2 \geq cM,$$

for some constant $C > 0$ independent of n and m. Thus $\inf_{n\neq m}|\lambda_n - \lambda_m| > 0$.

Choosing in equation (12.4) $x := \psi_n$ and $s := \overline{\lambda}_n$ we obtain in particular that $\inf_{n\in\mathbb{N}}|\langle b, \psi_n\rangle| > 0$. Moreover, $\sup_{n\in\mathbb{N}}|\langle b, \psi_n\rangle| < \infty$ is already implied by the admissibility of b (see Jacob and Zwart [9, Proof of Proposition 6.3]).

(iv)⇒(iii) Since $\{\langle b, \phi_n\rangle e^{-\overline{\lambda}_n \cdot}\}_{n\in\mathbb{N}}$ is a Riesz basis of the closure of its span in $\boldsymbol{L}_2(0, t_0)$ there exist constants $m, M > 0$ such that

$$m \leq \|\langle b, \phi_n\rangle e^{-\overline{\lambda}_n \cdot}\|_{\boldsymbol{L}_2(0,t_0)} \leq M, \quad n \in \mathbb{N} \tag{12.16}$$

(see Lemma 12.4). A simple calculation shows

$$\|\langle b, \phi_n\rangle e^{-\overline{\lambda}_n \cdot}\|_{L_2(0,t_0)} = |\langle b, \psi_n\rangle| \left(\frac{1 - e^{-2\operatorname{Re}\lambda_n t_0}}{2\operatorname{Re}\lambda_n}\right)^{1/2}.$$

Since $0 < \inf \operatorname{Re}\lambda_n \leq \sup \operatorname{Re}\lambda_n < \infty$, we have that

$$0 < \inf_{n\in\mathbb{N}} \left(\frac{1 - e^{-2\operatorname{Re}\lambda_n t_0}}{2\operatorname{Re}\lambda_n}\right)^{1/2} \leq \sup_{n\in\mathbb{N}} \left(\frac{1 - e^{-2\operatorname{Re}\lambda_n t_0}}{2\operatorname{Re}\lambda_n}\right)^{1/2} < \infty.$$

Thus (12.16) shows that $0 < \inf_{n\in\mathbb{N}}|\langle b, \psi_n\rangle| \leq \sup_{n\in\mathbb{N}}|\langle b, \psi_n\rangle| < \infty$. Now it is easy to see that $\{e^{-\overline{\lambda}_n \cdot}\}_{n\in\mathbb{N}}$ is a Riesz basis of the closure of its span in $\boldsymbol{L}_2(0, t_0)$. From Proposition 12.8 we conclude that *(iii)* holds.

(iii)⇒(iv) From Proposition 12.8 it follows that $\{e^{-\overline{\lambda}_n \cdot}\}_{n\in\mathbb{N}}$ is a Riesz basis of the closure of $\operatorname{span}_{n\in\mathbb{N}} e^{-\overline{\lambda}_n \cdot}$ in $\boldsymbol{L}_2(0, t_0)$ for some $t_0 > 0$. This together with $0 < \inf_{n\in\mathbb{N}}|\langle b, \psi_n\rangle|$ and $\sup_{n\in\mathbb{N}}|\langle b, \psi_n\rangle| < \infty$ implies that Part *(iv)* holds. □

We conclude this section with the proof of Theorem 12.4.

Proof: By Lemma 12.2 there exists a function $\omega \in \boldsymbol{H}_2(\mathbb{C}^+, \mathbb{C})$ such that the inverse Laplace transform of ω is an $\boldsymbol{L}_2(0,\infty)$ function with compact support and

$$\omega(\lambda_n) = -\frac{\langle \phi_1, \psi_n \rangle}{\langle b, \psi_n \rangle}, \qquad n \in \mathbb{N}.$$

Since the inverse Laplace transform of ω is an $\boldsymbol{L}_2(0,\infty)$ function with compact support it is easy to see that $\omega \in \boldsymbol{H}_\infty(\mathbb{C}^+;\mathbb{C})$. Thus ω is an entire function of order one and Proposition 12.7 implies

$$\omega^{(l)}(\lambda_n) = 0, \quad n \neq 1, \quad l \in \{0, \ldots, k_n - 1\},$$

and $\omega \neq 0$. Therefore, Lewin [10, Lehrsatz 6, p. 15] shows that for every $\varepsilon > 0$ we have

$$\sum_{n \in \mathbb{N}} \frac{k_n}{|\lambda_n|^{1+\varepsilon}} < \infty.$$

□

12.5 An Example

We now construct a system $\Sigma(A, b)$ that is exactly controllable, but the eigenvalues $\{\lambda_n\}_n$ do not satisfy $\inf_{n \neq m} |\lambda_n - \lambda_m| > 0$.

Example 12.1. Let $\{\rho_n\}_n \subset \mathbb{C}^+$ be given by

$$\rho_{2n} = 1 + in, \qquad n \in \mathbb{N},$$
$$\rho_{2n-1} = 1 + i\left(n + \frac{1}{n}\right), \qquad n \in \mathbb{N}.$$

Clearly $\inf_{n \neq m} |\rho_n - \rho_m| = 0$ and $\inf_{n \neq m} |\rho_{2n} - \rho_{2m}| = \inf_{n \neq m} |\rho_{2n-1} - \rho_{2m-1}| = 1$. Hence, Avdonin and Ivanov [1] prove that $\sum_{n \in \mathbb{N}} \delta_{\rho_{2n}}$ as well as $\sum_{n \in \mathbb{N}} \delta_{\rho_{2n-1}}$ are Carleson measures. For background information on Carleson measures we refer the reader to Garnett [4]. Since the sum of two Carleson measures again is a Carleson measure, we get that $\sum_{n \in \mathbb{N}} \delta_{\rho_n}$ is a Carleson measure.

Let $\Theta \in \boldsymbol{H}_\infty(\mathbb{C}^+)$ be the Blaschke product corresponding to the zeros $\{\rho_n\}_n$; that is,

$$\Theta(s) = \prod_{n \in \mathbb{N}} \frac{|1-\rho_n^2|}{1-\rho_n^2} \frac{s-\rho_n}{s+\bar{\rho}_n}, \qquad s \in \mathbb{C}^+.$$

Let $H = [\Theta \boldsymbol{H}_2(\mathbb{C}^+)]^{\perp}_{\boldsymbol{H}_2(\mathbb{C}^+)}$, let $P_{\boldsymbol{H}_2(\mathbb{C}^+)}$ be the orthogonal projection from $\boldsymbol{L}_2(i\mathbb{R})$ to $\boldsymbol{H}_2(\mathbb{C}^+)$, and let $\tilde{A}$ be the generator of the left shift semigroup $\tilde{T}(t)$ on V; that is,

$$\tilde{T}(t)x = P_{\boldsymbol{H}_2(\mathbb{C}^+)}[e^{t\cdot}x(\cdot)], \quad x \in H,$$

and let $b = \Theta$. In Jacob and Zwart [8] it is shown that b is an admissible control operator for $\tilde{T}(t)$, $\Sigma(\tilde{A}, b)$ is exactly controllable in finite time, $\tilde{A}$ generates an exponentially stable C_0-group on H, and every $-\overline{\rho}_n$, $n \in \mathbb{N}$, is an eigenvalue of $\tilde{A}$ with algebraic multiplicity one. Thus there exists a $\delta > 0$ such that $A := \tilde{A} + \delta I$ is completely unstable. Let $T(t)$ be the C_0-group generated by A. Is it easy to see that b also is an admissible control operator for A and Jacob and Zwart [9, Lemma 8.3] prove that $\Sigma(A, b)$ is exactly controllable in finite time. Since $\{-\bar{\rho}_n + \delta\}$ are eigenvalues of A, the eigenvalues of A are not properly spaced.

References

[1] S.A. Avdonin and S.A. Ivanov, *Families of Exponentials: The Method of Moments in Controllability Problems for Distributed Parameter Systems*, Cambridge University Press, Cambridge, 1995.

[2] R.F. Curtain and H. Zwart, *An Introduction to Infinite-Dimensional Linear Systems Theory*, Texts in Applied Mathematics 21, Springer Verlag, New York, 1995.

[3] P.L. Duren and D.L. Williams, Interpolation problems in function spaces, *J. Functional Analysis* 9:75–86, 1972.

[4] J.B. Garnett, *Bounded Analytic Functions*, Academic Press, New York, NY, 1981.

[5] P. Grabowski, On the spectral-Lyapunov approach to parametric optimization of distributed parameter systems, *IMA J. Math. Control Inf.* 7:317–338, 1990.

[6] P. Grabowski and F.M. Callier, Admissible observation operators, semigroup criteria of admissibility, *Integral Equation Operator Theory* 25:182–198, 1996.

[7] K. Hoffman, *Banach Spaces of Analytic Functions*, Prentice-Hall, Englewood Cliffs, NJ, 1962.

[8] B. Jacob and H. Zwart, Realization of inner functions, Technical Report 1998-17, School of Mathematics, University of Leeds, 1998.

[9] B. Jacob and H. Zwart, Equivalent conditions for stabilizability of infinite-dimensional systems with abmissible control operators, *SIAM J. Control Optimization* 37(5):1419-1455, 1999.

[10] B.J. Lewin, *Nullstellenverteilung ganzer Funktionen*, Akademie Verlag, Berlin, 1962.

[11] R. Rebarber and G. Weiss, Necessary conditions for exact controllability with a finite-dimensional input space, *Syst. Control Letters* 40:217–227, 2000.

[12] D.L. Russell and G. Weiss, A general necessary condition for exact observability, *SIAM J. Control Optimization* 32(1):1–23, 1994.

[13] G. Weiss, Admissibility of input elements for diagonal semigroups on l^2, *Syst. Control Letters* 10:79–82, 1988.

[14] G. Weiss, Admissibility of unbounded control operators, *SIAM J. Control Optimization* 27:527–545, 1989.

[15] G. Weiss, Representation of shift invariant operators on L_2 by H^∞ transfer functions: An elementary proof, a generalization to L^p and a counterexample for L^∞, *Math. Control, Signals Syst.* 4:193–203, 1991.

Chapter 13

Normalized Coprime Factorizations for Strongly Stabilizable Systems

Ruth F. Curtain
Job C. Oostveen

ABSTRACT *The existence of normalized coprime factorizations is established for the class of strongly stabilizable and detectable infinite-dimensional systems $\Sigma(A, B, C, D)$, where A has compact resolvent and generates a C_0- semigroup on a Hilbert space Z, and B, C, D are bounded, finite-rank operators. Explicit formulas are given for the normalized left- and right-coprime factors. With this result, the theory of robust stabilization with respect to normalized coprime factor perturbations can be generalized to this class of systems. An application to models of hybrid flexible structures is given.*

13.1 Introduction

The thesis "strongly stabilizable infinite-dimensional systems" by J.C. Oostveen [9] concerns the class of strongly stabilizable systems $\Sigma(A, B, C, D)$ on a Hilbert space Z, where B, C, and D are bounded operators and A generates a C_0-semigroup $T(t)$ on Z. The novel feature is that it is not assumed that (A, B) is exponentially stabilizable, but only strongly stabilizable.[1] The motivating example is systems of the form $\Sigma(A, B, B^*)$, where A generates a contraction semigroup, which arise in models of flexible structures with collocated actuators and sensors (See [9, Chapter 9] for several examples).

Despite the limitation of only assuming strong stabilizability it was possible to develop a fairly complete theory for strong stabilizability, strong detectability, the existence of solutions to Lyapunov and Riccati equations, and robust dynamic stabilization. One topic that was not satisfactorily resolved in [9] was that of normalized coprime factorizations. While it was possible to establish their existence under reasonable assumptions, it was

---

[1] $T(t)$ is exponentially stable if there exist constants $M \geq 1$ and $\alpha > 0$ such that $\|T(t)\| \leq Me^{\alpha t}$. $T(t)$ is strongly stable if for all $z \in Z$, $\lim_{t\to\infty} T(t)z = 0$.

only possible to give explicit formulae for the special class $\Sigma(A, B, B^*, D)$.

Explicit formulae for normalized coprime factorizations are required in the solution to the robust stabilization problem (see McFarlane and Glover [8] and Chapter 9.4 in Curtain and Zwart [1] for the case of exponentially stabilizable systems). This problem was solved for the strongly stabilizable class $\Sigma(A, B, B^*, D)$ in Oostveen and Curtain [11]. All that is needed to obtain a solution for the more general class $\Sigma(A, B, C, D)$ is explicit formulae for normalized coprime factorizations. In this chapter, we achieve this under reasonable conditions for the case of finite-dimensional input and output spaces. For the case of infinite-dimensional input and output spaces, the problem remains open.

In Section 13.2, we introduce the necessary mathematical background and formulate the problem. In Section 13.3, we propose explicit formulae for normalized coprime factorizations of $\Sigma(A, B, C, D)$ and prove that they are coprime. In Section 13.4, we present a class of hybrid flexible structures that are not in the special class $\Sigma(A, B, B^*, D)$, but which do satisfy the assumptions of our new results.

13.2 Problem Formulation and Mathematical Background

Let Z, U, Y be Hilbert spaces and consider the linear system $\Sigma(A, B, C, D)$, where A generates a C_0-semigroup $T(t)$ on Z, $B \in \mathcal{L}(U, Z)$, $C \in \mathcal{L}(Z, Y)$, and $D \in \mathcal{L}(U, Y)$. The transfer function $G(s) = D + C(sI - A)^{-1}B$ is well defined and holomorphic for $\mathrm{Re}(s) > \alpha$, for some $\alpha \in \mathbb{R}$. For more details on such systems, we refer to Curtain and Zwart [1]. We follow the definition of coprime factorizations from Curtain et al. [3].

Definition 13.1. *Let G be a well-posed transfer function (i.e. G is analytic on some right half-plane). A* left-coprime factorization *of G over $\mathbf{H}_\infty$ is a factorization*

$$G = \tilde{M}^{-1}\tilde{N},$$

where $\tilde{N} \in \mathbf{H}_\infty(\mathcal{L}(U, Y))$, $\tilde{M} \in \mathbf{H}_\infty(\mathcal{L}(Y))$, such that $\tilde{M}(s)$ is invertible on some right half-plane and for which there exist $Y \in \mathbf{H}_\infty(\mathcal{L}(Y, U))$ and $X \in \mathbf{H}_\infty(\mathcal{L}(Y))$ such that for $s \in \mathbb{C}_0^+$,

$$\tilde{M}(s)X(s) - \tilde{N}(s)Y(s) = I. \tag{13.1}$$

The left-coprime factorization is normalized *if, in addition,*

$$\tilde{M}(j\omega)\tilde{M}^*(j\omega) + \tilde{N}(j\omega)\tilde{N}^*(j\omega) = I \qquad \text{for almost all } \omega \in \mathbb{R}. \tag{13.2}$$

A right-coprime factorization *of G over $\mathbf{H}_\infty$ is a factorization*

$$G = NM^{-1},$$

where $N \in \mathbf{H}_\infty(\mathcal{L}(U,Y))$, $M \in \mathbf{H}_\infty(\mathcal{L}(U))$, such that $M(s)$ is invertible on some right half-plane and for which there exist $\tilde{Y} \in \mathbf{H}_\infty(\mathcal{L}(Y,U))$ and $\tilde{X} \in \mathbf{H}_\infty(\mathcal{L}(U))$ such that for $s \in \mathbb{C}_0^+$

$$\tilde{X}(s)M(s) - \tilde{Y}(s)N(s) = I. \tag{13.3}$$

The right-coprime factorization is normalized *if, in addition,*

$$M(j\omega)^* M(j\omega) + N(j\omega)^* N(j\omega) = I \qquad \text{for almost all } \omega \in \mathbb{R}. \tag{13.4}$$

A factorization

$$G = NM^{-1} = \tilde{M}^{-1}\tilde{N}$$

is a doubly coprime factorization *of G over $\mathbf{H}_\infty$ if (M,N) and $(\tilde{M},\tilde{N})$ are a right- and left-coprime factorization of G, respectively, and there exist X, Y, $\tilde{X}$, $\tilde{Y} \in \mathbf{H}_\infty$ such that*

$$\begin{bmatrix} M & Y \\ N & X \end{bmatrix}^{-1} = \begin{bmatrix} \tilde{X} & -\tilde{Y} \\ -\tilde{N} & \tilde{M} \end{bmatrix}. \tag{13.5}$$

If, in addition, (13.2) *and* (13.4) *are satisfied, then the factorization is a* normalized doubly coprime factorization.

As is well known, the candidate formulae for doubly coprime factorizations of $G(s) = D + C(sI - A)^{-1}B$ are

$$\begin{aligned} \begin{bmatrix} M_F & Y_F \\ N_F & X_F \end{bmatrix} &= \begin{bmatrix} I & 0 \\ D & I \end{bmatrix} \\ &\quad + \begin{bmatrix} F \\ DF + C \end{bmatrix} (sI - A_F)^{-1} \begin{bmatrix} B & L \end{bmatrix} \end{aligned} \tag{13.6}$$

$$\begin{aligned} \begin{bmatrix} \tilde{X}_L & -\tilde{Y}_L \\ -\tilde{N}_L & \tilde{M}_L \end{bmatrix} &= \begin{bmatrix} I & 0 \\ -D & I \end{bmatrix} \\ &\quad - \begin{bmatrix} F \\ C \end{bmatrix} (sI - A_L)^{-1} \begin{bmatrix} B + LD & -L \end{bmatrix}, \end{aligned} \tag{13.7}$$

where $A_F = A + BF$, $A_L = A + LC$, and $F \in \mathcal{L}(Z,U)$, $L \in \mathcal{L}(Y,Z)$ must be chosen such that all components in (13.6) and (13.7) are holomorphic and bounded in norm on $\mathrm{Re}(s) > 0$.

If A_F and A_L generate exponentially stable semigroups, this assumption is automatically satisfied. If, however, A_F and A_L generate strongly stable semigroups, this is not the case, and we need to check each component separately. To obtain normalized coprime factorizations, one chooses F and L through the control and filter Riccati equations. To this end, we quote the results on strongly stabilizing solutions of Riccati equations and the definitions of strong stabilizability and detectability from Oostveen and Curtain [10].

Definition 13.2. *$\Sigma(A,B,C,D)$ is* strongly stabilizable *if there exists an $F \in \mathcal{L}(Z,U)$ such that $\begin{bmatrix} C & F \end{bmatrix}(sI - A - BF)^{-1}z \in \mathbf{H}_2(Y \times U)$ for all $z \in Z$, and $A - BF$ generates a strongly stable semigroup.*

$\Sigma(A,B,C,D)$ is strongly detectable *if there exists an $L \in \mathcal{L}(Y,Z)$ such that $\begin{bmatrix} L & B \end{bmatrix}^*(sI - A^* - C^*L^*)^{-1}z \in \mathbf{H}_2(Y \times U)$ for all $z \in Z$, and $A - LC$ generates a strongly stable semigroup.*

Theorem 13.1.

(i) Suppose that $\Sigma(A,B,C,D)$ is such that there exists an $F \in \mathcal{L}(Z,U)$ such that $\begin{bmatrix} F^* & C^* \end{bmatrix}^*(sI - A_F)^{-1}z \in \mathbf{H}_2(U \times Y)$ for all $z \in Z$. Then there exists a self-adjoint nonnegative solution to the Riccati equation

$$A^*Qz + QAz - F_Q{}^*RF_Qz + C^*Cz \quad = \quad 0, \tag{13.8}$$

for $z \in \mathcal{D}(A)$, where

$$A_Q \quad = \quad A + BF_Q, \qquad R = I + D^*D, \tag{13.9}$$

$$F_Q \quad = \quad -R^{-1}(B^*Q + D^*C). \tag{13.10}$$

Moreover, $\begin{bmatrix} F_Q^* & C^* \end{bmatrix}^*(sI - A_Q)^{-1}z \in \mathbf{H}_2(U \times Y)$ and $\begin{bmatrix} F_Q^* & C^* \end{bmatrix}^*(sI - A_Q)^{-1}B \in \mathbf{H}_\infty(\mathcal{L}(U, U \times Y))$.

(ii) Suppose that $\Sigma(A,B,C,D)$ is such that there exists an $L \in \mathcal{L}(Y,Z)$ such that $\begin{bmatrix} L & B \end{bmatrix}^*(sI - A_L^*)^{-1}z \in \mathbf{H}_2(Y \times U)$ for all $z \in Z$; then there exists a self-adjoint nonnegative definite solution P to the Riccati equation

$$APz + PA^*z - L_PSL_P{}^*z + BB^*z \quad = \quad 0, \tag{13.11}$$

for $z \in \mathcal{D}(A^*)$ where

$$A_P \quad = \quad A + L_PC, \qquad S = I + DD^*, \tag{13.12}$$

$$L_P \quad = \quad -(PC^* + BD^*)S^{-1}. \tag{13.13}$$

Moreover, $\begin{bmatrix} L_P & B \end{bmatrix}^*(sI - A_P^*)^{-1}z \in \mathbf{H}_2(Y \times U)$ and furthermore $C(sI - A_P)^{-1}\begin{bmatrix} L_P & B \end{bmatrix} \in \mathbf{H}_\infty(\mathcal{L}(Y \times U, Y))$.

(*iii*) *Suppose that, in addition to the assumptions in (i),* $\Sigma(A, B, C, D)$ *is strongly detectable; then* (13.8) *has a unique self-adjoint nonnegative definite solution and* A_Q *generates a strongly stable semigroup.*

(*iv*) *Suppose that, in addition to the assumptions in (ii),* $\Sigma(A^*, C^*, B^*, D^*)$ *is strongly detectable; then* (13.11) *has a unique self-adjoint nonnegative definite solution and* A_P^* *generates a strongly stable semigroup.*

As in Curtain and Zwart [1], the obvious candidates for formulae for a normalized doubly coprime factorization together with the Bezout factors are given by

$$\begin{bmatrix} M & Y \\ N & X \end{bmatrix} = \begin{bmatrix} M_{F_Q} R^{-1/2} & Y_{F_Q} S^{1/2} \\ N_{F_Q} R^{-1/2} & X_{F_Q} S^{1/2} \end{bmatrix} \tag{13.14}$$

$$\begin{bmatrix} \tilde{X} & -\tilde{Y} \\ -\tilde{N} & \tilde{M} \end{bmatrix} = \begin{bmatrix} R^{1/2}\tilde{X}_{L_P} & -R^{1/2}\tilde{Y}_{L_P} \\ -S^{-1/2}\tilde{N}_{L_P} & S^{-1/2}\tilde{M}_{L_P} \end{bmatrix}, \tag{13.15}$$

where R, S are as in (13.9) and (13.12) respectively, M_{F_Q} and so forth are as in (13.6), with F_Q defined by (13.10), and $\tilde{X}_{L_P}$ and so forth are as in (13.7), with L_P defined by (13.13).

Furthermore, the proof that they satisfy (13.5), (13.2), and (13.4) follows exactly as in [1]. It remains to show that all the components in (13.6) and (13.7) are in $\mathbf{H}_\infty$. Theorem 13.1 gives conditions under which $M, N, \tilde{M}$, and $\tilde{N}$ are in $\mathbf{H}_\infty$, but whether $X, Y, \tilde{X}$, and $\tilde{Y}$ are in $\mathbf{H}_\infty$ is not at all clear. Theorem 13.1 only guarantees that they are in $\mathbf{H}_2$ (modulo a constant). For the special case of systems $\Sigma(A, B, B^*, D)$, explicit formulae for normalized doubly coprime factorizations are in Curtain and Zwart [2]. In Section 13.3, we use another approach to show that, for the general case, M, N (resp., $\tilde{M}, \tilde{N}$) are indeed coprime. We appeal to a matrix generalization of the Carleson Corona Theorem from Fuhrmann ([5], Theorem 14-10, p.204 or [4], Theorem 3.1). For easy reference, we abstract a special case of the results.

Theorem 13.2.

(*i*) *Given* $M \in \mathbf{H}_\infty(\mathbb{C}^{m\times m})$ *and* $N \in \mathbf{H}_\infty(\mathbb{C}^{k\times m})$, *a necessary and sufficient condition for* M *and* N *to be coprime over* $\mathbf{H}_\infty$ *is the existence of a* $\delta > 0$ *such that*

$$\inf_{s\in\mathbb{C}_0^+} \left\{\|M(s)u\|^2 + \|N(s)u\|^2,\ u \in \mathbb{C}^m,\ \|u\| = 1\right\} \geq \delta, \tag{13.16}$$

where $\|\cdot\|$ *denotes the Euclidean norm.*

(*ii*) *Given* $\tilde{M} \in \mathbf{H}_\infty(\mathbb{C}^{k\times k})$ *and* $\tilde{N} \in \mathbf{H}_\infty(\mathbb{C}^{k\times m})$, *a necessary and sufficient condition for* $\tilde{M}$ *and* $\tilde{N}$ *to be coprime over* $\mathbf{H}_\infty$ *is the existence of a* $\delta > 0$

such that

$$\inf_{s\in\mathbb{C}_0^+}\left\{\|\tilde{M}(s)^*y\|^2+\|\tilde{N}(s)^*y\|^2,\ y\in\mathbb{C}^k,\ \|y\|=1\right\}\geq\delta. \tag{13.17}$$

We remark that in (13.16) and (13.17), Fuhrmann used $\|M(s)u\| + \|N(s)u\|$ instead of $\|M(s)u\|^2 + \|N(s)u\|^2$. Of course, they are equivalent statements, but the version given above is more convenient for our purposes.

13.3 Formulae for Normalized Coprime Factorizations

Our main result is the following.

Theorem 13.3. *Suppose that $\Sigma(A,B,C,D)$ is such that A generates a C_0-semigroup on the Hilbert space Z, $B\in\mathcal{L}(\mathbb{C}^m,Z)$, $C\in\mathcal{L}(Z,\mathbb{C}^k)$, $D\in\mathcal{L}(\mathbb{C}^m,\mathbb{C}^k)$.*

(1) If

(i) A has no essential spectrum on the imaginary axis,

(ii) there exists an $F\in\mathcal{L}(Z,\mathbb{C}^m)$ such that $\begin{bmatrix} F^* & C^*\end{bmatrix}^*(sI-A_F)^{-1}z\in\mathbf{H}_2(U\times Y)$ for all $z\in Z$, and

(iii) $\Sigma(A,B,C,D)$ is strongly detectable

then (M,N) given by (13.14) represents a normalized right-coprime factorization of $G=D+C(sI-A)^{-1}B$.

(2) If

(i)′ A^* has no essential spectrum on the imaginary axis,

(ii)′ there exists an $L\in\mathcal{L}(\mathbb{C}^k,Z)$ such that $\begin{bmatrix} B & L\end{bmatrix}^*(sI-A_L^*)^{-1}z\in\mathbf{H}_2(U\times Y)$ for all $z\in Z$, and

(iii)′ $\Sigma(A^*,C^*,B^*,D^*)$ is strongly detectable

then $(\tilde{M},\tilde{N})$ given by (13.15) represents a normalized left-coprime factorization of $G=D+C(sI-A)^{-1}B$.

A more balanced, slightly stronger version of the above is the following.

Corollary 13.1. *Suppose that $\Sigma(A,B,C,D)$ is such that A generates a C_0-semigroup on the Hilbert space Z, $B\in\mathcal{L}(\mathbb{C}^m,Z)$, $C\in\mathcal{L}(Z,\mathbb{C}^k)$, and $D\in\mathcal{L}(\mathbb{C}^m,\mathbb{C}^k)$. If A has compact resolvent and $\Sigma(A,B,C,D)$ is strongly stabilizable* *and strongly detectable, then* (M, N) *and* $(\tilde{M}, \tilde{N})$ *given by* (13.14) *and* (13.15) *are normalized right-coprime and left-coprime factorizations of* $G(s)$*, respectively.*

Proof: (of Theorem 13.3) We prove only Part *(1)*, as Part *(2)* is its dual. (a) First we note that conditions *(ii)* and *(iii)* guarantee the existence of unique solutions Q and P to the algebraic Riccati equations (13.8) and (13.11), respectively. So all the components in (13.14) and (13.15) are well defined. Since A_Q is a finite-rank perturbation of A, their essential spectrum is identical (see Kato [7, Theorem 5.35, p. 244]). Moreover, *(iii)* implies that A_Q generates a strongly stable semigroup. So, A_Q has no spectrum in $\mathrm{Re}(s) \geq 0$.

(b) We now show that M and N satisfy (13.16). By direct calculation, following Curtain and Zwart [p. 372][1], we find that

$$\begin{aligned} R^{1/2}(M(s)^*M(s) + N(s)^*N(s))R^{1/2} &= M_{F_Q}(s)^*M_{F_Q}(s) + N_{F_Q}(s)^*N_{F_Q}(s) \\ = \quad & R + B^*(\bar{s}I - A_Q^*)^{-1}(F_Q^*F_Q + C_Q^*C_Q)(sI - A_Q)^{-1}B \\ & +(F_Q + D^*C_Q)(sI - A_Q)^{-1}B + B^*(\bar{s}I - A_Q^*)^{-1}(F_Q^* + C_Q^*D), \end{aligned}$$

where $C_Q = C + DF_Q$. Using the Riccati equation (13.8) and the identities $F_Q + D^*C_Q = -B^*Q$, $F_Q^*F_Q + C_Q^*C_Q = \tilde{C}^*\tilde{C} + Q\tilde{B}\tilde{B}^*Q$ from Curtain and Zwart [1, p. 372], (where $\tilde{C} = S^{-1/2}C$ and $\tilde{B} = BR^{-1/2}$), the above simplifies to

$$M(s)^*M(s) + N(s)^*N(s) \quad = \quad I - 2\mathrm{Re}(s)\tilde{B}^*(\bar{s}I - A_Q^*)^{-1}Q(sI - A_Q)^{-1}\tilde{B},$$

and so,

$$\|M(s)u\|^2 + \|N(s)u\|^2 \quad = \quad \|u\|^2 - 2\mathrm{Re}(s)\|Q^{1/2}(sI - A_Q)^{-1}\tilde{B}u\|^2 \tag{13.18}$$

which holds for $\mathrm{Re}(s) \geq 0$, since from Part (a) of the proof $\sigma(A_Q) \cap \overline{\mathbb{C}_0^+} = \emptyset$. In particular, there holds

$$\|M(i\omega)u\|^2 + \|N(i\omega)u\|^2 \quad = \quad \|u\|^2, \qquad \text{for all } \omega \in \mathbb{R}, \tag{13.19}$$

whence, $\left\| \begin{bmatrix} M \\ N \end{bmatrix} \right\|_\infty = 1$.

Recall that the candidate Bezout factors $\tilde{X}$ and $\tilde{Y}$ from (13.15) are in $\mathbf{H}_2$ (modulo a constant) and so the shifted factors $\tilde{X}(s - \varepsilon)$, $\tilde{Y}(s - \varepsilon)$ are in $\mathbf{H}_\infty$ for each $\varepsilon > 0$. In effect, we have a coprime factorization on $\mathrm{Re}(s) > \varepsilon$,

$$\tilde{X}M - \tilde{Y}N \quad = \quad I,$$

with M, N, $\tilde{X}$, $\tilde{Y}$ all holomorphic and bounded on $\mathrm{Re}(s) > \varepsilon$. Applying Theorem 13.2 to this situation, we conclude there exists a $\delta(\varepsilon) > 0$ such that

$$\inf_{s \in \mathbb{C}_\varepsilon^+} \left\{ \|M(s)u\|^2 + \|N(s)u\|^2, \; u \in \mathbb{C}^m, \; \|u\| = 1 \right\} \quad \geq \quad \delta(\varepsilon). \tag{13.20}$$

We need to show that $\delta(\varepsilon)$ has a nonzero lower limit as ε tends to zero. Suppose, on the contrary, that there exists a sequence $u_n \in \mathbb{C}^m$, $\|u_n\| = 1$, and $z_n \in \overline{\mathbb{C}_0^+}$, such that $\|M(z_n)u_n\| < 1/n$ and $\|N(z_n)u_n\| < 1/n$. Since u_n is a bounded sequence in $\mathbb{C}^m$, it has a convergent subsequence $u_{n_k} \to \xi$ as $k \to \infty$, and we have

$$\begin{aligned} \|M(z_{n_k})\xi\| &\leq \|M(z_{n_k})(\xi - u_{n_k})\| + \|M(z_{n_k})u_{n_k}\| \\ &\leq \|\xi - u_{n_k}\| + 1/n_k \qquad \text{since } \|M\|_\infty \leq 1 \\ &< 2/n_k \end{aligned}$$

for sufficiently large k. Similarly, $\|N(z_{n_k})\xi\| < 2/n_k$. Now (13.20) with $u = \xi$ implies that for sufficiently large n_k, $\text{Re}(z_{n_k}) \leq \varepsilon$. Since ε can be chosen arbitrarily small and (13.19) with $u = \xi$ holds on the imaginary axis, we can conclude that there exists no such sequence z_n in $\overline{\mathbb{C}_0^+}$.

(c) So, we have shown that (M, N) represents a normalized right-coprime factorization for G. □

Proof: **(of Corollary 13.1)** If A has compact resolvent, then *(i)* and *(i)'* hold and $A + BF$ generates a strongly stable semigroup if and only if $A^* + F^*B^*$ does. □

An important consequence of Theorem 13.3 is that it is now fairly routine to extend the arguments in Oostveen and Curtain [11] to solve the problem of robust stabilization for bounded linear systems $\Sigma(A, B, C, D)$ satisfying the assumptions of the above theorem.

A corollary to the above theorem is that there exist Bezout factors $\tilde{X}, \tilde{Y} \in \mathbf{H}_\infty$ satisfying (13.3). Note that Bezout factors are not unique, and we have not shown that the candidate Bezout factors $\tilde{X}, \tilde{Y}, X, Y$ in (13.14), and (13.15) are true Bezout factors. We only know that they are in $\mathbf{H}_2$ (modulo a constant), but not that they are in $\mathbf{H}_\infty$. While this is not essential for the robust stabilization problem with respect to coprime factor perturbations, it is needed for dynamic stabilization results, in particular, for the LQG controller (Oostveen [9], Chapter 6). This remains an interesting open question.

13.4 Application to Hybrid Flexible Structures

Here we give a general class that satisfies the conditions of Theorem 13.3, but is not of the form $\Sigma(A, B, B^*, D)$. Indeed A does not generate a contraction semigroup and the transfer function is not even square. The monograph [6] by Joshi is devoted to the study of modeling and control of various flexible space structures. Although it treats exclusively finite-dimensional models, modeling the flexible components as partial differential equations produces infinite-dimensional models. In Oostveen [9], Chapter 9.4 a detailed analysis was given of an infinite-dimensional version of the flexible structure model from [6], Chapters 2.1 and 2.2. We quote the results here without proofs.

The structure is equipped with a number (m_f) of force actuators and a number (m_T) of torque actuators. Let $\xi \in \mathbb{R}^3$ denote the position of the center of mass and let the force exerted by the ith actuator be $f_i \in \mathbb{R}^3$. The translation of the center of mass is given by

$$m\ddot{\xi} = \sum_{i=1}^{m_f} f_i. \tag{13.21}$$

The attitude of the structure is given by the vector $\alpha \in \mathbb{R}^3$ describing the rigid body Euler angle with respect to the x-, y-, and z-axes. It satisfies

$$J\ddot{\alpha} = \sum_{i=1}^{m_f} r_i \times f_i + \sum_{j=1}^{m_T} T_j, \tag{13.22}$$

where T_j denotes the torque exerted by the jth torque actuator and $r_i \in \mathbb{R}^3$ describes the location where the force f_i is applied. The elastic motion is described by

$$\begin{aligned} \ddot{q} + D\dot{q} + \Lambda q &= \sum_{i=1}^{m_f} \Psi_i^T f_i + \sum_{j=1}^{m_T} \Phi_j^T T_j \\ &= \Psi^T F + \Phi^T T. \end{aligned} \tag{13.23}$$

Here, q is the infinite sequence of modal amplitudes, which is assumed to be an element of a Hilbert space Q (for instance, $Q = l_2(\mathbb{N})$). $F = \text{col}(f_1, \ldots, f_{m_f})$ and $T = \text{col}(T_1, \ldots, T_{m_T})$. The infinite matrices $\Psi_i \in \mathcal{L}(Q, \mathbb{R}^3)$ and $\Phi_j \in \mathcal{L}(Q, \mathbb{R}^3)$ are called the *mode shape matrices* and *mode slope matrices*, respectively. Λ is the *stiffness operator* and it is assumed that $\Lambda = \Lambda^* > 0$. The *damping operator* D satisfies $D = D^* \geq 0$. Both D and Λ are closed and densely defined. In general, both are unbounded operators.

We call this a hybrid system because it consists of a finite-dimensional part (the dynamics of ξ and α) coupled with a distributed part (the elastic motion). The coupling consists of the fact that both parts are affected by the same control inputs f_i and T_j. The aim of this section is to show that under certain conditions this system has a statically stabilizable state space realization.

We define $p = \text{col}(m\xi, J\alpha, q) \in \mathbb{R}^3 \times \mathbb{R}^3 \times Q =: Z_1$. Then the dynamics is described as

$$\ddot{p} + \tilde{D}\dot{p} + \tilde{\Lambda} p = \Gamma^T u, \tag{13.24}$$

where

$$\tilde{D} = \begin{bmatrix} 0 & 0 & 0 \\ 0 & 0 & 0 \\ 0 & 0 & D \end{bmatrix}, \qquad \tilde{\Lambda} = \begin{bmatrix} 0 & 0 & 0 \\ 0 & 0 & 0 \\ 0 & 0 & \Lambda \end{bmatrix}, \tag{13.25}$$

$$\Gamma^T = \begin{bmatrix} I_3 & \cdots & I_3 & 0_3 & \cdots & 0_3 \\ R_1 & \cdots & R_{m_f} & I_3 & \cdots & I_3 \\ \Psi_1^T & \cdots & \Psi_{m_f}^T & \Phi_1^T & \cdots & \Phi_{m_T}^T \end{bmatrix},$$

$u = \mathrm{col}(f_1, \ldots, f_{m_f}, T_1, \ldots, T_{m_T}) \in \mathbb{R}^{3m_f+3m_T} =: U$ and the matrices R_i, $i = 1\ldots, m_f$ are given by

$$R_i = \begin{bmatrix} 0 & -r_z^i & r_x^i \\ -r_z^i & 0 & -r_y^i \\ -r_x^i & r_y^i & 0 \end{bmatrix}, \quad \text{where } r_i = \begin{pmatrix} r_x^i \\ r_y^i \\ r_z^i \end{pmatrix}.$$

If we have position sensors at the same locations as the force actuators and attitude sensors at the same locations as the torque actuators, then the observation is given by

$$y_p = \Gamma p. \tag{13.26}$$

On the other hand, the use of rate sensors, measuring translational and rotational velocities at those locations, leads to the observation

$$y_r = \Gamma \dot{p}. \tag{13.27}$$

Following Joshi [6], let us suppose that both types of measurements are available and the observation is $y = \mathrm{col}(y_p, y_r)$. To construct a first-order representation, we take as the state $z = \mathrm{col}(z, \dot{z})$ on the state space $Z = \{z = \mathrm{col}(z_1, z_2) \in Z_1 \times Z_1\}$ with inner product

$$\langle z, w\rangle_Z = \langle(\tilde{\Lambda} + \Gamma^T\Gamma)z_1, w_1\rangle_{Z_1} + \langle z_2, w_2\rangle_{Z_1}. \tag{13.28}$$

This leads to the following representation of the system (13.21) to (13.23), (13.26), and (13.27),

$$\begin{aligned} \dot{z}(t) &= Az(t) + Bu(t), \qquad z(0) = z_0, \\ y(t) &= Cz(t), \end{aligned} \tag{13.29}$$

where

$$A = \begin{bmatrix} 0 & I \\ -\tilde{\Lambda} & -\tilde{D} \end{bmatrix}, \quad B = \begin{bmatrix} 0 \\ \Gamma^T \end{bmatrix}, \quad C = \begin{bmatrix} C_p \\ C_r \end{bmatrix} = \begin{bmatrix} \Gamma & 0 \\ 0 & \Gamma \end{bmatrix}.$$

Note that this system is not square and $C \neq B^*$. The following was established in Oostveen [9].

Lemma 13.1. *Let the system $\Sigma(A, B, C)$ be defined as in* (13.29) *on the state space Z and assume that the following conditions are satisfied.*

(i) $\Lambda = \Lambda^* > 0$, $D = D^* \geq 0$ *are closed, densely defined operators with* $\mathcal{D}(D) \supset \mathcal{D}(\Lambda)$,

(ii) A *has compact resolvent.*

(iii) $\Sigma(A, B, C)$ *is approximately controllable.*

(iv) $\Sigma(A_2, B, C_r)$ *is approximately observable, where* $A_2 = A - \begin{bmatrix} \Gamma^T & 0 \end{bmatrix} \begin{bmatrix} 0 \\ \Gamma \end{bmatrix}$.

(v) $\tilde{\Lambda} + \Gamma^T\Gamma$ *is boundedly invertible and* $(\tilde{\Lambda} + \Gamma^T\Gamma)^{-1}\tilde{D}$ *is bounded.*

Then the system $\Sigma(A, B, C)$ *is strongly stabilized by the static output feedback law* $u = -G_p y_p - G_r y_r$ *for any positive-definite matrices* G_p *and* G_r. *In particular, the closed-loop system* $\Sigma(A_{cl}, B, C)$ *is such that* $A_{cl} = A - BG_rC_r - BG_pC_p$ *generates a strongly stable semigroup,* $B^*(sI - A_{cl}^*)^{-1}z \in \mathbf{H}_2$, $C(sI - A_{cl})^{-1}z \in \mathbf{H}_2$, *and* $C(sI - A_{cl})^{-1}B \in \mathbf{H}_\infty$.

We remark that *(v)* is not needed to establish that A_{cl} generates a strongly stable semigroup. The assumptions *(i)* and *(ii)* are satisfied by most, if not all, systems of interest. Assumptions *(iii)* and *(iv)* depend on the actuators and sensors implemented. Note that assumption *(iv)* concerns A_2, not A. The reason is that the pair (A, C_r) can never be observable; the position of the center of mass and the attitude of the structure can not be reconstructed from the measurements of the rate sensor, because the dynamics does not depend on the position and attitude. The coupling introduced in A_2 makes the dynamics dependent on them, and so it becomes possible to observe the position and the attitude from the rate sensors.

The above example satisfies all the conditions of Corollary 13.1 with $F = -\begin{bmatrix} G_r & G_p \end{bmatrix} C$ and $L = -B \begin{bmatrix} G_r & G_p \end{bmatrix}$. So, it possesses a normalized doubly coprime factorization and we can design robustly stabilizing controllers as Curtain and Zwart [1], Section 9.4 .

Acknowledgments

The authors thank Paul Fuhrmann and Luciano Pandolfi for their useful comments.

References

[1] R.F. Curtain and H.J. Zwart, *An Introduction to Infinite-Dimensional Linear Systems Theory*, Springer-Verlag, New York, 1995.

[2] R.F. Curtain and H.J. Zwart. Riccati equations and normalized coprime factorizations for strongly stabilizable infinite-dimensional systems. *Syst. Control Letters*, 28:11–22, 1996.

[3] R.F. Curtain, G. Weiss, and M. Weiss, Coprime factorizations for regular linear systems, *Automatica*, 32:1519–1532, 1996.

[4] P.A. Fuhrmann, On the Corona theorem and its application to spectral problems in Hilbert space, *Trans. Amer. Math. Soc.*, 132:493–520, 1968.

[5] P.A. Fuhrmann, *Linear Systems and Operators in Hilbert Space*, McGraw-Hill, New York, 1981.

[6] S.M. Joshi, *Control of Large Flexible Space Structures*, 131, Lecture Notes in Control and Information Sciences, Springer-Verlag, Berlin, 1989.

[7] T. Kato, *Perturbation Theory of Linear Operators*, Springer-Verlag, New York, 1966.

[8] D.C. McFarlane and K. Glover, *Robust Controller Design using Normalized Coprime Factor Plant Descriptions*, 138, Lecture Notes in Control and Information Sciences, Springer-Verlag, Berlin, 1989.

[9] J.C. Oostveen, *Strongly stabilizable infinite-dimensional systems*, Ph.D. thesis, Department of Mathematics, University of Groningen, The Netherlands, 1999.

[10] J.C. Oostveen and R.F. Curtain, Necessary and sufficient conditions for strong stability of distributed parameter systems, *Syst. Control Letters* 37:11–18, 1999.

[11] J.C. Oostveen and R.F. Curtain, Robustly stabilizing controllers for dissipative infinite-dimensional systems with collocated actuators and sensors, *Automatica* 36:337–348, 2000.

Chapter 14

Low-Gain Integral Control of Infinite-Dimensional Regular Linear Systems Subject to Input Hysteresis

Hartmut Logemann
Adam D. Mawby

ABSTRACT In the present chapter we introduce a general class of causal dynamic nonlinearities with certain monotonicity and Lipschitz continuity properties. It is shown that closing the loop around an exponentially stable, single-input, single-output, infinite-dimensional, regular, linear system, subject to an input nonlinearity from this class and compensated by an integral controller, guarantees asymptotic tracking of constant reference signals, provided that (i) the steady-state gain of the linear part of the plant is positive, (ii) the positive integrator gain is smaller than a certain constant given by a positive-real condition in terms of the linear part of the plant, and (iii) the reference value is feasible in a very natural sense. The class of nonlinearities under consideration contains in particular relay hysteresis, backlash, and hysteresis operators of Prandtl and Preisach types.

14.1 Introduction

This chapter extends a sequence [12, 13] of recent results pertaining to integral control of infinite-dimensional systems subject to static input nonlinearities. Underpinning these results are generalizations of the well-known principle (see, e.g., [6, 15, 17]) that closing the loop around a stable, linear, finite-dimensional, continuous-time, single-input, single-output plant, with transfer function $\mathbf{G}$ compensated by a pure integral controller k/s, will result in a stable closed-loop system that achieves asymptotic tracking of arbitrary constant reference signals, provided that $|k|$ is sufficiently small and $\mathbf{G}(0)k > 0$.[1] This result has been extended to various classes of infinite-dimensional systems; see [11] and the references therein. In par-

[1] Therefore, under the above assumptions on the plant, the problem of tracking constant reference signals reduces to that of tuning the gain parameter k. This so-called "tuning regulator theory" [6] has been successfully applied in process control (see [5, 14]).

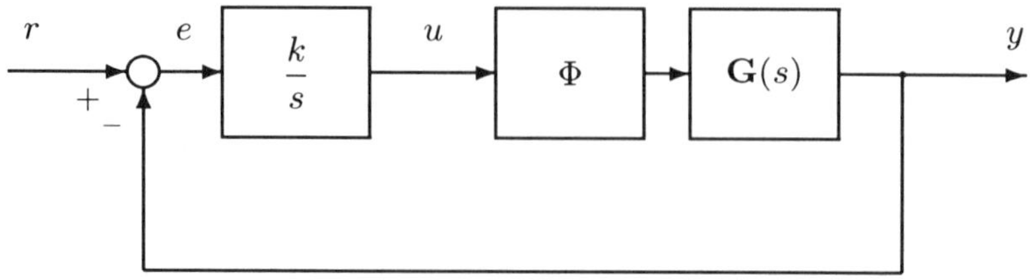

FIGURE 14.1. Low-gain control with input nonlinearity.

ticular, Logemann et al. [13] have shown that the above principle remains true for a single-input, single-output, linear, regular, infinite-dimensional system subject to a static, nondecreasing, globally Lipschitz input nonlinearity Φ (such as, e.g., saturation), provided the reference value r is feasible in the sense that $r/\mathbf{G}(0)$ is in the image of the nonlinearity; see Figure 14.1. Here we consider the same problem as in [13], but for a wider class of causal dynamic nonlinearities that satisfy a certain Lipschitz condition. This class encompasses, in particular, a large number of hysteresis nonlinearities important in applications such as relay (or passive), backlash, and plastic-elastic hysteresis. Generally speaking, hysteresis is a special type of memory-based relation between a scalar input signal $u(\cdot)$ and a scalar output signal $v(\cdot)$ that cannot be expressed in terms of a single-valued function, but takes the form of "hysteresis" loops; in particular, the operator $u(\cdot) \mapsto v(\cdot)$ is causal and rate independent. This type of behavior arises in mechanical plays, thermostats, elastoplasticity, ferromagnetism, and in smart material structures such as piezoelectric elements and magnetostrictive transducers (see Banks et al. [1] for hysteresis phenomena in smart materials). There exists a substantial literature on mathematical modeling and mathematical theory of hysteresis phenomena; see, for example, Brokate [3], Brokate and Sprekels [4], Krasnosel'skiĭ and Pokrovskiĭ [9], Macki et al. [16], and Visintin [19]. Of particular importance in a systems and control context is the pioneering work [9].

As in [13] we assume that the linear part of the system to be controlled (described in Figure 14.1 by the transfer function $\mathbf{G}(s)$) is an exponentially stable, single-output, single-input, regular, infinite-dimensional system. This class, introduced by Weiss (see [23] through [22]), is rather general and allows for highly unbounded control and observation operators. It includes most distributed parameter systems and time-delay systems of interest in control engineeering.

The main result in this chapter shows that for the class of dynamic nonlinearities under consideration, the output $y(t)$ of the closed-loop system, shown in Figure 14.1, converges to r as $t \to \infty$, provided that $\mathbf{G}(0) > 0$, r is feasible in some natural sense, and $k \in (0, K/\lambda)$, where $\lambda > 0$ is a Lipschitz constant for the nonlinearity and K is the supremum of the set of all numbers $k > 0$ such that the function

$$1 + k \operatorname{Re} \frac{\mathbf{G}(s)}{s}$$

is positive real.

The chapter is organized as follows. In Section 14.2 we briefly discuss regular linear infinite-dimensional systems. In Section 14.3 we define a class of dynamic nonlinear operators for which, in Section 14.4, we show that the output $y(t)$ of the closed-loop system (shown in Figure 14.1) converges to the reference r as $t \to \infty$. In Section 14.5 we introduce various hysteresis operators, such as relay, backlash, and elastic-plastic as well as hysteresis operators of Prandtl and Preisach types; we show that under a few natural assumptions they are contained within the class of nonlinearities introduced in Section 14.3. Section 14.6 contains simulations of two controlled diffusion process examples with input hysteresis nonlinearities. Finally, a number of technicalities have been relegated to the Appendix (Section 14.7), which, in particular, contains an existence and uniqueness result for the solutions of the nonlinear abstract Cauchy problem describing the closed-loop system shown in Figure 14.1.

Notation: If $I \subset \mathbb{R}$ is a compact interval, then $AC(I, \mathbb{R})$ denotes the space of absolutely continuous real-valued functions defined on I; $AC(\mathbb{R}_+, \mathbb{R})$ denotes the space of real-valued functions defined on $\mathbb{R}_+$ that are absolutely continuous on any compact interval $I \subset \mathbb{R}_+$; that is, a function $f : \mathbb{R}_+ \to \mathbb{R}$ is in $AC(\mathbb{R}_+, \mathbb{R})$ if and only if there exists a function $g \in L^1_{\text{loc}}(\mathbb{R}_+, \mathbb{R})$ such that

$$f(t) = f(0) + \int_0^t g(\tau)\, d\tau\,, \quad \forall\, t \geq 0\,.$$

We call a function $f : [a, b] \to \mathbb{R}$ piecewise monotone if there exist numbers $a = t_0 < t_1 < \ldots < t_n = b$ such that f is monotone on $[t_{i-1}, t_i]$ for $i = 1, 2, \ldots, n$. A function $f : \mathbb{R}_+ \to \mathbb{R}$ is called piecewise monotone if f is piecewise monotone on each compact interval $I \subset \mathbb{R}_+$. We denote the space of piecewise monotone continuous functions $f : \mathbb{R}_+ \to \mathbb{R}$ by $C_{\text{pm}}(\mathbb{R}_+, \mathbb{R})$. It is straightforward to show that $C_{\text{pm}}(\mathbb{R}_+, \mathbb{R})$ is dense in $C(\mathbb{R}_+, \mathbb{R})$ in the sense that for all $f \in C(\mathbb{R}_+, \mathbb{R})$ and all $\varepsilon > 0$, there exists $g \in C_{\text{pm}}(\mathbb{R}_+, \mathbb{R})$ such that

$$|f(t) - g(t)| \leq \varepsilon\,, \quad \forall\, t \in \mathbb{R}_+\,.$$

For $\alpha \in \mathbb{R}$, we define the exponentially weighted L^p-space

$$L^p_\alpha(\mathbb{R}_+, \mathbb{R}) := \{f \in L^p_{\text{loc}}(\mathbb{R}_+, \mathbb{R}) |\ f(\cdot) \exp(-\alpha\, \cdot) \in L^p(\mathbb{R}_+, \mathbb{R})\}\,.$$

$L(X, Y)$ denotes the space of bounded linear operators from a Banach space X to a Banach space Y. Let $\mathbb{N}$ denote the nonnegative integers. For $\alpha \in \mathbb{R}$, we define $\mathbb{C}_\alpha := \{s \in \mathbb{C} \,|\, \operatorname{Re} s > \alpha\}$. The Laplace transform is denoted by $\mathfrak{L}$.

14.2 Preliminaries on Regular Linear Systems

In Figure 14.1 the underlying linear system (i.e., the system with transfer function $\mathbf{G}(s)$) is assumed to be a single-input, single-output, continuous-time, regular system Σ with state space X (a Hilbert space) and with generating operators (A, B, C, D). This means in particular that A generates a strongly continuous semigroup $\mathbf{T} = (\mathbf{T}_t)_{t \geq 0}$, $C \in L(X_1, \mathbb{R})$ is an admissible observation operator for $\mathbf{T}$, $B \in L(\mathbb{R}, X_{-1})$ is an admissible control operator for $\mathbf{T}$, and $D \in \mathbb{R}$ is the feedthrough of the system. Here X_1 denotes the space $\mathrm{dom}(A)$ (the domain of A) endowed with the graph norm and X_{-1} denotes the completion of X with respect to the norm $\|x\|_{-1} = \|(s_0 I - A)^{-1}x\|$, where s_0 is any fixed element in the resolvent set of A. The norm on X is denoted by $\|\cdot\|$, while $\|\cdot\|_1$ and $\|\cdot\|_{-1}$ denote the norms on X_1 and X_{-1}, respectively. Then $X_1 \hookrightarrow X \hookrightarrow X_{-1}$ and $\mathbf{T}$ restricts (resp., extends) to a strongly continuous semigroup on X_1 (resp., X_{-1}). The exponential growth constant

$$\omega(\mathbf{T}) := \lim_{t \to \infty} \frac{1}{t} \ln \|\mathbf{T}_t\|$$

is the same on all three spaces. The generator of $\mathbf{T}$ on X_{-1} is an extension of A to X (which is bounded as an operator from X to X_{-1}). We use the same symbol $\mathbf{T}$ (resp., A) for the original semigroup (resp., its generator) and the associated restrictions and extensions. With this convention, we may write $A \in L(X, X_{-1})$. Considered as a generator on X_{-1}, the domain of A is X.

We regard a regular system Σ as synonymous with its generating operators and simply write $\Sigma = (A, B, C, D)$. The regular system is said to be *exponentially stable* if the semigroup $\mathbf{T}$ is exponentially stable; that is, $\omega(\mathbf{T}) < 0$. The control operator B (resp., observation operator C) is said to be *bounded* if $B \in L(\mathbb{R}, X)$ (resp., $C \in L(X, \mathbb{R})$); otherwise, B (resp., C) is said to be *unbounded*. In terms of the generating operators (A, B, C, D), the transfer function $\mathbf{G}(s)$ can be expressed as

$$\mathbf{G}(s) = C_L(sI - A)^{-1}B + D\,,$$

where C_L denotes the so-called Lebesgue extension of C. The transfer function $\mathbf{G}(s)$ is bounded and holomorphic in any half-plane $\mathrm{Re}\, s > \alpha$ with $\alpha > \omega(\mathbf{T})$. Moreover,

$$\lim_{s \to \infty,\, s \in \mathbb{R}} \mathbf{G}(s) = D\,.$$

For any $x_0 \in X$ and $u \in L^2_{\mathrm{loc}}(\mathbb{R}_+, \mathbb{R})$, the state and output functions $x(\cdot)$ and $y(\cdot)$, respectively, satisfy the equations

$$\dot{x}(t) = Ax(t) + Bu(t)\,, \qquad x(0) = x_0\,, \tag{14.1a}$$

$$y(t) = C_L x(t) + Du(t)\,, \tag{14.1b}$$

for almost all $t \geq 0$ (in particular, $x(t) \in \mathrm{dom}(C_L)$ for almost all $t \geq 0$). The derivative on the left-hand side of (14.1a) has, of course, to be understood in X_{-1}. In other words, if we consider the initial-value problem (14.1a) in the space X_{-1}, then for any $x_0 \in X$ and $u \in L^2_{\mathrm{loc}}(\mathbb{R}_+, \mathbb{R})$, (14.1a) has a unique strong solution (in the sense of Pazy [18, p. 109]) given by the variation of parameters formula

$$x(t) = \mathbf{T}_t x_0 + \int_0^t \mathbf{T}_{t-\tau} B u(\tau)\, d\tau \,. \tag{14.2}$$

For any $x_0 \in X$ and any $u \in L^2_{\mathrm{loc}}(\mathbb{R}_+, \mathbb{R})$, (14.2) defines a continuous X-valued function.

Denoting the input-output operator of (14.1) by $\mathbf{F}$, we have that $\mathbf{F}$ is a shift-invariant (and thus causal) operator from $L^2_{\mathrm{loc}}(\mathbb{R}_+, \mathbb{R})$ into $L^2_{\mathrm{loc}}(\mathbb{R}_+, \mathbb{R})$. For $\alpha \in \mathbb{R}$ and $u \in L^2_\alpha(\mathbb{R}_+, \mathbb{R})$,

$$(\mathfrak{L}(\mathbf{F}u))(s) = \mathbf{G}(s)(\mathfrak{L}(u))(s)\,, \quad \mathrm{Re}\, s > \max(\omega(\mathbf{T}), \alpha)\,.$$

Finally, we introduce the state-to-output map $\mathbf{\Psi} : X \to L^2_{\mathrm{loc}}(\mathbb{R}_+, \mathbb{R})$ defined by

$$(\mathbf{\Psi} x_0)(t) = C_L \mathbf{T}_t x_0\,, \quad \text{a.e. } t \in \mathbb{R}_+\,.$$

For more details on regular systems see Weiss [23] through [22]. For details on regular systems in the context of low-gain control the reader is referred to [11] and [13].

For future reference we state the following lemma, the proof of which can be found in [13].

Lemma 14.1. *Assume that $\mathbf{T}$ is exponentially stable and that $B \in L(\mathbb{R}, X_{-1})$ is an admissible control operator for $\mathbf{T}$.*

If $u \in L^\infty(\mathbb{R}_+, \mathbb{R})$ is such that $\lim_{t\to\infty} u(t) = u_\infty$ exists, then, for all $x_0 \in X$, the state $x(\cdot)$ given by (14.2) satisfies

$$\lim_{t\to\infty} \|x(t) + A^{-1} B u_\infty\| = 0\,.$$

14.3 A Class of Causal Monotone Nonlinear Operators

Let $a \in (0, \infty]$ and let $J \subset \mathbb{R}_+$ be an interval of the form $[0, a)$ or $[0, a]$. For $\tau \in J$, we define the operator $\mathbf{Q}_\tau : C(J, \mathbb{R}) \to C(\mathbb{R}_+, \mathbb{R})$ by

$$(\mathbf{Q}_\tau u)(t) = \begin{cases} u(t) & \text{for } 0 \leq t \leq \tau\,, \\ u(\tau) & \text{for } t > \tau\,. \end{cases}$$

If the domain space of $\mathbf{Q}_\tau$ is $C(\mathbb{R}_+, \mathbb{R})$ (i.e., $J = [0, \infty)$), then $\mathbf{Q}_\tau$ is a projection operator. Let $\mathcal{C} \subset C(\mathbb{R}_+, \mathbb{R})$, $\mathcal{C} \neq \emptyset$. Recall that an operator

$\Phi : \mathcal{C} \to C(\mathbb{R}_+, \mathbb{R})$ is called *causal* if for all $u, v \in \mathcal{C}$ and all $\tau \geq 0$ with $u(t) = v(t)$ for all $t \in [0, \tau]$ it follows that $(\Phi(u))(t) = (\Phi(v))(t)$ for all $t \in [0, \tau]$. If $\mathcal{C}$ is invariant under $\mathbf{Q}_t$ for all $t \in \mathbb{R}_+$ (i.e., $\mathbf{Q}_t(\mathcal{C}) \subset \mathcal{C}$ for all $t \in \mathbb{R}_+$) then it is easy to show that Φ is causal if and only if for all $u \in \mathcal{C}$,

$$(\Phi(\mathbf{Q}_t u))(t) = (\Phi(u))(t)\,, \quad \forall\, t \in \mathbb{R}_+\,.$$

Given an operator $\Phi : C(\mathbb{R}_+, \mathbb{R}) \to C(\mathbb{R}_+, \mathbb{R})$ and a number $a > 0$ and denoting the space of all functions $f : [0, a) \to \mathbb{R}$ by $\mathcal{F}([0, a), \mathbb{R})$, we define an operator $\tilde{\Phi} : C([0, a), \mathbb{R}) \to \mathcal{F}([0, a), \mathbb{R})$ by setting

$$(\tilde{\Phi}(u))(t) = (\Phi(\mathbf{Q}_t u))(t)\,, \quad \forall\, t \in [0, a)\,.$$

If Φ is causal, then for each $\tau \in [0, a)$ we have

$$(\tilde{\Phi}(u))(t) = (\Phi(\mathbf{Q}_\tau u))(t)\,, \quad \forall\, t \in [0, \tau]\,,$$

implying in particular that $\tilde{\Phi}(C([0, a), \mathbb{R})) \subset C([0, a), \mathbb{R})$. In the following, we use the same symbol Φ to denote the original operator acting on $C(\mathbb{R}_+, \mathbb{R})$ and the associated operator $\tilde{\Phi}$ acting on $C([0, a), \mathbb{R})$.

Let $u \in C(\mathbb{R}_+, \mathbb{R})$. The function u is called *ultimately nondecreasing* if there exists $T \in \mathbb{R}_+$ such that u is nondecreasing on $[T, \infty)$; u is said to be *approximately ultimately nondecreasing*, if for all $\varepsilon > 0$, there exists an ultimately nondecreasing function $v \in C(\mathbb{R}_+, \mathbb{R})$ such that

$$|u(t) - v(t)| \leq \varepsilon\,, \quad \forall\, t \in \mathbb{R}_+\,.$$

The *numerical value set* $\operatorname{NVS}\Phi$ of an operator $\Phi : C(\mathbb{R}_+, \mathbb{R}) \to C(\mathbb{R}_+, \mathbb{R})$ is defined by

$$\operatorname{NVS}\Phi := \{(\Phi(u))(t) \mid u \in C(\mathbb{R}_+, \mathbb{R}),\, t \in \mathbb{R}_+\}\,.$$

For $\alpha \geq 0$, $w \in C([0, \alpha], \mathbb{R})$, and $\delta_1, \delta_2 > 0$, we define $\mathcal{C}(w; \delta_1, \delta_2)$ to be the set of all $u \in C(\mathbb{R}_+, \mathbb{R})$ such that

$$u(t) = w(t)\,, \ \forall\, t \in [0, \alpha] \qquad \text{and} \qquad |u(t) - w(\alpha)| \leq \delta_1\,, \ \forall\, t \in [\alpha, \alpha + \delta_2]\,.$$

We introduce the following assumptions on the nonlinear operator $\Phi : C(\mathbb{R}_+, \mathbb{R}) \to C(\mathbb{R}_+, \mathbb{R})$.

(N1) Φ is causal.

(N2) For all $u \in C(\mathbb{R}_+, \mathbb{R})$ and all $\tau \in \mathbb{R}_+$, $(\Phi(\mathbf{Q}_\tau u))(t) = (\Phi(\mathbf{Q}_\tau u))(\tau)$ for all $t \geq \tau$.

(N3) $\Phi(AC(\mathbb{R}_+, \mathbb{R})) \subset AC(\mathbb{R}_+, \mathbb{R})$.

(N4) Φ is monotone in the sense that for all $u \in AC(\mathbb{R}_+, \mathbb{R})$ with $\Phi(u) \in AC(\mathbb{R}_+, \mathbb{R})$,

$$\frac{d}{dt}(\Phi(u))(t)\,\dot{u}(t) \geq 0\,, \quad \text{a.e. } t \in \mathbb{R}_+\,.$$

(N5) There exists $\lambda > 0$ such that for all $\alpha \in \mathbb{R}_+$, $w \in C([0,\alpha], \mathbb{R})$, there exist numbers $\delta_1, \delta_2 > 0$ such that for all $u, v \in \mathcal{C}(w; \delta_1, \delta_2)$,

$$\sup_{t\in[\alpha,\alpha+\delta_2]} |(\Phi(u))(t) - (\Phi(v))(t)| \leq \lambda \sup_{t\in[\alpha,\alpha+\delta_2]} |u(t) - v(t)|\,.$$

(N6) If $u \in C(\mathbb{R}_+, \mathbb{R})$ is approximately ultimately nondecreasing and furthermore $\lim_{t\to\infty} u(t) = \infty$, then $\Phi(u)(t)$ and $\Phi(-u)(t)$ converge to $\sup \mathrm{NVS}\,\Phi$ and $\inf \mathrm{NVS}\,\Phi$, respectively, as $t \to \infty$.

(N7) If $u \in C(\mathbb{R}_+, \mathbb{R})$ is such that $\lim_{t\to\infty}(\Phi(u))(t) \in \mathrm{int}\,\mathrm{NVS}\,\Phi$, then u is bounded.

(N8) For all $a > 0$ and all $u \in C([0,a), \mathbb{R})$, there exist $\alpha, \beta > 0$ such that

$$\sup_{t\in[0,\tau]} |(\Phi(u))(t)| \leq \alpha + \beta \sup_{t\in[0,\tau]} |u(t)|\,, \quad \forall \tau \in [0,a)\,.$$

Remark 14.1. (i) Assumption (N2) says that if the input u of the nonlinearity Φ is constant on $[\tau,\infty)$, then the output $\Phi(u)$ is constant and equal to $(\Phi(u))(\tau)$ on $[\tau,\infty)$.

(ii) Hysteresis operators as defined in [4] are simply causal and rate independent operators defined on $C_{\mathrm{pm}}(\mathbb{R}_+, \mathbb{R})$ (see [4, Definition 2.2.8 and Proposition 2.2.9]). Most hysteresis operators admit causal extensions to $C(\mathbb{R}_+, \mathbb{R})$ satisfying (N2). In this sense, a large class of hysteresis operators satisfies the assumptions (N1) and (N2). This is in particular true for hysteresis operators that are Lipschitz continuous in the sense of Definition 14.2. We mention that none of the assumptions (N1) through (N8) imply rate independence. However, most of the operators satisfying (N1) and (N2) and which are of interest in the modeling of dynamic actuator nonlinearities will be hysteresis operators in the sense of [4].

(iii) Assumptions (N1) and (N5) ensure local existence and uniqueness of solutions to the nonlinear closed-loop system shown in Figure 14.1; if (N1), (N5), and (N8) hold, then the solution of this closed-loop system exists and is unique on the time interval $\mathbb{R}_+$ (see Section 14.4).

(iv) If (N1) and (N6) hold, then $\mathrm{NVS}\,\Phi$ is an interval.

We show in Section 14.5 that the assumptions (N1) through (N8) are satisfied by a large class of hysteresis operators. Some of the implications of the assumptions (N1) through (N5) are described in the following lemma.

Lemma 14.2. *For an operator $\Phi : C(\mathbb{R}_+, \mathbb{R}) \to C(\mathbb{R}_+, \mathbb{R})$ the following statements hold.*

(i) If Φ satisfies (N1), (N2), and (N5), then for all $u \in C(\mathbb{R}_+, \mathbb{R})$ and all $\alpha \in \mathbb{R}_+$, there exists $\delta > 0$ such that for all $t \in [\alpha, \alpha + \delta]$,

$$|(\Phi(u))(t) - (\Phi(u))(\alpha)| \leq \lambda \sup_{\tau \in [\alpha, t]} |u(\tau) - u(\alpha)| . \tag{14.3}$$

(ii) If Φ satisfies (N1) through (N3) and (N5), then for all $u \in AC(\mathbb{R}_+, \mathbb{R})$,

$$\left| \frac{d}{dt}(\Phi(u))(t) \right| \leq \lambda |\dot{u}(t)| , \quad \text{a.e. } t \in \mathbb{R}_+ . \tag{14.4}$$

(iii) If Φ satisfies (N1) through (N5), then for every $u \in AC(\mathbb{R}_+, \mathbb{R})$, there exists a measurable function $d_u : \mathbb{R}_+ \to [0, \lambda]$ such that

$$\frac{d}{dt}(\Phi(u))(t) = d_u(t)\dot{u}(t) , \quad \text{a.e. } t \in \mathbb{R}_+ . \tag{14.5}$$

Proof: To prove statement (i), let $u \in C(\mathbb{R}_+, \mathbb{R})$ and $\alpha \in \mathbb{R}_+$ and define $w \in C([0, \alpha], \mathbb{R})$ by $w(t) = u(t)$ for all $t \in [0, \alpha]$. By (N5), there exist numbers $\delta_1, \delta_2 > 0$ such that for all $v_1, v_2 \in \mathcal{C}(w; \delta_1, \delta_2)$,

$$\sup_{t \in [\alpha, \alpha+\delta_2]} |(\Phi(v_1))(t) - (\Phi(v_2))(t)| \leq \lambda \sup_{t \in [\alpha, \alpha+\delta_2]} |v_1(t) - v_2(t)| .$$

By continuity of u, there exists $\delta \in (0, \delta_2)$ such that $\mathbf{Q}_t u \in \mathcal{C}(w; \delta_1, \delta_2)$ for all $t \in [\alpha, \alpha + \delta]$. Thus, using (N1) and (N2), we may conclude that for $t \in [\alpha, \alpha + \delta]$,

$$\begin{aligned} |(\Phi(u))(t) - (\Phi(u))(\alpha)| &\leq \sup_{\tau \in [\alpha, t]} |(\Phi(u))(\tau) - (\Phi(u))(\alpha)| \\ &= \sup_{\tau \in [\alpha, \alpha+\delta_2]} |(\Phi(\mathbf{Q}_t u))(\tau) - (\Phi(\mathbf{Q}_\alpha u))(\tau)| \\ &\leq \lambda \sup_{\tau \in [\alpha, \alpha+\delta_2]} |(\mathbf{Q}_t u)(\tau) - (\mathbf{Q}_\alpha u)(\tau)| \\ &= \lambda \sup_{\tau \in [\alpha, t]} |u(\tau) - u(\alpha)| , \end{aligned}$$

which is (14.3).

To prove statements (ii) and (iii), let $u \in AC(\mathbb{R}_+, \mathbb{R})$. Let E be the set of all $t \in \mathbb{R}_+$ such that u or $\Phi(u)$ is not differentiable at t. By (N3), E is of measure zero. Using statement (i), we obtain for all $t \in \mathbb{R}_+ \setminus E$,

$$\begin{aligned} \left| \frac{d}{dt}(\Phi(u))(t) \right| &= \lim_{\varepsilon \downarrow 0} \frac{|(\Phi(u))(t+\varepsilon) - (\Phi(u))(t)|}{\varepsilon} \\ &\leq \lambda \lim_{\varepsilon \downarrow 0} \frac{\sup_{\tau \in [t, t+\varepsilon]} |u(\tau) - u(t)|}{\varepsilon} \\ &\leq \lambda \lim_{\varepsilon \downarrow 0} \left(\sup_{\tau \in (t, t+\varepsilon]} \left| \frac{u(\tau) - u(t)}{\tau - t} \right| \right) = \lambda |\dot{u}(t)| , \end{aligned}$$

which is (14.4). Finally, to prove statement (iii), let $E' \subset \mathbb{R}_+$ be of measure zero and such that $E \subset E'$ and

$$\left|\frac{d}{dt}(\Phi(u))(t)\right| \leq \lambda|\dot{u}(t)|, \quad \frac{d}{dt}(\Phi(u))(t)\,\dot{u}(t) \geq 0; \qquad \forall\, t \in \mathbb{R}_+ \setminus E'.$$

The existence of such a set E' is guaranteed by statement (ii) and (N4). Set $F = \{t \in \mathbb{R}_+ \setminus E \mid \dot{u}(t) = 0\}$ and define

$$d_u(t) = \begin{cases} \frac{d}{dt}(\Phi(u))(t)/\dot{u}(t) & \text{if } t \in \mathbb{R}_+ \setminus (E' \cup F), \\ 0 & \text{if } t \in E' \cup F. \end{cases}$$

By construction the function d_u is measurable, $d_u(t) \in [0, \lambda]$ for all $t \in \mathbb{R}_+$, and (14.5) holds. □

The following remark proves useful in Section 14.5.

Remark 14.2. Consider the following assumption which is slightly stronger than Assumption (N4).

(N4′) Φ is monotone in the sense that for all $u \in AC(\mathbb{R}_+, \mathbb{R})$ with $\Phi(u) \in AC(\mathbb{R}_+, \mathbb{R})$,

$$\frac{d}{dt}(\Phi(u))(t)\,\dot{u}(t) \geq 0, \quad \forall\, t \in \mathbb{R}_+ \setminus E_u,$$

where E_u is the set of all $t \in \mathbb{R}_+$ such that u or $\Phi(u)$ is not differentiable at t.

If in statement (iii) of Lemma 14.2, (N4) is replaced by (N4′), then for every $u \in AC(\mathbb{R}_+, \mathbb{R})$, there exists a measurable function $d_u : \mathbb{R}_+ \to [0, \lambda]$ such that

$$\frac{d}{dt}(\Phi(u))(t) = d_u(t)\dot{u}(t), \quad \forall\, t \in \mathbb{R}_+ \setminus E_u.$$

This follows from the observation that $E' = E$ in the proof of Lemma 14.2.

We are now in the position to define the class of nonlinear operators we consider in the context of the low-gain integral control problem in Section 14.4. If $\Phi : C(\mathbb{R}_+, \mathbb{R}) \to C(\mathbb{R}_+, \mathbb{R})$ satisfies (N5), then any number $l > 0$ such that (N5) holds for $\lambda = l$, is called a *Lipschitz constant* of Φ.

Definition 14.1. *Let $\lambda > 0$. The set of all operators $\Phi : C(\mathbb{R}_+, \mathbb{R}) \to C(\mathbb{R}_+, \mathbb{R})$ satisfying (N1) through (N8) and having Lipschitz constant λ is denoted by $\mathcal{N}(\lambda)$.*

We next introduce a concept of Lipschitz continuity for operators from $C(\mathbb{R}_+, \mathbb{R})$ to $C(\mathbb{R}_+, \mathbb{R})$ and show that if $\Phi : C(\mathbb{R}_+, \mathbb{R}) \to C(\mathbb{R}_+, \mathbb{R})$ is Lipschitz continuous and satisfies (N1) and (N2), then Φ also satisfies (N3), (N5), and (N8).

Definition 14.2. *Let* $\mathcal{C} = C(\mathbb{R}_+, \mathbb{R})$ *or* $\mathcal{C} = C_{\mathrm{pm}}(\mathbb{R}_+, \mathbb{R})$. *An operator* $\Phi : \mathcal{C} \to C(\mathbb{R}_+, \mathbb{R})$ *is called* Lipschitz continuous with Lipschitz continuity constant $l > 0$ *if*

$$\sup_{t\in\mathbb{R}_+} |(\Phi(u))(t) - (\Phi(v))(t)| \leq l \sup_{t\in\mathbb{R}_+} |u(t) - v(t)|\,, \quad \forall\, u, v \in \mathcal{C}\,.$$

For later convenience we define for every $\tau \geq 0$ a seminorm

$$\sigma_\tau : C(\mathbb{R}_+, \mathbb{R}) \to \mathbb{R}_+\,, \quad u \mapsto \sup_{t\in[0,\tau]} |u(t)|\,.$$

Remark 14.3. Let $\mathcal{C} = C(\mathbb{R}_+, \mathbb{R})$ or $\mathcal{C} = C_{\mathrm{pm}}(\mathbb{R}_+, \mathbb{R})$. Then $\Phi : \mathcal{C} \to C(\mathbb{R}_+, \mathbb{R})$ is causal and Lipschitz continuous with Lipschitz continuity constant $l > 0$ if and only if

$$\sigma_\tau(\Phi(u) - \Phi(v)) \leq l\sigma_\tau(u - v)\,, \quad \forall\, u, v \in \mathcal{C}\,, \ \forall\, \tau \in \mathbb{R}_+\,.$$

Lemma 14.3. *If* $\Phi : C(\mathbb{R}_+, \mathbb{R}) \to C(\mathbb{R}_+, \mathbb{R})$ *is Lipschitz continuous with Lipschitz continuity constant* $l > 0$ *and satisfies (N1) and (N2), then assumptions (N3), (N5) (with Lipschitz constant* $\lambda = l$*), and (N8) hold.*

Proof: Let $\Phi : C(\mathbb{R}_+, \mathbb{R}) \to C(\mathbb{R}_+, \mathbb{R})$ be Lipschitz continuous with Lipschitz continuity constant $l > 0$ and let Φ satisfy assumptions (N1) and (N2). To show that (N3) holds, let $u \in AC(\mathbb{R}_+, \mathbb{R})$, $\varepsilon > 0$, and $b > a \geq 0$. Then there exists $\delta > 0$ such that

$$\sum_{k=1}^{n} |u(b_k) - u(a_k)| \leq \frac{\varepsilon}{l}\,,$$

for every finite family of pairwise disjoint subintervals $(a_k, b_k) \subset [a, b]$ of total length

$$\sum_{k=1}^{n} (b_k - a_k) \leq \delta\,. \tag{14.6}$$

Since u is continuous, there exists $c_k \in [a_k, b_k]$ such that

$$|u(c_k) - u(a_k)| = \max_{t\in[a_k,b_k]} |u(t) - u(a_k)|\,.$$

Using (N1), (N2), Lipschitz continuity, and Remark 14.3, we obtain for any $\tau_1, \tau_2 \in \mathbb{R}_+$ with $\tau_2 \geq \tau_1$

$$\begin{aligned} |(\Phi(u))(\tau_2) - (\Phi(u))(\tau_1)| &= |(\Phi(\mathbf{Q}_{\tau_2} u))(\tau_2) - (\Phi(\mathbf{Q}_{\tau_1} u))(\tau_2)| \\ &\leq l\sigma_{\tau_2}(\mathbf{Q}_{\tau_2} u - \mathbf{Q}_{\tau_1} u) \\ &= l \max_{t\in[\tau_1,\tau_2]} |u(t) - u(\tau_1)|\,. \end{aligned} \tag{14.7}$$

Now suppose that the family of intervals (a_k, b_k) satisfies (14.6). Then

$$\sum_{k=1}^{n}(c_k - a_k) \le \delta\,,$$

and so

$$\sum_{k=1}^{n} \max_{t\in[a_k,b_k]} |u(t) - u(a_k)| = \sum_{k=1}^{n} |u(c_k) - u(a_k)| \le \frac{\varepsilon}{l}\,. \tag{14.8}$$

Using (14.7) and (14.8), we may conclude

$$\sum_{k=1}^{n} |(\Phi(u))(b_k) - (\Phi(u))(a_k)| \le l \sum_{k=1}^{n} \max_{t\in[a_k,b_k]} |u(t) - u(a_k)| \le \varepsilon\,,$$

showing that $\Phi(u) \in AC(\mathbb{R}_+, \mathbb{R})$.

By (N1) and Lipschitz continuity (with Lipschitz continuity constant l), it is clear that (N5) holds with Lipschitz constant $\lambda = l$. Finally, we show that (N8) is satisfied. To this end let $a > 0$ and $u \in C([0, a), \mathbb{R})$, then by (N1), Lipschitz continuity, and Remark 14.3,

$$\sigma_\tau(\Phi(\mathbf{Q}_\tau u) - \Phi(0)) \le l\sigma_\tau(\mathbf{Q}_\tau u)\,, \quad \forall \tau \in [0, a)\,.$$

Therefore, by (N1) and (N2),

$$\sup_{t\in[0,\tau]} |(\Phi(u))(t)| \le l \sup_{t\in[0,\tau]} |u(t)| + |(\Phi(0))(0)|\,, \quad \forall \tau \in [0, a)\,,$$

showing that assumption (N8) is satisfied with $\alpha = (\Phi(0))(0)$ and $\beta = l$. □

For future reference we state the following lemma.

Lemma 14.4. *Let $\Phi : C(\mathbb{R}_+, \mathbb{R}) \to C(\mathbb{R}_+, \mathbb{R})$ be Lipschitz continuous. If for any ultimately nondecreasing $u \in C(\mathbb{R}_+, \mathbb{R})$ with $\lim_{t\to\infty} u(t) = \infty$,*

$$\lim_{t\to\infty} (\Phi(u))(t) = \sup \mathrm{NVS}\,\Phi \quad \textit{and} \quad \lim_{t\to\infty} (\Phi(-u))(t) = \inf \mathrm{NVS}\,\Phi\,,$$

then Φ satisfies (N6).

The proof of Lemma 14.4 is straightforward and is therefore omitted.

14.4 Integral Control in the Presence of Input Nonlinearities Satisfying (N1) to (N8)

In the following, let (A, B, C, D) be the generating operators of a linear single-input single-output regular system Σ with state space X and transfer function $\mathbf{G}$, and let $\Phi : C(\mathbb{R}_+, \mathbb{R}) \to C(\mathbb{R}_+, \mathbb{R})$ be a dynamic input nonlinearity. Denoting the constant reference signal by r, an application of the integrator

$$u(t) = u_0 + k \int_0^t [r - C_L x(\tau) - D(\Phi(u))(\tau)]\, d\tau\,,$$

where k is a real parameter (see Figure 14.1), leads to the following nonlinear system of differential equations

$$\dot{x} = Ax + B\Phi(u)\,, \qquad x(0) = x_0 \in X\,, \tag{14.9a}$$

$$\dot{u} = k[r - C_L x - D\Phi(u)]\,, \qquad u(0) = u_0 \in \mathbb{R}\,. \tag{14.9b}$$

A continuous function

$$[0, \tau) \to X \times \mathbb{R}\,, \quad t \mapsto (x(t), u(t))$$

is called a *solution* of (14.9) if $(x(\cdot), u(\cdot))$ is absolutely continuous as a $(X_{-1} \times \mathbb{R})$-valued function, $x(t) \in \operatorname{dom}(C_L)$ for a.e. $t \in [0, \tau)$, $(x(0), u(0)) = (x_0, u_0)$, and the differential equations in (14.9) are satisfied a.e. on $[0, \tau)$. Of course, the derivative on the left-hand side of (14.9a) has to be understood in X_{-1}.[2]

An application of a well-known result on abstract Cauchy problems (see Pazy [18, Theorem 2.4, p. 107]) shows that a continuous $(X \times \mathbb{R})$-valued function $(x(\cdot), u(\cdot))$ is a solution of (14.9) if and only if it satisfies the following integrated version of (14.9),

$$x(t) = \mathbf{T}_t x_0 + \int_0^t \mathbf{T}_{t-\tau} B(\Phi(u))(\tau)\, d\tau\,,$$

$$u(t) = u_0 + k \int_0^t [r - C_L x(\tau) - D(\Phi(u))(\tau)]\, d\tau\,.$$

The next result shows that (14.9) has a unique solution.

Proposition 14.1. *For any $\Phi : C(\mathbb{R}_+, \mathbb{R}) \to C(\mathbb{R}_+, \mathbb{R})$ satisfying (N1), (N5), and (N8) and any pair $(x_0, u_0) \in X \times \mathbb{R}$ of initial conditions, there exists a unique solution $(x(\cdot), u(\cdot))$ of (14.9) defined on $\mathbb{R}_+$.*

For the proof of the above result it is useful to consider the following initial-value problem for u,

$$\dot{u} = k[r - \mathbf{\Psi} x_0 - \mathbf{F}(\Phi(u))]\,, \qquad u(0) = u_0\,, \tag{14.10}$$

---

[2] Being a Hilbert space, $X_{-1} \times \mathbb{R}$ is reflexive. Hence any absolutely continuous $(X_{-1} \times \mathbb{R})$-valued function is a.e. differentiable and can be recovered from its derivative by integration; see [2, Theorem 3.1, p. 10].

where $\mathbf{\Psi}$ and $\mathbf{F}$ are the state-to-output and input-to-output operators of Σ, respectively (see Section 14.2). Clearly, (14.10) is obtained from (14.9b) on noting that $C_L x(t) + D(\Phi(u))(t) = (\mathbf{\Psi} x_0)(t) + [\mathbf{F}(\Phi(u))](t)$. An absolutely continuous function $u : [0, \tau) \to \mathbb{R}$ is a *solution* of (14.10) if $u(0) = u_0$ and the differential equation in (14.10) is satisfied a.e. on $[0, \tau)$.

Lemma 14.5. *Let $x_0 \in X$. For any $\Phi : C(\mathbb{R}_+, \mathbb{R}) \to C(\mathbb{R}_+, \mathbb{R})$ satisfying (N1), (N5) and (N8) and any initial condition $u_0 \in \mathbb{R}$, there exists a unique solution $u(\cdot)$ of (14.10) defined on $\mathbb{R}_+$.*

The proof of this lemma is relegated to the Appendix (Section 14.7).

Proof: **(of Proposition 14.1)** Let $u : \mathbb{R}_+ \to \mathbb{R}$ be the unique solution of (14.10) (the existence of such a solution is guaranteed by Lemma 14.5) and *define* $x(\cdot)$ to be the unique solution of

$$\dot{x} = Ax + B\Phi(u), \qquad x(0) = x_0 .$$

Then $(x(\cdot), u(\cdot))$ is the unique solution of (14.9) defined on $\mathbb{R}_+$. □

If $\mathbf{G}$ is holomorphic and bounded on $\mathbb{C}_\alpha$ for some $\alpha < 0$ (which is the case if $\mathbf{T}_t$ is exponentially stable) and $\mathbf{G}(0) > 0$, then it is easy to show that

$$1 + k \operatorname{Re} \frac{\mathbf{G}(s)}{s} \geq 0, \quad \forall\, s \in \mathbb{C}_0, \tag{14.11}$$

for all sufficiently small $k > 0$; see [11, Lemma 3.10]. We define

$$K := \sup\{k > 0 \,|\, (14.11) \text{ holds}\}. \tag{14.12}$$

Henceforth, let $\mathcal{M}_f(\mathbb{R}_+)$ denote the space of all finite signed Borel measures on $\mathbb{R}_+$. Recall that a signed measure μ on $\mathbb{R}_+$ is called finite if $|\mu|(\mathbb{R}_+) < \infty$, where $|\mu|$ denotes the total variation of μ.

The main result of this section is the following theorem.

Theorem 14.1. *Let $\lambda > 0$. Assume that $\Phi \in \mathcal{N}(\lambda)$, $\mathfrak{L}^{-1}(\mathbf{G}) \in \mathcal{M}_f(\mathbb{R}_+)$, $\mathbf{T}_t$ is exponentially stable, $\mathbf{G}(0) > 0$, $k \in (0, K/\lambda)$, and $r \in \mathbb{R}$ is such that*

$$\Phi_r := r/\mathbf{G}(0) \in \operatorname{clos}(\mathrm{NVS}\, \Phi). \tag{14.13}$$

Then, for all $(x_0, u_0) \in X \times \mathbb{R}$, a unique solution $(x(\cdot), u(\cdot))$ of (14.9) exists on $\mathbb{R}_+$ and satisfies

(i) $\lim_{t\to\infty} (\Phi(u))(t) = \Phi_r$,

(ii) $\lim_{t\to\infty} \|x(t) + A^{-1}B\Phi_r\| = 0$,

(iii) $\lim_{t\to\infty}[r - y(t) + (\Psi x_0)(t)] = 0$, *where* $y(t) = C_L x(t) + D(\Phi(u))(t)$,

(iv) *if* $\Phi_r \in \operatorname{int}(\mathrm{NVS}\,\Phi)$, *then* $u(\cdot)$ *is bounded.*

Remark 14.4. (i) Since $(\Psi x_0)(t)$ converges exponentially to 0 as $t \to \infty$ for all $x_0 \in X_1 = \operatorname{dom}(A)$, it follows from (iii) that the error $e(t) = r - y(t)$ converges to 0 for all $x_0 \in \operatorname{dom}(A)$. If C is bounded, then this statement is true for all $x_0 \in X$. If C is unbounded and $x_0 \notin \operatorname{dom}(A)$, then $e(t)$ does not necessarily converge to 0 as $t \to \infty$. However, the proof of Theorem 14.1 shows that $e(t)$ is small for large t in the sense that $e(t) = e_1(t) + e_2(t)$, where the function e_1 is bounded with $\lim_{t\to\infty} e_1(t) = 0$ and $e_2 \in L^2_\alpha(\mathbb{R}_+, \mathbb{R})$ for some $\alpha < 0$.

(ii) The assumption that $\mathfrak{L}^{-1}(\mathbf{G}) \in \mathcal{M}_f(\mathbb{R}_+)$ is not very restrictive and seems to be satisfied in all practical examples of exponentially stable systems. In particular, this assumption is satisfied if B or C is bounded (see [10, Lemma 2.3]).

(iii) In applying Theorem 14.1 it is important to know the constant K or at least a lower bound for K. In principle, K can be obtained from frequency/step response experiments performed on the linear part of the plant; see [12] for details.

Proof: **(of Theorem 14.1)** By Proposition 14.1, there exists a unique solution of (14.9) on $\mathbb{R}_+$. We denote this solution by $(x(\cdot), u(\cdot))$ and introduce new variables by defining

$$z(t) := x(t) + A^{-1}B(\Phi(u))(t), \qquad v(t) := (\Phi(u))(t) - \Phi_r; \qquad \forall\, t \geq 0.$$

By regularity it follows that $z(t) \in \operatorname{dom}(C_L)$ for a.e. $t \in \mathbb{R}_+$. Moreover, by Lemma 14.2 (iii), there exists a measurable function $d_u : \mathbb{R}_+ \to [0, \lambda]$ such that $\dot{v}(t) = d_u(t)\dot{u}(t)$ for a.e. $t \in \mathbb{R}_+$. Therefore an easy calculation yields that for a.e. $t \in \mathbb{R}_+$,

$$\dot{z}(t) = Az(t) - kd_u(t)A^{-1}B(C_L z(t) + \mathbf{G}(0)v(t)), \qquad z(0) = z_0, \quad (14.14a)$$

$$\dot{v}(t) = -kd_u(t)(C_L z(t) + \mathbf{G}(0)v(t)), \qquad v(0) = v_0, \quad (14.14b)$$

where

$$z_0 := x_0 + A^{-1}B(\Phi(u))(0), \qquad v_0 := (\Phi(u))(0) - \Phi_r.$$

The derivative on the left-hand side of (14.14a) has to be understood in X_{-1}. We observe that, while in these new variables we still have an unbounded operator $A^{-1}BC_L$, the operator $A^{-1}B$ is in $L(\mathbb{R}, X)$. Since d_u is a measurable function satisfying $d_u(t) \in [0, \lambda]$ for all $t \in \mathbb{R}_+$ and $k \in (0, K/\lambda)$, it follows from the Lyapunov argument developed in [13] (see the proof of Theorem 3.3 in [13]) that the limit of $v(t)$ as $t \to \infty$ exists and is finite, and hence there exists a number $L \in \mathbb{R}$ such that

$$\lim_{t\to\infty} (\Phi(u))(t) = L.$$

The essence of the proof is to show that $L = \Phi_r$. Setting

$$y_0(t) = (\Psi x_0)(t), \qquad y_1(t) = [\mathfrak{L}^{-1}(\mathbf{G}) \star (\Phi(u))](t),$$

where $\star$ denotes convolution, we have

$$\dot{u}(t) = k[r - y_0(t) - y_1(t)], \quad \text{a.e. } t \in \mathbb{R}. \tag{14.15}$$

Since $\lim_{t\to\infty}(\Phi(u))(t) = L$ and $\mathfrak{L}^{-1}(\mathbf{G}) \in \mathcal{M}_f(\mathbb{R}_+)$, it follows that

$$\lim_{t\to\infty} y_1(t) = \mathbf{G}(0)L; \tag{14.16}$$

see [8, Theorem 6.1 (ii), p. 96]. Define a function $\tilde{y}_1 : \mathbb{R}_+ \to \mathbb{R}$ by setting

$$\tilde{y}_1(t) = r - y_1(t) = \mathbf{G}(0)\Phi_r - y_1(t).$$

Seeking a contradiction, suppose that $L \neq \Phi_r$. Then, either $\Phi_r > L$ or $\Phi_r < L$. If $\Phi_r > L$, then by (14.16), there exists a number $\tau_0 \geq 0$ such that

$$\tilde{y}_1(t) \geq \frac{1}{2}\mathbf{G}(0)(\Phi_r - L) > 0, \quad \forall\, t \geq \tau_0. \tag{14.17}$$

Hence, integrating (14.15) yields

$$u(t) = u(\tau) + k\left(\int_\tau^t \tilde{y}_1(s)\,ds - \int_\tau^t y_0(s)\,ds\right), \quad t \geq \tau \geq \tau_0. \tag{14.18}$$

By exponential stability, $y_0 \in L^2_\alpha(\mathbb{R}_+, \mathbb{R})$ for some $\alpha < 0$, and thus $y_0 \in L^1(\mathbb{R}_+, \mathbb{R})$. Therefore, for given $\varepsilon > 0$, there exists $\tau_\varepsilon \geq \tau_0$ such that

$$\int_{\tau_\varepsilon}^\infty |y_0(s)|\,ds \leq \frac{\varepsilon}{k}. \tag{14.19}$$

Defining $u_\varepsilon \in C(\mathbb{R}_+, \mathbb{R})$ by

$$u_\varepsilon(t) = \begin{cases} u(t) & \text{for } 0 \leq t \leq \tau_\varepsilon, \\ u(\tau_\varepsilon) + k\int_{\tau_\varepsilon}^t \tilde{y}_1(s)\,ds & \text{for } t > \tau_\varepsilon, \end{cases}$$

it follows from (14.17) that u_ε is ultimately nondecreasing, and moreover, by (14.18) and (14.19),

$$|u(t) - u_\varepsilon(t)| \leq \varepsilon, \quad \forall\, t \in \mathbb{R}_+,$$

showing that u is approximately ultimately nondecreasing. Since $u(t) \to \infty$ as $t \to \infty$, we may invoke (N6) to conclude that

$$\Phi_r > L = \lim_{t\to\infty}(\Phi(u))(t) = \sup \mathrm{NVS}\,\Phi,$$

which is in contradiction to (14.13). If $\Phi_r < L$, then a very similar argument shows that $-u$ is approximately ultimately nondecreasing and $\lim_{t\to\infty}(-u)(t) = \infty$. Invoking (N6) gives

$$\Phi_r < L = \lim_{t\to\infty}(\Phi(u))(t) = \inf \mathrm{NVS}\,\Phi,$$

which again is in contradiction to (14.13). Therefore, we may conclude that $L = \Phi_r$ and thus $\lim_{t\to\infty}(\Phi(u))(t) = \Phi_r$, which is statement (i). Statement (ii) follows from statement (i) and Lemma 14.1. Statement (iii) is a consequence of statement (i), together with the identity

$$r - y(t) + (\mathbf{\Psi} x_0)(t) = \mathbf{G}(0)\Phi_r - [\mathfrak{L}^{-1}(\mathbf{G}) \star (\Phi(u))](t)\,,$$

and the fact that $\lim_{t\to\infty}[\mathfrak{L}^{-1}(\mathbf{G}) \star (\Phi(u))](t) = \mathbf{G}(0)\Phi_r$. Finally, to prove statement (iv), let $\Phi_r \in \operatorname{int} \mathrm{NVS}\,\Phi$. Then, boundedness of u follows immediately from statement (i) and (N7). □

Remark 14.5. (i) For $\lambda, \tilde{\lambda} > 0$, define $\mathcal{N}(\lambda, \tilde{\lambda})$ to be the set of all $\Phi \in \mathcal{N}(\lambda)$ such that for all $u \in AC(\mathbb{R}_+, \mathbb{R})$, there exists a measurable function $d_u : \mathbb{R}_+ \to [0, \tilde{\lambda}]$ such that (14.5) holds. Of course, by Lemma 14.2 (iii), $\mathcal{N}(\lambda, \tilde{\lambda}) = \mathcal{N}(\lambda)$ if $\tilde{\lambda} \geq \lambda$. The proof of Theorem 14.1 shows that if $\Phi \in \mathcal{N}(\lambda, \tilde{\lambda})$, then statements (i) through (iv) of Theorem 14.1 are true for all $k \in (0, K/\tilde{\lambda})$. If $\tilde{\lambda} < \lambda$, then $K/\tilde{\lambda} > K/\lambda$, which means that in this case a wider range of gain parameters k may be used. This observation is relevant for Section 14.5, where we show that there exist $\lambda, \tilde{\lambda} > 0$ with $\tilde{\lambda} < \lambda$ and hysteresis operators $\Phi \in \mathcal{N}(\lambda, \tilde{\lambda})$ such that $\Phi \notin \mathcal{N}(l)$ for all $l \in (0, \lambda)$.

(ii) We see from the proof of Theorem 14.1 that (N7) is only needed for statement (iv).

14.5 Hysteresis Nonlinearities Satisfying (N1) to (N8)

In this section we consider various classes of hysteresis operators and we show that under certain conditions these operators satisfy (N1) to (N8).

Static nonlinearities

Although static nonlinearities do not describe hysteresis phenomena, we include them here because (i) they form a special subclass of hysteresis operators as defined in [4] and (ii) we would like to recover the main result in [13] whose content is essentially Theorem 14.1 for *static* nondecreasing globally Lipschitz nonlinearities.

For a continuous function $\phi : \mathbb{R} \to \mathbb{R}$, define the corresponding static nonlinearity by

$$\mathcal{S}_\phi : C(\mathbb{R}_+, \mathbb{R}) \to C(\mathbb{R}_+, \mathbb{R})\,, \quad u \mapsto \phi \circ u\,.$$

The proof of the following proposition is straightforward and is therefore left to the reader.

Proposition 14.2. *If $\phi : \mathbb{R} \to \mathbb{R}$ is nondecreasing and globally Lipschitz with Lipschitz constant $\lambda > 0$, then $\mathcal{S}_\phi \in \mathcal{N}(\lambda)$.*

By a combination of Theorem 14.1 and Proposition 14.2 we recover the main result of [13]. Note that saturation and deadzone nonlinearities satisfy the assumptions of Proposition 14.2.

Relay hysteresis

In *relay* (also called *passive* or *positive*) hysteresis, the relationship between input and output is determined by two threshold values $a_1 < a_2$ for the input. The output $v(t) = (\mathcal{R}_\xi(u))(t)$ moves, for a given continuous input $u(t)$, on one of two fixed curves $\rho_1 : [a_1, \infty) \to \mathbb{R}$ and $\rho_2 : (-\infty, a_2] \to \mathbb{R}$ (see Figure 14.2), depending on which threshold, a_1 or a_2, was last attained. In the following we restrict our attention to "continuous" relay hysteresis nonlinearities; that is, the two curves ρ_1 and ρ_2 join at a_1 and a_2.

More formally, let $a_1, a_2 \in \mathbb{R}$ with $a_1 < a_2$ and let $\rho_1 : [a_1, \infty) \to \mathbb{R}$ and $\rho_2 : (-\infty, a_2] \to \mathbb{R}$ be continuous and such that $\rho_1(a_1) = \rho_2(a_1)$ and $\rho_1(a_2) = \rho_2(a_2)$. For $u \in C(\mathbb{R}_+, \mathbb{R})$ and $t \geq 0$ define

$$S(u,t) := u^{-1}(\{a_1, a_2\}) \cap [0,t]\,, \quad \tau(u,t) := \begin{cases} \max S(u,t) & \text{if } S(u,t) \neq \emptyset\,, \\ -1 & \text{if } S(u,t) = \emptyset\,. \end{cases}$$

Following Macki et al. [16], for each $\xi \in \mathbb{R}$, we define an operator $\mathcal{R}_\xi : C(\mathbb{R}_+, \mathbb{R}) \to C(\mathbb{R}_+, \mathbb{R})$ by

$$(\mathcal{R}_\xi(u))(t) = \begin{cases} \rho_2(u(t)) & \text{if } u(t) \leq a_1, \\ \rho_1(u(t)) & \text{if } u(t) \geq a_2, \\ \rho_2(u(t)) & \text{if } u(t) \in (a_1, a_2),\ \tau(u,t) \neq -1,\ u(\tau(u,t)) = a_1, \\ \rho_1(u(t)) & \text{if } u(t) \in (a_1, a_2),\ \tau(u,t) \neq -1,\ u(\tau(u,t)) = a_2, \\ \rho_1(u(t)) & \text{if } u(t) \in (a_1, a_2),\ \tau(u,t) = -1,\ \xi > 0, \\ \rho_2(u(t)) & \text{if } u(t) \in (a_1, a_2),\ \tau(u,t) = -1,\ \xi \leq 0. \end{cases} \tag{14.20}$$

The number ξ plays the role of an "initial state" that determines the output value $(\mathcal{R}_\xi(u))(t)$ if $u(s) \in (a_1, a_2)$ for all $s \in [0,t]$. The operator $\mathcal{R}_\xi$ is called a *relay hysteresis* operator and is illustrated in Figure 14.2.

Proposition 14.3. *If ρ_1 and ρ_2 are both nondecreasing and globally Lipschitz with Lipschitz constant $\lambda > 0$, then for each $\xi \in \mathbb{R}$, the operator $\mathcal{R}_\xi$ defined by* (14.20) *is in $\mathcal{N}(\lambda)$.*

Proof: A straightforward consequence of the definition of the relay hysteresis operator is that $\mathcal{R}_\xi$ satisfies conditions (N1), (N2), (N5), (N6), and (N8). To show that (N3) and (N4) hold, let $u \in AC(\mathbb{R}_+, \mathbb{R})$. For any compact interval $J \subset \mathbb{R}_+$,

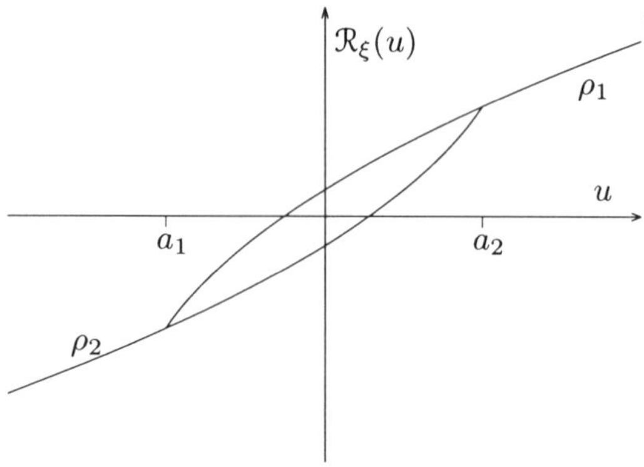

FIGURE 14.2. Relay hysteresis.

u is uniformly continuous on J, and therefore, using that $a_1 \neq a_2$, there exists $\delta > 0$, such that for all $t_1, t_2 \in J$,

$$u(t_1) = a_1 , \quad u(t_2) = a_2 \quad \Longrightarrow \quad |t_2 - t_1| \geq \delta .$$

As a consequence, there exist numbers $t_i \uparrow \infty$ as $i \to \infty$, with $t_0 = 0$ and a map $j : \mathbb{N} \to \{1, 2\}$ such that for all $i \in \mathbb{N}$,

$$(\mathcal{R}_\xi(u))(t) = \rho_{j(i)}(u(t)) , \quad \forall\, t \in [t_i, t_{i+1}] . \tag{14.21}$$

It follows that $\mathcal{R}_\xi(u)$ is absolutely continuous on $[t_i, t_{i+1}]$ for each $i \in \mathbb{N}$. Hence, by continuity of $\mathcal{R}_\xi(u)$, we may conclude that $\mathcal{R}_\xi(u) \in AC(\mathbb{R}_+, \mathbb{R})$, showing that (N3) holds. Furthermore, since ρ_1 and ρ_2 are nondecreasing and Lipschitz, (14.21) yields that for all $i \in \mathbb{N}$,

$$\frac{d}{dt}(\mathcal{R}_\xi(u))(t)\dot{u}(t) \geq 0 , \quad \text{a.e. } t \in [t_i, t_{i+1}] ,$$

which implies that (N4) holds.

Finally, to show that (N7) is satisfied, note first that $\mathrm{NVS}\,\mathcal{R}_\xi = \mathrm{im}\,\rho_1 \cup \mathrm{im}\,\rho_2$. Let $u \in C(\mathbb{R}_+, \mathbb{R})$ and suppose $\lim_{t\to\infty}(\mathcal{R}_\xi(u))(t) = l \in \mathrm{int}\,\mathrm{NVS}\,\mathcal{R}_\xi$. Then there exist $\varepsilon > 0$ and $T \geq 0$ such that $I_\varepsilon := (l - \varepsilon, l + \varepsilon) \subset \mathrm{int}\,\mathrm{NVS}\,\mathcal{R}_\xi$ and $(\mathcal{R}_\xi(u))(t) \in I_\varepsilon$ for all $t \geq T$, which implies

$$u(t) \in \rho_1^{-1}(I_\varepsilon) \cup \rho_2^{-1}(I_\varepsilon) =: U , \quad \forall\, t \geq T . \tag{14.22}$$

But the set U is bounded since $\sup \rho_1, \inf \rho_2 \notin I_\varepsilon$, and ρ_1 and ρ_2 are nondecreasing. Combining this with (14.22) shows that u is bounded. □

We remark that although the relay hysteresis operator $\mathcal{R}_\xi$ as defined in (14.20) satisfies the Lipschitz condition (N5), $\mathcal{R}_\xi$ is not Lipschitz continuous in the sense of Definition 14.2. In fact, it is easy to show that $\mathcal{R}_\xi$ is

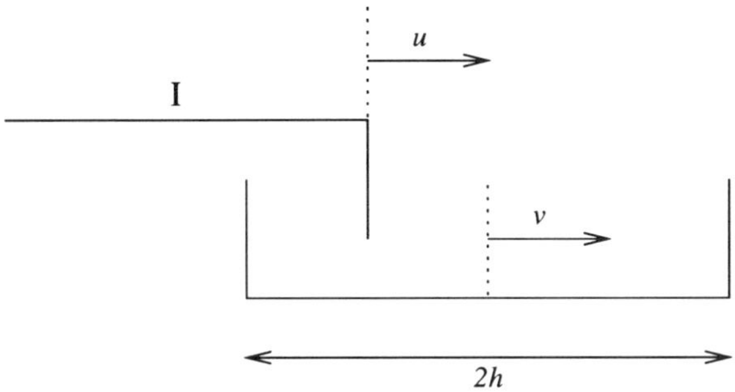

FIGURE 14.3. Schematic representation of backlash.

not even continuous with respect to the topology on $C(\mathbb{R}_+,\mathbb{R})$ given by the family of seminorms $\{\sigma_n \mid n \in \mathbb{N}\}$. In particular, when we talk about "continuous" relay hysteresis, we simply mean that the output corresponding to a continuous input is continuous, but not that the relay hysteresis operator is continuous with respect to any natural topology on $C(\mathbb{R}_+,\mathbb{R})$.

Backlash hysteresis

The *backlash* operator (also called *play* operator) has been discussed in a mathematically rigorous context in a number of references; see, for example [3, 4, 9, 19]. Intuitively, the backlash operator describes the input-output behavior of a simple mechanical play between two mechanical elements I and II shown in Figure 14.3. The position of element I at time t is denoted by $u(t)$. The position $v(t)$ of the middle point of element II at time t will remain constant as long as $u(t)$ moves in the interior and it will change at the rate $\dot{v}=\dot{u}$ as long as $u(t)$ hits the boundary of element II with an outward directed velocity.

To give a formal definition of backlash, define for each $h \in \mathbb{R}_+$ the function $b_h : \mathbb{R}^2 \to \mathbb{R}$ by

$$b_h(v,w) = \max\{v-h, \min\{v+h, w\}\}\,.$$

The proof of the following semigroup property can be found in the Appendix (Section 14.7).

Lemma 14.6. *Let $t_1 < t_2$, $u : [t_1,t_2] \to \mathbb{R}$ be monotone and $w \in [u(t_1) - h, u(t_1)+h]$. Then, for all $t,\tau \in [t_1,t_2]$ with $t \geq \tau$,*

$$b_h(u(t),w) = b_h(u(t), b_h(u(\tau),w))\,.$$

For all $h \in \mathbb{R}_+$ and all $\xi \in \mathbb{R}$ we introduce an operator $\mathcal{B}_{h,\xi}$ on $C_{\text{pm}}(\mathbb{R}_+,\mathbb{R})$

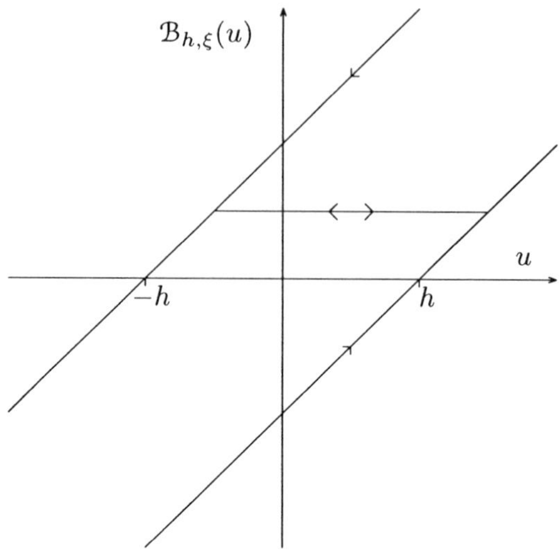

FIGURE 14.4. Backlash hysteresis.

by defining recursively for every $u \in C_{\rm pm}(\mathbb{R}_+, \mathbb{R})$,

$$(\mathcal{B}_{h,\,\xi}(u))(t) = \begin{cases} b_h(u(0), \xi) & \text{for } t = 0\,, \\ b_h(u(t), (\mathcal{B}_{h,\,\xi}(u))(t_i)) & \text{for } t_i < t \leq t_{i+1},\ i \in \mathbb{N}\,, \end{cases} \tag{14.23}$$

where $0 = t_0 < t_1 < t_2 < \ldots$ is a partition of $\mathbb{R}_+$, such that u is monotone on each of the intervals $[t_i, t_{i+1}]$. Again, ξ plays the role of an "initial state." Using Lemma 14.6 it is not difficult to show that the definition of $\mathcal{B}_{h,\,\xi}(u)$ is independent of the choice of partition. Clearly, $\mathcal{B}_{h,\,\xi}(u)$ is continuous at each $t \in \mathbb{R}$, $t \neq t_i$ for all $i \in \mathbb{N}$ and is left-continuous at t_i for all $i \in \mathbb{N}\backslash\{0\}$. Moreover, an application of Lemma 14.6 shows that $\mathcal{B}_{h,\,\xi}(u)$ is right-continuous at t_i for all $i \in \mathbb{N}$. Consequently, $\mathcal{B}_{h,\,\xi}(u)$ is continuous for all $u \in C_{\rm pm}(\mathbb{R}_+, \mathbb{R})$. The backlash operator $\mathcal{B}_{h,\,\xi}$ is illustrated in Figure 14.4.

Proposition 14.4. *Let $(h, \xi) \in \mathbb{R}_+ \times \mathbb{R}$. The backlash operator $\mathcal{B}_{h,\,\xi}$ has the following properties.*

(i) $\mathcal{B}_{h,\,\xi} : C_{\rm pm}(\mathbb{R}_+, \mathbb{R}) \to C(\mathbb{R}_+, \mathbb{R})$ is Lipschitz continuous with Lipschitz continuity constant $l = 1$ and uniquely extends to a Lipschitz continuous operator $\mathcal{B}_{h,\,\xi} : C(\mathbb{R}_+, \mathbb{R}) \to C(\mathbb{R}_+, \mathbb{R})$ with Lipschitz continuity constant $l = 1$.

(ii) $\mathcal{B}_{h,\,\xi} : C(\mathbb{R}_+, \mathbb{R}) \to C(\mathbb{R}_+, \mathbb{R})$ satisfies (N1) to (N3), (N4′), and (N5) to (N8). In particular, $\mathcal{B}_{h,\,\xi} \in \mathcal{N}(1)$.

Proof: Statement (i) follows from [4, p. 42]. For statement (ii), we first note that as an immediate consequence of the definition of $\mathcal{B}_{h,\,\xi}$ and statement (i),

(N1) and (N2) are satisfied. Combining statement (i) and Lemma 14.3, $\mathcal{B}_{h,\xi}$ also satisfies conditions (N3), (N5), and (N8). To show that (N4′) holds, let $u \in AC(\mathbb{R}_+, \mathbb{R})$ and let E be the set of all $t \in \mathbb{R}_+$ such that u or $\mathcal{B}_{h,\xi}(u)$ is not differentiable at t. Clearly, E has zero measure. We need to show that

$$\frac{d}{dt}(\mathcal{B}_{h,\xi}(u))(t)\dot{u}(t) \geq 0\,, \quad \forall\, t \in \mathbb{R}_+ \setminus E\,. \tag{14.24}$$

Let $t \in \mathbb{R}_+ \setminus E$. If $\dot{u}(t) = 0$, then (14.24) holds trivially. If $\dot{u}(t) > 0$, then there exist $t_1 > t$ and $u_n \in C_{\rm pm}(\mathbb{R}_+, \mathbb{R})$ such that $\sigma_{t_1}(u_n - u) \to 0$ as $n \to \infty$ and $u_n(\tau) \geq u_n(t)$ for all $\tau \in (t, t_1)$ and all $n \in \mathbb{N}$. It follows that $(\mathcal{B}_{h,\xi}(u_n))(\tau) \geq (\mathcal{B}_{h,\xi}(u_n))(t)$ for all $\tau \in (t, t_1)$, which in turn implies $(\mathcal{B}_{h,\xi}(u))(\tau) \geq (\mathcal{B}_{h,\xi}(u))(t)$ for all $\tau \in (t, t_1)$. Therefore

$$\frac{d}{dt}(\mathcal{B}_{h,\xi}(u))(t) = \lim_{\varepsilon \downarrow 0} \frac{(\mathcal{B}_{h,\xi}(u))(t+\varepsilon) - (\mathcal{B}_{h,\xi}(u))(t)}{\varepsilon} \geq 0\,,$$

and so (14.24) holds. If $\dot{u}(t) < 0$, then (14.24) can be obtained by a very similar argument.

To show that (N6) is satisfied, note first that NVS $\mathcal{B}_{h,\xi} = \mathbb{R}$. Let $u \in C(\mathbb{R}_+, \mathbb{R})$ be ultimately nondecreasing with $\lim_{t\to\infty} u(t) = \infty$. Then there exists $T \in \mathbb{R}_+$ such that $(\mathcal{B}_{h,\xi}(u))(t) = u(t) - h$ for all $t \geq T$. Thus, $\lim_{t\to\infty}(\mathcal{B}_{h,\xi}(u))(t) = \infty$. Similarly, $\lim_{t\to\infty}(\mathcal{B}_{h,\xi}(-u))(t) = -\infty$. It follows from statement (i) and Lemma 14.4 that (N6) holds.

For (N7), let $u \in C(\mathbb{R}_+, \mathbb{R})$ and suppose $\lim_{t\to\infty}(\mathcal{B}_{h,\xi}(u))(t) = l \in \mathbb{R}$. Then there exist $\varepsilon > 0$, $T \in \mathbb{R}_+$ such that $(\mathcal{B}_{h,\xi}(u))(t) \in (l - \varepsilon, l + \varepsilon)$ for all $t \geq T$. Consequently, $u(t) \in (l - \varepsilon - h, l + \varepsilon + h)$ for all $t \geq T$, and hence, u is bounded.
□

Remark 14.6. Let $\beta_1, \beta_2 : \mathbb{R} \to \mathbb{R}$ be continuous and such that $\operatorname{im} \beta_1 = \operatorname{im} \beta_2$ and $\beta_1(v) \leq \beta_2(v)$ for all $v \in \mathbb{R}$. Setting

$$b(v, w) = \max\{\beta_1(v), \min\{\beta_2(v), w\}\}\,,$$

we can define for each $\xi \in \mathbb{R}$ a *generalized backlash* operator $\mathcal{B}_\xi : C_{\rm pm}(\mathbb{R}_+, \mathbb{R}) \to C(\mathbb{R}_+, \mathbb{R})$ by (14.23) with b_h replaced by b. Generalized backlash (also called *generalized play*) was introduced in [9]. If β_1 and β_2 are both nondecreasing and globally Lipschitz with Lipschitz constant $\lambda > 0$, then a suitably modified version of Proposition 14.23 holds for generalized backlash, in particular $\mathcal{B}_\xi \in \mathcal{N}(\lambda)$. An example of generalized backlash is illustrated in Figure 14.5.

Elastic-plastic hysteresis

The *elastic-plastic* operator (also called *stop* operator) models the stress-strain relationship in a one-dimensional elastic-plastic element. As long as the modulus of the stress v is smaller than the yield stress h, the strain u is related to v through the linear Hooke's Law. Once the stress exceeds

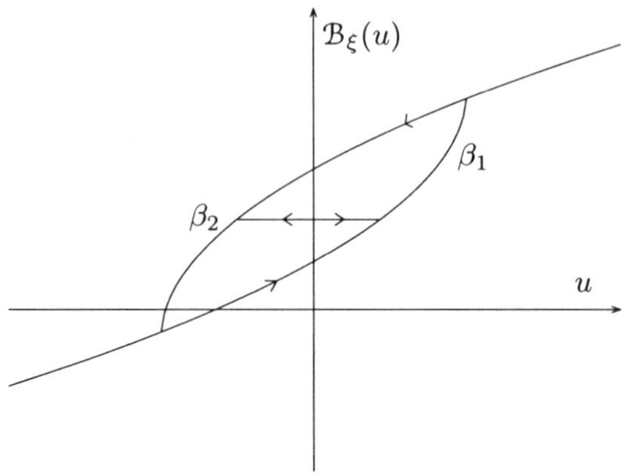

FIGURE 14.5. Generalized backlash hysteresis.

the yield value it remains constant under further increasing of the strain; however, the elastic behavior is instantly recovered when the strain is again decreased. As we show, elastic-plastic hysteresis is closely related to backlash hysteresis.

To give a formal definition of the elastic-plastic operator, define for each $h \in \mathbb{R}_+$ the function $e_h : \mathbb{R} \to \mathbb{R}$ by

$$e_h(u) = \min\{h, \max\{-h, u\}\}.$$

Following [4], for all $h \in \mathbb{R}_+$ and all $\xi \in \mathbb{R}$, we introduce an operator $\mathcal{E}_{h,\xi}$ on $C_{\mathrm{pm}}(\mathbb{R}_+, \mathbb{R})$ by defining recursively for every $u \in C_{\mathrm{pm}}(\mathbb{R}_+, \mathbb{R})$,

$$(\mathcal{E}_{h,\xi}(u))(t) = \begin{cases} e_h(u(0) - \xi) & \text{for } t = 0, \\ e_h(u(t) - u(t_i) + (\mathcal{E}_{h,\xi}(u))(t_i)) & \text{for } t_i < t \le t_{i+1},\ i \in \mathbb{N}, \end{cases} \tag{14.25}$$

where $0 = t_0 < t_1 < t_2 < \ldots$ is a partition of $\mathbb{R}_+$, such that u is monotone on each of the intervals $[t_i, t_{i+1}]$. As with backlash we note that the definition is independent of the choice of partition and $\mathcal{E}_{h,\xi}(C_{\mathrm{pm}}(\mathbb{R}_+, \mathbb{R})) \subset C(\mathbb{R}_+, \mathbb{R})$. The elastic-plastic operator $\mathcal{E}_{h,\xi}$ is illustrated in Figure 14.6.

To show that elastic-plastic hysteresis and the Preisach operator (introduced later) satisfy (N1) through (N8), we need the following lemma. The proof is routine and is therefore omitted.

Lemma 14.7. *Let $u \in C(\mathbb{R}_+, \mathbb{R})$ be unbounded. Then there exists an increasing sequence $(t_n) \subset \mathbb{R}_+$ with $\lim_{n\to\infty} t_n = \infty$ such that either*

$$u(t_n) = \sup_{t\in[0,t_n]} |u(t)|, \quad \forall n \in \mathbb{N} \quad \text{or} \quad u(t_n) = -\sup_{t\in[0,t_n]} |u(t)|, \quad \forall n \in \mathbb{N}.$$

We recall the definition of $\mathcal{N}(\lambda, \tilde{\lambda})$ from Remark 14.5.

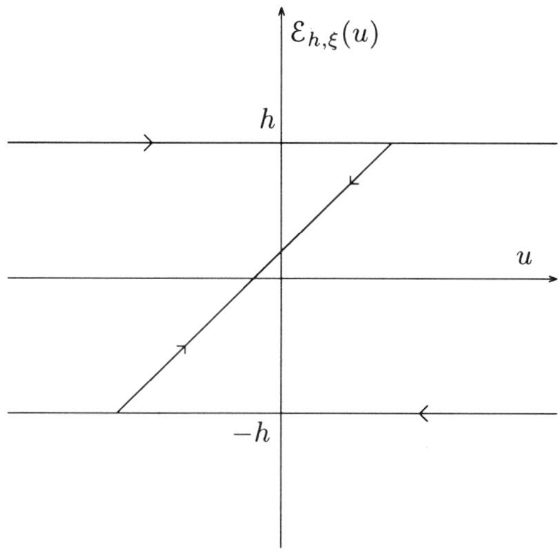

FIGURE 14.6. Elastic-plastic hysteresis.

Proposition 14.5. *Let $(h, \xi) \in \mathbb{R}_+ \times \mathbb{R}$. The elastic-plastic operator $\mathcal{E}_{h,\,\xi}$ has the following properties.*

(i) $\mathcal{E}_{h,\,\xi} : C_{\mathrm{pm}}(\mathbb{R}_+, \mathbb{R}) \to C(\mathbb{R}_+, \mathbb{R})$ *is Lipschitz continuous with Lipschitz continuity constant $l = 2$ and uniquely extends to a Lipschitz continuous operator $\mathcal{E}_{h,\,\xi} : C(\mathbb{R}_+, \mathbb{R}) \to C(\mathbb{R}_+, \mathbb{R})$ with Lipschitz continuity constant $l = 2$.*

(ii) for $H \in \mathbb{R}_+$, globally Lipschitz $\zeta : \mathbb{R}_+ \to \mathbb{R}$ with Lipschitz constant 1, $u \in C(\mathbb{R}_+, \mathbb{R})$, and $t \in \mathbb{R}_+$,

$$(\mathcal{E}_{H,\,\zeta(H)}(u))(t) = H \quad \Longrightarrow \quad (\mathcal{E}_{h,\,\zeta(h)}(u))(t) = h\,, \quad \forall\, h \in [0, H]\,,$$

and

$$(\mathcal{E}_{H,\,\zeta(H)}(u))(t) = -H \quad \Longrightarrow \quad (\mathcal{E}_{h,\,\zeta(h)}(u))(t) = -h\,, \quad \forall\, h \in [0, H]\,.$$

(iii) $\mathcal{E}_{h,\,\xi} : C(\mathbb{R}_+, \mathbb{R}) \to C(\mathbb{R}_+, \mathbb{R})$ *satisfies (N1) to (N3), (N4′), and (N5) to (N8). Furthermore, $\mathcal{E}_{h,\,\xi} \in \mathcal{N}(2,1) \subset \mathcal{N}(2)$.*

Remark 14.7. In statement (i), $l = 2$ is the smallest possible Lipschitz continuity constant for $\mathcal{E}_{h,\,\xi}$. To illustrate this, consider $u, v \in C_{\mathrm{pm}}(\mathbb{R}_+, \mathbb{R})$ defined by

$$u(t) = \begin{cases} t + \xi & \text{for } t \in [0, h], \\ h + \xi & \text{for } t > h, \end{cases}$$

$$v(t) = \begin{cases} t + \xi & \text{for } t \in [0, 3h/2], \\ 3h - t + \xi & \text{for } t \in (3h/2, 5h/2], \\ h/2 + \xi & \text{for } t > 5h/2. \end{cases}$$

Then $\sigma_\tau(u-v) = h/2$ and $\sigma_\tau(\mathcal{E}_{h,\xi}(u) - \mathcal{E}_{h,\xi}(v)) = h$ for all $\tau \geq 5h/2$.

Proof: (of Proposition 14.5) Statement (i) follows from [4, p. 44]. To prove statement (ii), note that by [4, p. 42] and the Lipschitz continuity of $\mathcal{B}_{h,\xi}$, we have for every $u \in C(\mathbb{R}_+,\mathbb{R})$, $\xi_1,\xi_2 \in \mathbb{R}$, and $t, h_1, h_2 \in \mathbb{R}_+$,

$$|(\mathcal{B}_{h_1,\xi_1}(u))(t) - (\mathcal{B}_{h_2,\xi_2}(u))(t)| \leq \max(|h_1-h_2|, |\xi_1-\xi_2|)\,. \tag{14.26}$$

Also from [4, p. 44],

$$\mathcal{E}_{h,\xi}(u) + \mathcal{B}_{h,\xi}(u) = u\,, \quad \forall\, u \in C(\mathbb{R}_+,\mathbb{R})\,. \tag{14.27}$$

Now let $H \in \mathbb{R}_+$, $\zeta : \mathbb{R}_+ \to \mathbb{R}$ be globally Lipschitz with Lipschitz constant 1, $u \in C(\mathbb{R}_+,\mathbb{R})$, $t \in \mathbb{R}_+$ and suppose $(\mathcal{E}_{H,\zeta(H)}(u))(t) = H$. Then using (14.26) and (14.27), we have for all $h \in [0,H]$,

$$H - (\mathcal{E}_{h,\zeta(h)}(u))(t) = (\mathcal{E}_{H,\zeta(H)}(u))(t) - (\mathcal{E}_{h,\zeta(h)}(u))(t) \leq H - h\,,$$

and so since $(\mathcal{E}_{h,\zeta(h)}(u))(t) \leq h$, we obtain $(\mathcal{E}_{h,\zeta(h)}(u))(t) = h$ for all $h \in [0,H]$. The second implication in statement (ii) can be proved in a similar way.

To prove statement (iii), we first note that as an immediate consequence of the definition of $\mathcal{E}_{h,\xi}$ and statement (i), (N1) and (N2) are satisfied. Therefore, combining statement (i) and Lemma 14.3, $\mathcal{E}_{h,\xi}$ also satisfies conditions (N3), (N5), and (N8). To show (N4′) holds, let $u \in AC(\mathbb{R}_+,\mathbb{R})$ and E be the set of all $t \in \mathbb{R}_+$ such that u or $\mathcal{E}_{h,\xi}(u)$ is not differentiable at t. By (N3), E has zero measure. We need to show that

$$\frac{d}{dt}(\mathcal{E}_{h,\xi}(u))(t)\dot{u}(t) \geq 0\,, \quad \forall\, t \in \mathbb{R}_+ \setminus E\,. \tag{14.28}$$

Let $t \in \mathbb{R}_+ \setminus E$, then by (14.27), u, $\mathcal{E}_{h,\xi}(u)$ and $\mathcal{B}_{h,\xi}(u)$ are all differentiable at t and

$$\frac{d}{dt}(\mathcal{E}_{h,\xi}(u))(t) = -\frac{d}{dt}(\mathcal{B}_{h,\xi}(u))(t) + \dot{u}(t)\,.$$

Therefore, since $\mathcal{B}_{h,\xi} \in \mathcal{N}(1)$ and $\mathcal{B}_{h,\xi}$ satisfies (N4′), it follows from Lemma 14.2 (i), and Remark 14.2 that there exists a measurable function $d_u : \mathbb{R}_+ \to [0,1]$ such that

$$\frac{d}{dt}(\mathcal{E}_{h,\xi}(u))(t) = (1 - d_u(t))\dot{u}(t)\,, \quad \forall\, t \in \mathbb{R}_+ \setminus E\,,$$

and thus (14.28) holds. We note that $1 - d_u(t) \in [0,1]$ for all $t \in \mathbb{R}_+$ (although the smallest possible Lipschitz constant of $\mathcal{E}_{h,\xi}$ is $\lambda = 2$; see Remark 14.7). It follows that $\mathcal{E}_{h,\xi} \in \mathcal{N}(2,1) \subset \mathcal{N}(2)$ (cf. Remark 14.5 (i)) once we have shown that (N6) and (N7) hold.

To show that (N6) is satisfied, let $u \in C(\mathbb{R}_+,\mathbb{R})$ be ultimately non-decreasing with $\lim_{t\to\infty} u(t) = \infty$; then

$$\lim_{t\to\infty} (\mathcal{E}_{h,\xi}(u))(t) = h = \sup \mathrm{NVS}\, \mathcal{E}_{h,\xi}$$

and, similarly, $\lim_{t\to\infty}(\mathcal{E}_{h,\xi}(-u))(t) = -h = \inf \mathrm{NVS}\, \mathcal{E}_{h,\xi}$. It follows from statement (i) and Lemma 14.4 that (N6) holds.

For (N7), let $u \in C(\mathbb{R}_+, \mathbb{R})$ and suppose

$$\lim_{t\to\infty} (\mathcal{E}_{h,\xi}(u))(t) \in \mathrm{int}\, \mathrm{NVS}\, \mathcal{E}_{h,\xi} = (-h, h)\,.$$

Seeking a contradiction, assume that u is unbounded. Then, by Lemma 14.7, without loss of generality, we may assume that there exists an increasing sequence $(t_n) \subset \mathbb{R}_+$ such that $\lim_{n\to\infty} t_n = \infty$ and $u(t_n) - \xi = \sup_{t\in[0,t_n]} |u(t) - \xi|$. Moreover, again without loss of generality, we may assume that $u(t_n) > h + \xi$ for all $n \in \mathbb{N}$. Define for each $n \in \mathbb{N}$, $H_n := u(t_n) - \xi > h$; then $(\mathcal{E}_{H_n,\xi}(u))(t_n) = H_n$ for all $n \in \mathbb{N}$. By statement (ii), $(\mathcal{E}_{h,\xi}(u))(t_n) = h$ for all $n \in \mathbb{N}$, which is in contradiction to the assumption that $\lim_{t\to\infty}(\mathcal{E}_{h,\xi}(u))(t) \in (-h, h)$. □

Preisach Operators

All the hysteresis operators considered so far model relatively simple hysteresis loops. The Preisach operator, introduced below, represents a far more general type of hysteresis which for certain input functions exhibits nested loops in the corresponding input-output graphs.

In the following, let $\mathcal{M}_c(\mathbb{R}_+)$ denote the set of all signed Borel measures μ on $\mathbb{R}_+$ such that $|\mu|(S) < \infty$ for all compact sets $S \subset \mathbb{R}_+$.[3] Clearly, $\mathcal{M}_f(\mathbb{R}_+) \subset \mathcal{M}_c(\mathbb{R}_+)$. We denote the Lebesgue measure on $\mathbb{R}$ by μ_L.

Let $C_0(\mathbb{R}_+, \mathbb{R})$ be the set of all continuous functions $\zeta : \mathbb{R}_+ \to \mathbb{R}$ with compact support. We define the set of Preisach memory curves

$$P := \{\zeta \in C_0(\mathbb{R}_+, \mathbb{R}) \mid |\zeta(h_1) - \zeta(h_2)| \le |h_1 - h_2| \;\forall\, h_1, h_2 \in \mathbb{R}_+\}\,.$$

For given $\zeta \in P$, the *Preisach operator*, $\mathcal{P}_\zeta : C(\mathbb{R}_+, \mathbb{R}) \to C(\mathbb{R}_+, \mathbb{R})$, is defined by

$$(\mathcal{P}_\zeta(u))(t) = \int_0^\infty \int_0^{(\mathcal{B}_{h,\zeta(h)}(u))(t)} w(h,s)\, ds\, d\mu(h) + w_0\,, \tag{14.29}$$

where $\mu \in \mathcal{M}_c(\mathbb{R}_+)$, $w \in L^1_{\mathrm{loc}}(\mathbb{R}_+ \times \mathbb{R}; \mu \otimes \mu_L)$, and $w_0 \in \mathbb{R}$. It is clear that for fixed $\zeta \in P$, $u \in C(\mathbb{R}_+, \mathbb{R})$, and $t \in \mathbb{R}_+$, the map

$$\psi : \mathbb{R}_+ \to \mathbb{R}\,, \quad h \mapsto (\mathcal{B}_{h,\zeta(h)}(u))(t)\,,$$

is in P: by (14.26), ψ is globally Lipschitz with Lipschitz constant 1, and as a direct consequence of the definition of the backlash operator, ψ also has compact support. Consequently, the right-hand side of (14.29) is finite for all $u \in C(\mathbb{R}_+, \mathbb{R})$ and all $t \in \mathbb{R}_+$.

---

[3] If $\mu \in \mathcal{M}_c(\mathbb{R}_+)$, then it follows that the measure $|\mu|$ is regular, and hence that μ is a signed Radon measure; see [7, pp. 205–216].

The following two lemmas are useful for the verification of (N1) through (N8) for a large class of Preisach operators.

Lemma 14.8. *Suppose that $\mu \in \mathcal{M}_c(\mathbb{R}_+)$, $w \in L^1_{\rm loc}(\mathbb{R}_+ \times \mathbb{R}; \mu \otimes \mu_L)$, $w_0 \in \mathbb{R}$, $\zeta \in P$, and that $\mathcal{P}_\zeta$ is defined by (14.29). Let $u \in C(\mathbb{R}_+, \mathbb{R})$ and $t \in \mathbb{R}_+$. If $u(t) = \sup_{\tau\in[0,t]} |u(\tau)|$ and $\zeta = 0$ on $[u(t), \infty)$, then*

$$(\mathcal{P}_\zeta(u))(t) = \int_0^{u(t)} \int_0^{u(t)-h} w(h,s)\, ds\, d\mu(h) + w_0 .$$

Proof: Let $u \in C(\mathbb{R}_+, \mathbb{R})$, $t \in \mathbb{R}_+$, and suppose that $u(t) = \sup_{\tau\in[0,t]} |u(\tau)|$ and $\zeta = 0$ on $[u(t), \infty)$. Setting $H := u(t)$, we have $(\mathcal{B}_{H,\zeta(H)}(u))(t) = 0 = u(t) - H$ and $(\mathcal{B}_{h,\zeta(h)}(u))(t) = 0$ for all $h > H$. Combining Proposition 14.5 (ii) and (14.27) shows that $(\mathcal{B}_{h,\zeta(h)}(u))(t) = u(t) - h$ for all $h \in [0, H]$ and therefore by (14.29),

$$(\mathcal{P}_\zeta(u))(t) = \int_0^{u(t)} \int_0^{u(t)-h} w(h,s)\, ds\, d\mu(h) + w_0 .$$

□

The proof of the following lemma follows immediately from [4, pp. 58–60].

Lemma 14.9. *Let $\mu \in \mathcal{M}_c(\mathbb{R}_+)$, $w \in L^1_{\rm loc}(\mathbb{R}_+ \times \mathbb{R}; \mu \otimes \mu_L)$, and $w_0 \in \mathbb{R}$. Suppose that $\lambda := \int_0^\infty \sup_{s\in\mathbb{R}} |w(h,s)|\, d|\mu|(h) < \infty$. Then for all $\zeta \in P$, the Preisach operator $\mathcal{P}_\zeta : C(\mathbb{R}_+, \mathbb{R}) \to C(\mathbb{R}_+, \mathbb{R})$, defined by (14.29), is Lipschitz continuous with Lipschitz continuity constant λ and for $u \in AC(\mathbb{R}_+, \mathbb{R})$,*

$$(\mathcal{P}_\zeta(u))'(t) = \int_0^\infty w(h, (\mathcal{B}_{h,\zeta(h)}(u))(t))(\mathcal{B}_{h,\zeta(h)}(u))'(t)\, d\mu(h)\,, \quad \text{a.e. } t \in \mathbb{R}_+ ,$$

where $'$ denotes differentiation with respect to t.

Remark 14.8. Let $\zeta \in \mathcal{P}$ and $u \in AC(\mathbb{R}_+, \mathbb{R})$. It is implicit in Lemma 14.9 that for μ_L-almost every $t \in \mathbb{R}_+$, $(\mathcal{B}_{h,\zeta(h)}(u))'(t)$ exists for $|\mu|$-almost every $h \in \mathbb{R}_+$. This result is proved in [4, Lemma 2.4.8].

Proposition 14.6. *Let $\mu \in \mathcal{M}_c(\mathbb{R}_+)$ be positive, let $w \in L^1_{\rm loc}(\mathbb{R}_+ \times \mathbb{R}; \mu \otimes \mu_L)$ be nonnegative, and let $w_0 \in \mathbb{R}$. Suppose that $\lambda := \int_0^\infty \sup_{s\in\mathbb{R}} w(h,s)\, d\mu(h) < \infty$. Then, for all $\zeta \in P$, the Preisach operator $\mathcal{P}_\zeta$, defined by (14.29), is in $\mathcal{N}(\lambda)$.*

Proof: By Lemma 14.9, $\mathcal{P}_\zeta : C(\mathbb{R}_+, \mathbb{R}) \to C(\mathbb{R}_+, \mathbb{R})$ is Lipschitz continuous with Lipschitz continuity constant λ and, by its definition, satisfies conditions (N1) and (N2). Therefore, by Lemma 14.3, $\mathcal{P}_\zeta$ also satisfies conditions (N3), (N5), and (N8).

To show that (N4) holds, let $u \in AC(\mathbb{R}_+, \mathbb{R})$. By (N3) and Lemma 14.9 (see also Remark 14.8) there exists $E \subset \mathbb{R}_+$ with $\mu_L(E) = 0$ and such that for all

$t \in \mathbb{R}_+ \setminus E$, $u'(t)$ and $(\mathcal{P}_\varsigma(u))'(t)$ exist, $(\mathcal{B}_{h,\varsigma(h)}(u))'(t)$ exists for μ-almost every $h \in \mathbb{R}_+$, and

$$(\mathcal{P}_\varsigma(u))'(t) = \int_0^\infty w(h, (\mathcal{B}_{h,\varsigma(h)}(u))(t))(\mathcal{B}_{h,\varsigma(h)}(u))'(t)\, d\mu(h)\,. \tag{14.30}$$

Let $t \in \mathbb{R}_+ \setminus E$. If $u'(t) = 0$, (N4) immediately follows. If $u'(t) > 0$, then, since (N4′) holds for $\mathcal{B}_{h,\,\varsigma(h)}$, we have $(\mathcal{B}_{h,\varsigma(h)}(u))'(t) \geq 0$, whenever this derivative exists (which is the case for μ-almost every $h \in \mathbb{R}_+$). Since w and μ are nonnegative, we obtain from (14.30) that $(\mathcal{P}_\varsigma(u))'(t) \geq 0$. If $u'(t) < 0$, then (N4) can be shown to hold by a similar argument.

To show that (N6) is satisfied, let $u \in C(\mathbb{R}_+, \mathbb{R})$ be ultimately non-decreasing with $\lim_{t\to\infty} u(t) = \infty$. Then there exists $T \in \mathbb{R}_+$ such that for all $t \geq T$, $\sup_{\tau\in[0,t]} |u(\tau)| = u(t)$ and $\varsigma = 0$ on $[u(t), \infty)$. So by Lemma 14.8,

$$(\mathcal{P}_\varsigma(u))(t) = \int_0^{u(t)} \int_0^{u(t)-h} w(h,s)\, ds\, d\mu(h) + w_0\,, \quad \forall\, t \geq T\,,$$

and since $\lim_{t\to\infty} u(t) = \infty$,

$$\lim_{t\to\infty} (\mathcal{P}_\varsigma(u))(t) = \int_0^\infty \int_0^\infty w(h,s)\, ds\, d\mu(h) + w_0 \in [w_0, \infty]\,. \tag{14.31}$$

We note that because μ and w are nonnegative

$$\sup \mathrm{NVS}\, \mathcal{P}_\varsigma \leq \int_0^\infty \int_0^\infty w(h,s)\, ds\, d\mu(h) + w_0\,,$$

and therefore, by (14.31),

$$\lim_{t\to\infty} (\mathcal{P}_\varsigma(u))(t) = \sup \mathrm{NVS}\, \mathcal{P}_\varsigma = \int_0^\infty \int_0^\infty w(h,s)\, ds\, d\mu(h) + w_0\,. \tag{14.32}$$

Similarly, $\lim_{t\to\infty}(\mathcal{P}_\varsigma(-u))(t) = \inf \mathrm{NVS}\, \mathcal{P}_\varsigma$. It follows from Lipschitz continuity and Lemma 14.4 that (N6) holds.

For (N7), let $u \in C(\mathbb{R}_+, \mathbb{R})$ and suppose that

$$\lim_{t\to\infty} (\mathcal{P}_\varsigma(u))(t) \in \operatorname{int} \mathrm{NVS}\, \mathcal{P}_\varsigma\,.$$

Let $H \in \mathbb{R}_+$ be such that $\varsigma = 0$ on $[H, \infty)$. Seeking a contradiction, suppose that u is unbounded. Then, by Lemma 14.7, without loss of generality, we may assume that there exists an increasing sequence $(t_n) \subset \mathbb{R}_+$ such that $\lim_{n\to\infty} t_n = \infty$ and $u(t_n) = \sup_{t\in[0,t_n]} |u(t)|$. Moreover, again without loss of generality, we may assume that $u(t_n) \geq H$ for all $n \in \mathbb{N}$. By Lemma 14.8

$$(\mathcal{P}_\varsigma(u))(t_n) = \int_0^{u(t_n)} \int_0^{u(t_n)-h} w(h,s)\, ds\, d\mu(h) + w_0\,, \quad \forall\, n \in \mathbb{N}\,.$$

Since $\lim_{n\to\infty} u(t_n) = \infty$, it follows from the second equation in (14.32) that

$$\lim_{t\to\infty} (\mathcal{P}_\varsigma(u))(t) = \lim_{n\to\infty} (\mathcal{P}_\varsigma(u))(t_n) = \sup \mathrm{NVS}\, \mathcal{P}_\varsigma\,,$$

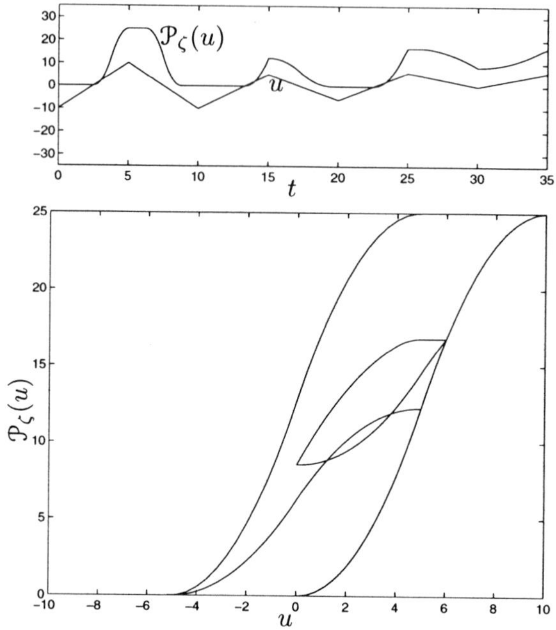

FIGURE 14.7. Example of Preisach hysteresis.

which is in contradiction to $\lim_{t\to\infty}(\mathcal{P}_\zeta(u))(t) \in \operatorname{int} \mathrm{NVS}\, \mathcal{P}_\zeta$. □

As an example, we consider the operator $\mathcal{P}_\zeta$ obtained by setting $\zeta \equiv 0$, $\mu = \mu_L$, $w_0 = 0$, and $w \equiv 2 \cdot \chi_{[0,5]\times[0,5]}$, where χ_S denotes the indicator function of the set S. This operator is illustrated in Figure 14.7.

If we set $w \equiv 1$ and $w_0 = 0$ in (14.29), we obtain the *Prandtl operator*

$$(\mathcal{P}_\zeta(u))(t) = \int_0^\infty (\mathcal{B}_{h,\zeta(h)}(u))(t)\, d\mu(h)\,, \quad \forall\, u \in C(\mathbb{R}_+, \mathbb{R})\,, \ \forall\, t \in \mathbb{R}_+\,, \tag{14.33}$$

where $\zeta \in P$ and $\mu \in \mathcal{M}_c(\mathbb{R}_+)$ (cf. [4, pp. 54]). The following corollary is a special case of Proposition 14.6.

Corollary 14.1. *Let μ be a finite positive Borel measure on $\mathbb{R}_+$. Then for all $\zeta \in P$, the Prandtl operator $\mathcal{P}_\zeta$, defined by (14.33), is in $\mathcal{N}(\lambda)$, where $\lambda := \mu(\mathbb{R}_+)$.*

For example, defining the measure μ by $\mu(E) = \int_E (\sin(\pi h) + 1)\chi_{[0,10]} dh$ and setting $\zeta \equiv 0$ yields the operator illustrated in Figure 14.8.

Another example covered by Corollary 14.1 is backlash hysteresis. Indeed, the backlash operator $\mathcal{B}_{h,\xi}$ can be obtained from (14.33) by setting $\mu = \delta_h$ (where δ_h is the unit point mass at h) and by letting $\zeta : \mathbb{R}_+ \to \mathbb{R}$ be any continuous function with compact support and such that $\zeta(h) = \xi$.

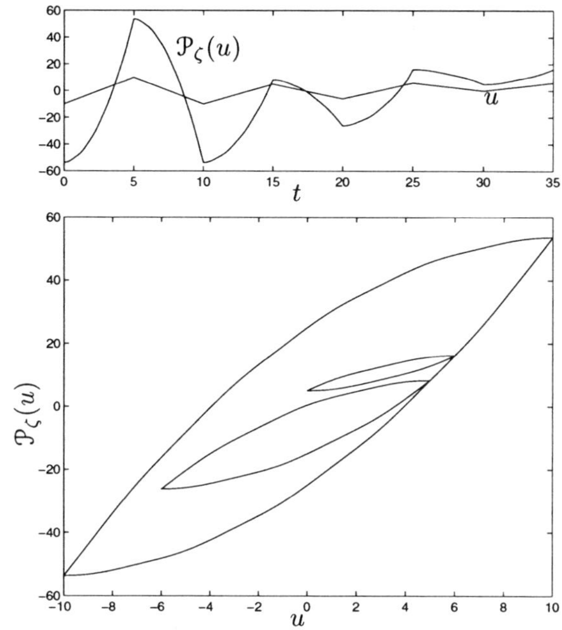

FIGURE 14.8. Example of Prandtl hysteresis.

Let $p \in L^1(\mathbb{R}_+, \mathbb{R})$; then setting $\mu = \left(\int_0^\infty p(h)\,dh\right)\delta_0 - p\mu_L$ in (14.33), we obtain, for $\zeta \in P$,

$$(\mathcal{P}_\zeta(u))(t) = \int_0^\infty p(h)(\mathcal{E}_{h,\zeta(h)}(u))(t)\,dh\,, \quad \forall\, u \in C(\mathbb{R}_+, \mathbb{R})\,, \ \forall\, t \in \mathbb{R}_+\,, \tag{14.34}$$

where we have used (14.27) and the fact that for all $\xi \in \mathbb{R}$ and $u \in C(\mathbb{R}_+, \mathbb{R})$, $\mathcal{B}_{0,\xi}(u) = u$.

Proposition 14.7. *Let $p \in L^1(\mathbb{R}_+, \mathbb{R})$ be nonnegative. Then for all $\zeta \in P$, the Prandtl operator $\mathcal{P}_\zeta : C(\mathbb{R}_+, \mathbb{R}) \to C(\mathbb{R}_+, \mathbb{R})$, given by (14.34), is in $\mathcal{N}(\lambda)$, where $\lambda := 2\int_0^\infty p(h)\,dh$.*

Proof: By Lemma 14.9, $\mathcal{P}_\zeta : C(\mathbb{R}_+, \mathbb{R}) \to C(\mathbb{R}_+, \mathbb{R})$ is Lipschitz continuous with Lipschitz continuity constant $\lambda = 2\int_0^\infty p(h)\,dh$ and clearly satisfies conditions (N1) and (N2) and therefore by Lemma 14.3, also satisfies conditions (N3), (N5), and (N8).

To show that (N4) holds, fix $u \in AC(\mathbb{R}_+, \mathbb{R})$. By (N3), (14.27), and Lemma 14.9 (see also Remark 14.8) there exists $E \subset \mathbb{R}_+$ with $\mu_L(E) = 0$ and such that for all $t \in \mathbb{R}_+ \setminus E$, $u'(t)$ and $(\mathcal{P}_\zeta(u))'(t)$ exist, $(\mathcal{E}_{h,\zeta(h)}(u))'(t)$ exists for $|\mu|$-almost every $h \in \mathbb{R}_+$, and

$$(\mathcal{P}_\zeta(u))'(t) = \int_0^\infty p(h)(\mathcal{E}_{h,\zeta(h)}(u))'(t)\,dh\,.$$

Let $t \in \mathbb{R}_+ \setminus E$. If $u'(t) = 0$, (N4) immediately follows. If $u'(t) > 0$, then, since (N4′) holds for $\mathcal{E}_{h,\zeta(h)}$, we have $(\mathcal{E}_{h,\zeta(h)}(u))'(t) \geq 0$ whenever this derivative exists (which is the case for $|\mu|$-almost every $h \in \mathbb{R}_+$). Since p is nonnegative, we may conclude that $(\mathcal{P}_\zeta(u))'(t) \geq 0$. If $u'(t) < 0$, then (N4) can be shown to hold by a similar argument.

To prove that (N6) is satisfied, let $u \in C(\mathbb{R}_+, \mathbb{R})$ be ultimately non-decreasing with $\lim_{t\to\infty} u(t) = \infty$. Then there exists $T \in \mathbb{R}_+$ such that for all $t \geq T$, $\sup_{\tau\in[0,t]} |u(\tau)| = u(t)$ and $\zeta = 0$ on $[u(t), \infty)$. So, by Lemma 14.8, with $w_0 = 0$, $w \equiv 1$, and $\mu = \left(\int_0^\infty p(h)\,dh\right)\delta_0 - p\mu_L$, we obtain

$$(\mathcal{P}_\zeta(u))(t) = \int_0^{u(t)} p(h)h\,dh + u(t)\int_{u(t)}^\infty p(h)\,dh\,, \quad \forall\, t \geq T\,. \tag{14.35}$$

We note that because p is nonnegative, $\sup \mathrm{NVS}\,\mathcal{P}_\zeta \leq \int_0^\infty p(h)h\,dh \in [0,\infty]$. Now using (14.35) and the fact that p is nonnegative

$$\int_0^\infty p(h)h\,dh \geq \sup \mathrm{NVS}\,\mathcal{P}_\zeta \geq (\mathcal{P}_\zeta(u))(t) \geq \int_0^{u(t)} p(h)h\,dh\,, \quad \forall\, t \geq T\,.$$

Since $\lim_{t\to\infty} u(t) = \infty$, it follows that

$$\lim_{t\to\infty} (\mathcal{P}_\zeta(u))(t) = \int_0^\infty p(h)h\,dh = \sup \mathrm{NVS}\,\mathcal{P}_\zeta\,.$$

Similarly, $\lim_{t\to\infty}(\mathcal{P}_\zeta(-u))(t) = \inf \mathrm{NVS}\,\mathcal{P}_\zeta$. Consequently, (N6) follows from the Lipschitz continuity of $\mathcal{P}_\zeta$ and an application of Lemma 14.4.

For (N7), let $u \in C(\mathbb{R}_+, \mathbb{R})$ and suppose $\lim_{t\to\infty}(\mathcal{P}_\zeta(u))(t) \in \mathrm{int}\,\mathrm{NVS}\,\mathcal{P}_\zeta$. Let $H \in \mathbb{R}_+$ be such that $\zeta = 0$ on $[H, \infty)$. Seeking a contradiction, suppose that u is unbounded. Then, by Lemma 14.7, without loss of generality, we may assume that there exists an increasing sequence $(t_n) \subset \mathbb{R}_+$ such that $\lim_{n\to\infty} t_n = \infty$ and $u(t_n) = \sup_{t\in[0,t_n]} |u(t)|$. Moreover, again without loss of generality, we may assume that $u(t_n) \geq H$ for all $n \in \mathbb{N}$. Then, by Lemma 14.8,

$$(\mathcal{P}_\zeta(u))(t_n) = \int_0^{u(t_n)} p(h)h\,dh + u(t_n)\int_{u(t_n)}^\infty p(h)\,dh\,, \quad \forall\, n \in \mathbb{N}\,.$$

Combining this with $\lim_{n\to\infty} u(t_n) = \infty$, we may conclude as in the proof of (N6) that $\lim_{t\to\infty}(\mathcal{P}_\zeta(u))(t) = \lim_{n\to\infty}(\mathcal{P}_\zeta(u))(t_n) = \sup \mathrm{NVS}\,\mathcal{P}_\zeta$, which is in contradiction to $\lim_{t\to\infty}(\mathcal{P}_\zeta(u))(t) \in \mathrm{int}\,\mathrm{NVS}\,\mathcal{P}_\zeta$. □

14.6 Example: Controlled Diffusion Process with Output Delay

Consider a diffusion process (with diffusion coefficient $\kappa > 0$ and with Dirichlet boundary conditions) on the one-dimensional spatial domain $[0, 1]$, with scalar nonlinear pointwise control action (applied at point $x_b \in (0,1)$, via an operator $\Phi : C(\mathbb{R}_+, \mathbb{R}) \to C(\mathbb{R}_+, \mathbb{R})$, as defined below) and delayed (delay $T \geq 0$) pointwise scalar observation (output at point $x_c \in (0,1)$, $x_c \geq x_b$).

We formally write this single-input, single-output system as

$$\begin{aligned} z_t(t,x) &= \kappa z_{xx}(t,x) + \delta(x - x_b)(\Phi(u))(t)\,, \\ y(t) &= z(t-T, x_c)\,, \end{aligned}$$

with boundary conditions

$$z(t,0) = 0 = z(t,1)\,, \qquad \forall\, t > 0\,.$$

For simplicity, we assume zero initial conditions

$$z(t,x) = 0\,, \qquad \forall\,(t,x) \in [-T,0] \times [0,1]\,.$$

With input $(\Phi(u))(\cdot)$ and output $y(\cdot)$, this example qualifies as a regular linear system with transfer function given by

$$\mathbf{G}(s) = \frac{e^{-sT}\sinh\left(x_b\sqrt{(s/\kappa)}\right)\sinh\left((1-x_c)\sqrt{(s/\kappa)}\right)}{\kappa\sqrt{(s/\kappa)}\sinh\sqrt{(s/\kappa)}}\,.$$

In this case, a detailed analysis (see [12] for related investigations) shows that K defined by (14.12) satisfies

$$K = \frac{1}{|\mathbf{G}'(0)|} = \frac{6\kappa^2}{x_b(1-x_c)(6T\kappa + 1 - x_b^2 - (1-x_c)^2)}\,.$$

Therefore, by Theorem 14.1, if $\Phi \in \mathcal{N}(\lambda)$ for some $\lambda > 0$ and $k \in (0, K/\lambda)$, the integral control, $\dot{u}(t) = k[r - y(t)]$, with $u(0) = 0$, guarantees asymptotic tracking of all feasible constant reference signals r. For purposes of illustration, we adopt the following values

$$\kappa = 0.1, \qquad x_b = \frac{1}{3}, \qquad x_c = \frac{2}{3}, \qquad T = 1\,,$$

and so $K = 243/620 \approx 0.3919$.

We consider relay and Prandtl hysteresis operators.

(i) Let $\Phi = \mathcal{R}_\xi$ be a relay hysteresis operator defined by (14.20), where $\xi = 0$, $a_1 = -1$, $a_2 = 1$, $\rho_1(u) = \sqrt{u + 1.1}$, and $\rho_2(u) = \sqrt{0.1} + \sqrt{2.1} - \sqrt{1.1 - u}$. Then $\Phi \in \mathcal{N}(\lambda)$, where $\lambda = 1.6$ and $\mathrm{NVS}\,\Phi = \mathrm{im}\,\rho_1 \cup \mathrm{im}\,\rho_2 = \mathbb{R}$. Hence $K/\lambda \approx 0.245$ and taking $r = 1.42$ gives

$$\Phi_r = \frac{r}{\mathbf{G}(0)} = \frac{r\kappa}{x_b(1-x_c)} = 1.278 \in \mathrm{int}\,(\mathrm{NVS}\,\Phi)\,.$$

In each of the following cases of admissible controller gains

$$\text{(i) } k = 0.24, \qquad \text{(ii) } k = 0.17, \qquad \text{(iii) } k = 0.1,$$

Figure 14.9 depicts the output behavior of the system under integral control, Figure 14.10 depicts the corresponding control input, and Figure 14.11 shows the input to the hysteresis nonlinearity. We see from Figure 14.11 that for (i), $\lim_{t\to\infty} u(t) = \rho_1^{-1}(\Phi_r)$ and for (ii) and (iii), $\lim_{t\to\infty} u(t) = \rho_2^{-1}(\Phi_r)$.

(ii) Let $\Phi = \mathcal{P}_\zeta$ be a Prandtl operator, as defined in (14.33), where $\mu = (1/8)\chi_{[1,9]}\mu_L$ and $\zeta \equiv 0$. Then $\Phi \in \mathcal{N}(\lambda)$, where $\lambda = 1$, and $\mathrm{NVS}\,\Phi = \mathbb{R}$. Then $K/\lambda = K \approx 0.3919$. For $r = 1$, we have

$$\Phi_r = \frac{r}{\mathbf{G}(0)} = \frac{r\kappa}{x_b(1 - x_c)} = 0.9 \in \mathrm{int}\,(\mathrm{NVS}\,\Phi)\,.$$

In each of the following cases of admissible controller gains

$$\text{(i) } k = 0.39, \qquad \text{(ii) } k = 0.2, \qquad \text{(iii) } k = 0.1,$$

Figure 14.12 depicts the output behavior of the system under integral control, Figure 14.13 depicts the corresponding control input, and Figure 14.14 shows the input to the Prandtl operator. We see from Figure 14.12 that for (i) (the largest gain) the output exhibits a small overshoot, which does not occur in (ii) and (iii). The overshoot leads to the formation of a hysteresis loop. Hence in Figure 14.14, the hysteresis input converges to a different value as compared to the other two cases.

Figures 14.9 through 14.14 were generated using SIMULINK Simulation Software within MATLAB wherein a truncated eigenfunction expansion, of order 10, was adopted to model the diffusion process.

14.7 Appendix

Proof of Lemma 14.5

Let $\alpha \geq 0$ and $w \in C([0,\alpha],\mathbb{R})$ and consider the following initial-value problem

$$\dot{u}(t) = k[r - (\mathbf{\Psi} x_0)(t) - (\mathbf{F}\Phi(u))(t)]\,, \quad t > \alpha\,, \tag{14.36a}$$
$$u(t) = w(t)\,, \quad t \in [0,\alpha]\,, \tag{14.36b}$$

where $\Phi : C(\mathbb{R}_+,\mathbb{R}) \to C(\mathbb{R}_+,\mathbb{R})$.

Lemma 14.10. *Assume that Φ satisfies (N1) and (N5) and let $x_0 \in X$. For any initial function $w \in C([0,\alpha],\mathbb{R})$, there exists $\varepsilon > 0$ and a unique function $u \in C([0,\alpha+\varepsilon],\mathbb{R})$ with $u(t) = w(t)$ for all $t \in [0,\alpha]$ and such that $u|_{[\alpha,\alpha+\varepsilon]}$ is absolutely continuous and* (14.36a) *is satisfied for a.e. $t \in [\alpha,\alpha+\varepsilon]$.*

Remark 14.9. For $\alpha = 0$, the above initial-value problem is identical to the initial-value problem (14.10). Note that Steps 2 and 3 in the proof of Lemma 14.5

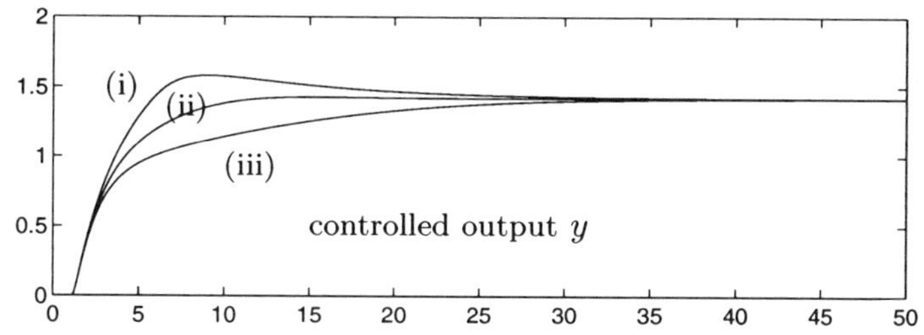

FIGURE 14.9. Controlled output.

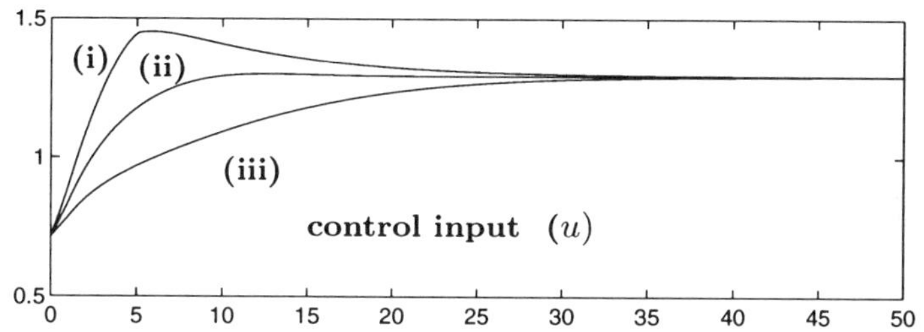

FIGURE 14.10. Control input.

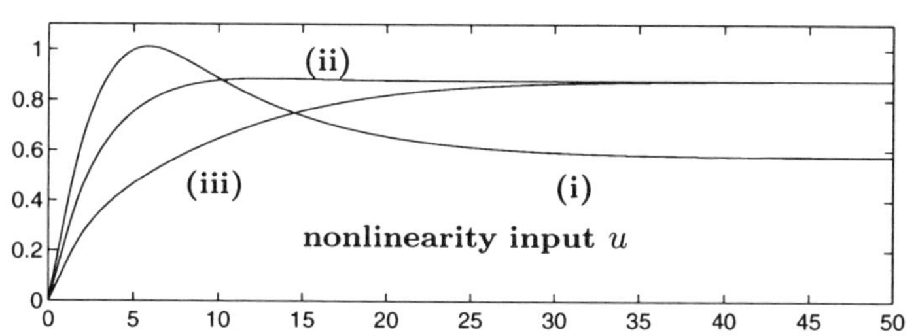

FIGURE 14.11. Input to relay nonlinearity.

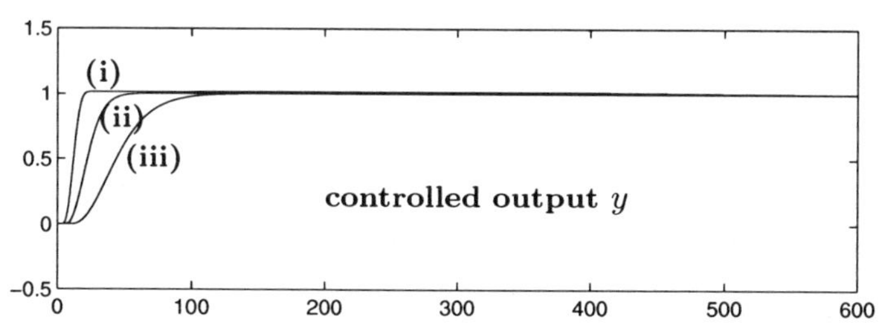

FIGURE 14.12. Controlled output.

(see below) require the existence and uniqueness result given in Lemma 14.10 in the more general context of $\alpha \geq 0$.

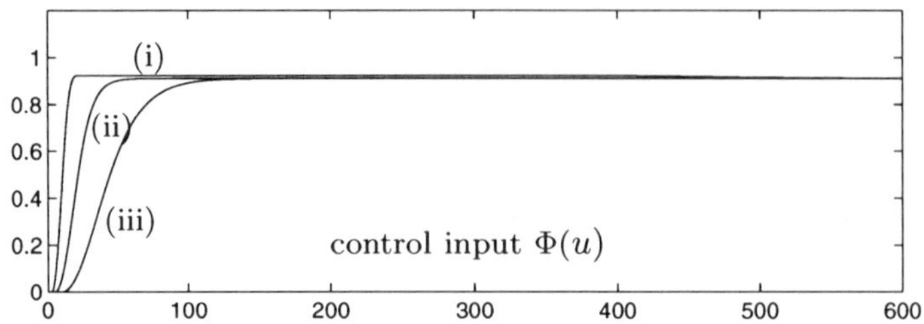

FIGURE 14.13. Control input.

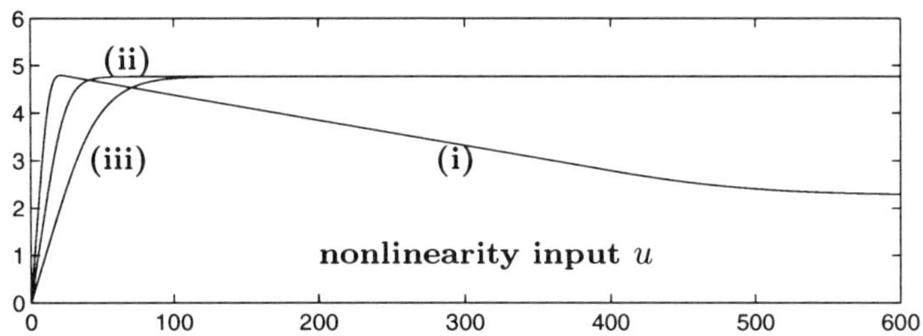

FIGURE 14.14. Input to Prandtl operator.

For the proof of Lemma 14.10 it is convenient to introduce a "truncation" of the operator $\mathbf{F}$ acting on functions defined on a finite time interval. To this end let $\tau > 0$ and define $\pi_\tau : L^2([0,\tau],\mathbb{R}) \to L^2(\mathbb{R}_+,\mathbb{R})$ by

$$(\pi_\tau v)(t) = \begin{cases} v(t) & \text{for } 0 \le t \le \tau\,, \\ 0 & \text{for } t > \tau\,. \end{cases}$$

Defining the linear operator $\mathbf{F}_\tau : L^2([0,\tau],\mathbb{R}) \to L^2([0,\tau],\mathbb{R})$ by

$$(\mathbf{F}_\tau v)(t) = (\mathbf{F}(\pi_\tau v))(t)\,, \quad \forall\, t \in [0,\tau]\,,$$

we have that $\mathbf{F}_\tau$ is bounded and $\|\mathbf{F}_{\tau_1}\| \le \|\mathbf{F}_{\tau_2}\|$ for all $\tau_2 \ge \tau_1 > 0$.

Proof: (of Lemma 14.10) Let $w \in C([0,\alpha],\mathbb{R})$. Then by (N5), there exist numbers $\delta_1, \delta_2 > 0$ such that for all $u, v \in \mathcal{C}(w;\delta_1,\delta_2)$,

$$\sup_{t\in[\alpha,\alpha+\delta_2]} |(\Phi(u))(t) - (\Phi(v))(t)| \le \lambda \sup_{t\in[\alpha,\alpha+\delta_2]} |u(t) - v(t)|\,. \tag{14.37}$$

For $\varepsilon, \eta > 0$ set

$$\mathcal{C}_{\eta,\varepsilon} := \{u \in C([0,\alpha+\varepsilon],\mathbb{R}) \mid u(t) = w(t) \text{ if } t \in [0,\alpha]\,; \\ |u(t) - w(\alpha)| \le \eta \text{ if } t \in [\alpha,\alpha+\varepsilon]\}\,. \tag{14.38}$$

Let $\eta \in (0, \delta_1)$, $\varepsilon > 0$, and $u, v \in \mathcal{C}_{\eta,\varepsilon}$. Then

$$\mathbf{Q}_{\alpha+\varepsilon}u,\ \mathbf{Q}_{\alpha+\varepsilon}v \in \mathcal{C}(w; \delta_1, \delta_2)\,,$$

and hence, by (14.37), we obtain for every $\eta \in (0, \delta_1)$ and every $\varepsilon \in (0, \delta_2)$,

$$\begin{aligned}
&\sup\nolimits_{t\in[\alpha,\alpha+\varepsilon]} |(\Phi(u))(t) - (\Phi(v))(t)| \\
&\qquad \le \sup\nolimits_{t\in[\alpha,\alpha+\delta_2]} |(\Phi(\mathbf{Q}_{\alpha+\varepsilon}u))(t) - (\Phi(\mathbf{Q}_{\alpha+\varepsilon}v))(t)| \\
&\qquad \le \lambda \sup\nolimits_{t\in[\alpha,\alpha+\delta_2]} |(\mathbf{Q}_{\alpha+\varepsilon}u)(t) - (\mathbf{Q}_{\alpha+\varepsilon}v)(t)| \\
&\qquad = \lambda \sup\nolimits_{t\in[\alpha,\alpha+\varepsilon]} |u(t) - v(t)|\,, \quad \forall\, u, v \in \mathcal{C}_{\eta,\varepsilon}\,.
\end{aligned}$$

Using the causality of $\mathbf{F}$ and Φ (Φ is causal by (N1)), the boundedness of $\mathbf{F}_\tau$ for every $\tau \ge 0$, Hölder's inequality, and (14.7), we conclude that there exists $L > 0$ such that, for every $\varepsilon \in (0, \delta_2)$,

$$\int_\alpha^{\alpha+\varepsilon} |\mathbf{F}\Phi(u) - \mathbf{F}\Phi(v)| \le \varepsilon L \sup_{t\in[\alpha,\alpha+\varepsilon]} |u(t) - v(t)|\,, \quad \forall\, u, v \in \mathcal{C}_{\eta,\varepsilon}\,. \tag{14.39}$$

Moreover, an application of (14.39) for $v = \mathbf{Q}_\alpha u$ shows that, for every $\varepsilon \in (0, \delta_2)$, we have for all $u \in \mathcal{C}_{\eta,\varepsilon}$,

$$\int_\alpha^{\alpha+\varepsilon} |\mathbf{F}\Phi(u)| \le \int_\alpha^{\alpha+\varepsilon} |(\mathbf{F}\Phi(\mathbf{Q}_\alpha u))(\tau)|\, d\tau + \varepsilon L \sup_{t\in[\alpha,\alpha+\varepsilon]} |u - w(\alpha)|\,. \tag{14.40}$$

Set $f(t) = r - (\mathbf{\Psi} x_0)(t)$ and choose $\rho > 0$ such that

$$\int_\alpha^{\alpha+\rho} (|f(\tau)| + |(\mathbf{F}\Phi(\mathbf{Q}_\alpha u))(\tau)|)\, d\tau \le \frac{\eta}{2|k|}\,. \tag{14.41}$$

In the following let $\eta \in (0, \delta_1)$ and choose $\varepsilon > 0$ such that

$$\varepsilon < \delta_2\,, \quad \varepsilon < \rho\,, \quad \varepsilon < \frac{1}{2|k|L}\,. \tag{14.42}$$

Define the operator Γ by

$$\begin{aligned}
(\Gamma u)(t) &= w(\alpha) + k\left(\int_\alpha^t f(\tau)\, d\tau - \int_\alpha^t (\mathbf{F}\Phi(u))(\tau)\, d\tau\right), \quad t \ge \alpha\,, \\
(\Gamma u)(t) &= w(t)\,, \quad 0 \le t \le \alpha\,.
\end{aligned}$$

Clearly, $\mathcal{C}_{\eta,\varepsilon}$ is a complete metric space, and the claim follows if we can show that Γ is a contraction on $\mathcal{C}_{\eta,\varepsilon}$.

We first show that $\Gamma(\mathcal{C}_{\eta,\varepsilon}) \subset \mathcal{C}_{\eta,\varepsilon}$. Using (14.40) through (14.42) we obtain, for all $u \in \mathcal{C}_{\eta,\varepsilon}$ and all $t \in [\alpha, \alpha+\varepsilon]$,

$$|(\Gamma u)(t) - w(\alpha)| \le \frac{\eta}{2} + \varepsilon|k|L \sup_{t\in[\alpha,\alpha+\varepsilon]} |u(t) - w(\alpha)| \le \frac{\eta}{2} + \varepsilon|k|L\eta \le \eta\,,$$

which shows that $\Gamma(\mathcal{C}_{\eta,\varepsilon}) \subset \mathcal{C}_{\eta,\varepsilon}$. It remains to show that Γ is a contraction on $\mathcal{C}_{\eta,\varepsilon}$. To this end, let $u, v \in \mathcal{C}_{\eta,\varepsilon}$. Using (14.39), we obtain

$$\sup_{\tau\in[\alpha,\alpha+\varepsilon]} |(\Gamma u)(\tau) - (\Gamma v)(\tau)| \le \varepsilon|k|L \sup_{\tau\in[\alpha,\alpha+\varepsilon]} |u(\tau) - v(\tau)| \,.$$

By (14.42) we have that $\varepsilon|k|L < 1$, showing that Γ is a contraction on $\mathcal{C}_{\varepsilon,\eta}$. □

Proof: **(of Lemma 14.5)** We proceed in several steps.

STEP 1. Existence and uniqueness on a small interval.

An application of Lemma 14.10 with $\alpha = 0$ shows that there exists an $\varepsilon > 0$ such that (14.10) has a unique solution on the interval $[0, \varepsilon)$.

STEP 2. Extended uniqueness.

Let u_i be a solution of (14.10) on the interval $[0, a_i)$, $i = 1, 2$. We claim that $u_1(t) = u_2(t)$ for all $t \in [0, a)$, where $a = \min(a_1, a_2)$. Seeking a contradiction, assume that there exists $t \in (0, a)$ such that $u_1(t) \neq u_2(t)$. Defining

$$a^* = \inf\{t \in (0, a) \,|\, u_1(t) \neq u_2(t)\}$$

it follows that $a^* > 0$ (by STEP 1), $a^* < a$ (by assumption), and $u_1(a^*) = u_2(a^*)$ (by continuity of u_1 and u_2). Clearly, the initial-value problem

$$\begin{aligned} \dot{u}(t) &= k[r - (\mathbf{\Psi} x_0)(t) - (\mathbf{F}\Phi(u))(t)]\,, \quad t \ge a^*\,, \\ u(t) &= u_1(t)\,, \quad t \in [0, a^*]\,, \end{aligned}$$

is solved by u_1 and u_2 on [0,a). This implies (by Lemma 14.10) that there exists an $\varepsilon > 0$ such that $u_1(t) = u_2(t)$ for all $t \in [0, a^* + \varepsilon)$, which contradicts the definition of a^*.

STEP 3. Global existence.

Let $\mathcal{T} \subset \mathbb{R}_+$ be the set of all $\tau > 0$ such that there exists a solution u^τ of (14.10) on the interval $[0, \tau)$. Set $t^* := \sup \mathcal{T}$ and define a function $u : [0, t^*) \to \mathbb{R}$ by setting

$$u(t) = u^\tau(t)\,, \quad \text{for } t \in [0, \tau)\,, \text{ where } \tau \in \mathcal{T}\,.$$

By STEP 2 the function u is well defined (i.e., the definition of $u(t)$ for a particular value $t \in [0, t^*)$ does not depend on the choice of $\tau \in \mathcal{T} \cap (t, \infty)$) and u is the unique solution of (14.10) on the interval $[0, t^*)$.

We claim that $t^* = \infty$. Seeking a contradiction, assume that $t^* < \infty$. Multiplying $\dot{u}$ by u, we obtain using (14.10),

$$u(t)\dot{u}(t) = ku(t)\,[f(t) - (\mathbf{F}\Phi(u))(t)]\,, \quad \forall\, t \in [0, t^*)\,,$$

where again $f(t) = r - (\mathbf{\Psi} x_0)(t)$. Integration yields

$$\frac{1}{2}u^2(t) = \frac{1}{2}u^2(0) + k\left(\int_0^t fu - \int_0^t \mathbf{F}\Phi(u)u\right), \quad \forall\, t \in [0, t^*)\,. \tag{14.43}$$

For $v \in C([0,t^*),\mathbb{R})$ and $t \in [0,t^*)$, we define

$$\sigma_t(v) = \sup_{\tau\in[0,t]} |v(\tau)|\,.$$

Using (14.43), the boundedness of $\mathbf{F}_{t^*}$, and applying Hölder's inequality, shows that there exists $\gamma_1 > 0$ such that for all $t \in [0,t^*)$,

$$\frac{1}{2}\sigma_t(u^2) \le \frac{1}{2}u^2(0) + |k|\left(\int_0^{t^*} f^2\right)^{1/2}\left(\int_0^t u^2\right)^{1/2} + |k|\gamma_1\left(\int_0^t [\Phi(u)]^2\right)^{1/2}\left(\int_0^t u^2\right)^{1/2}.$$

Denoting the map

$$C([0,t^*),\mathbb{R}) \to \mathbb{R}_+\,, \quad v \mapsto [\sigma_t(v)]^2$$

by σ_t^2, we see that there exist suitable constants $\gamma_2,\gamma_3,\gamma_4 > 0$ such that for all $t \in [0,t^*)$

$$\sigma_t^2(u) \le \gamma_2 + \gamma_3\left(\int_0^t \sigma_\tau^2(u)\,d\tau\right)^{1/2} + \gamma_4\left(\int_0^t \sigma_\tau^2(\Phi(u))\,d\tau\right)^{1/2}\left(\int_0^t \sigma_\tau^2(u)\,d\tau\right)^{1/2}.$$

Using Assumption (N8), we may conclude that there exist numbers $\alpha,\beta > 0$ such that for all $t \in [0,t^*)$,

$$\sigma_t^2(u) \le \gamma_2 + \gamma_3\left(\int_0^t \sigma_\tau^2(u)\,d\tau\right)^{1/2} + \gamma_4\left(\int_0^t [\alpha+\beta\sigma_\tau(u)]^2\,d\tau\right)^{1/2}\left(\int_0^t \sigma_\tau^2(u)\,d\tau\right)^{1/2}.$$

From this we obtain that there exist numbers $\gamma_5,\gamma_6 > 0$ such that

$$\sigma_t^2(u) \le \gamma_5 + \gamma_6\int_0^t \sigma_\tau^2(u)\,d\tau\,, \quad \forall\, t \in [0,t^*)\,,$$

and an application of Gronwall's lemma then shows that

$$\sigma_t^2(u) \le \gamma_5 e^{\gamma_6 t}\,, \quad \forall\, t \in [0,t^*)\,.$$

Since, by assumption, $t^* < \infty$, it follows that u is bounded on $[0,t^*)$. Consequently, the right-hand side of (14.43) converges to a finite limit as $t \uparrow t^*$, and so $\lim_{t\uparrow t^*} u^2(t)$ exists and is finite. By continuity of u, this in turn implies that there exists $\gamma \in \mathbb{R}$ such that $\lim_{t\uparrow t^*} u(t) = \gamma$, and hence setting $u(t^*) = \gamma$ makes u into a continuous function on $[0,t^*]$. Finally, Lemma 14.10 shows that the initial value problem

$$\begin{aligned}
\dot{v} &= k[r - \mathbf{\Psi}x_0 - \mathbf{F}\Phi(v)]\,, \quad t \ge a\,,\\
v(t) &= u(t)\,, \quad t \in [0,t^*]\,,
\end{aligned}$$

has a unique solution $\tilde{u}$ on $[0, t^* + \varepsilon)$ for some $\varepsilon > 0$. By the causality of the map $\mathbf{F}\Phi$, the function $\tilde{u}$ is a solution of (14.10) on $[0, t^* + \varepsilon)$; that is, $\tilde{u}$ is a continuation of u. But this means that $t^* + \varepsilon \in \mathcal{T}$, which is in contradiction to the definition of t^*. □

Proof: **(of Lemma 14.6)** Let $t_1 < t_2$, $u : [t_1, t_2] \to \mathbb{R}$ be monotone and $w \in [u(t_1) - h, u(t_1) + h]$. Fix $t, \tau \in [t_1, t_2]$ with $t \geq \tau$. We first note that $w = b_h(u(t_1), w)$ since $w \in [u(t_1) - h, u(t_1) + h]$. Without loss of generality we may assume that u is nondecreasing and so $w = b_h(u(t_1), w) \leq b_h(u(\tau), w)$. If $w = b_h(u(\tau), w)$, then, trivially, $b_h(u(t), w) = b_h(u(t), b_h(u(\tau), w))$. If $w < b_h(u(\tau), w)$, then $b_h(u(\tau), w) = u(\tau) - h$ and thus $w < b_h(u(\tau), w) \leq u(t) - h$. Consequently,

$$b_h(u(t), w) = u(t) - h = b_h(u(t), b_h(u(\tau), w)) \,.$$

□

Acknowledgment: This work was supported by the UK EPSRC Council (Grant GR/L78086).

References

[1] H.T. Banks, R.C. Smith and Y. Wang, *Smart Material Structures: Modeling, Estimation and Control*, Masson, Paris, 1996.

[2] V. Barbu, *Analysis and Control of Nonlinear Infinite-Dimensional Systems*, Academic Press, Boston, 1993.

[3] M. Brokate, Hysteresis operators, in *Phase Transitions and Hysteresis*, A. Visintin ed., Springer-Verlag, Berlin, 1–38, 1994.

[4] M. Brokate and J. Sprekels, *Hysteresis and Phase Transitions*, Springer-Verlag, New York, 1996.

[5] G.W.M. Coppus, S.L. Sha, and R.K. Wood, Robust multivariable control of a binary distillation column, *IEE Proc., D* 130:201–208, 1983.

[6] E.J. Davison, Multivariable tuning regulators: The feedforward and robust control of a general servomechanism problem, *IEEE Trans. Auto. Control* 21:35–47, 1976.

[7] G.B. Folland, *Real Analysis*, Wiley, New York, 1984.

[8] G. Gripenberg, S.-O. Londen, and O. Staffans, *Volterra Integral and Functional Equations*, Cambridge University Press, Cambridge, 1990.

[9] M.A. Krasnosel'skiĭ and A.V. Pokrovskiĭ, *Systems with Hysteresis*, Springer-Verlag, Berlin, 1989.

[10] H. Logemann and E.P. Ryan, Time-varying and adaptive integral control of infinite-dimensional regular linear systems with input nonlinearities, *SIAM J. Control Optimization* 38:1120–1144, 2000.

[11] H. Logemann and S. Townley, Low-gain control of uncertain regular linear systems, *SIAM J. Control & Optimization* 35:78–116, 1997.

[12] H. Logemann, E.P. Ryan, and S. Townley, Integral control of linear systems with actuator nonlinearities: lower bounds for the maximal regulating gain, *IEEE Trans. Auto. Control* 44:1315–1319, 1999.

[13] H. Logemann, E.P. Ryan, and S. Townley, Integral control of infinite-dimensional linear systems subject to input saturation, *SIAM J. Control & Optimization* 36:1940–1961, 1998.

[14] J. Lunze, Experimentelle Erprobung einer Einstellregel für PI-Mehrgrößenregler bei der Herstellung von Ammoniumnitrat-Harnstoff-Lösung, *Messen Steuern Regeln* 30:2–6, 1987.

[15] J. Lunze, *Robust Multivariable Feedback Control*, Prentice-Hall, London, 1988.

[16] J.W. Macki, P. Nistri, and P. Zecca, Mathematical models for hysteresis, *SIAM Review* 35:94–123, 1993.

[17] M. Morari, Robust stability of systems with integral control, *IEEE Trans. Auto. Control* 30:574–577, 1985.

[18] A. Pazy, *Semigroups of Linear Operators and Applications to Partial Differential Equations*, Springer-Verlag, New York, 1983.

[19] A. Visintin, *Differential Models of Hysteresis*, Springer-Verlag, Berlin, 1994.

[20] G. Weiss, Admissibility of unbounded control operators, *SIAM J. Control & Optimization* 27:527–545, 1989.

[21] G. Weiss, Admissible observation operators for linear semigroups, *Israel J. Math.* 65:17–43, 1989.

[22] G. Weiss, The representation of regular linear systems on Hilbert spaces, in *Distributed Parameter System*, F. Kappel, K. Kunisch, and W. Schappacher eds., Birkhäuser Verlag, Basel, 401–416, 1989.

[23] G. Weiss, Transfer functions of regular linear systems, part I: Characterization of regularity, *Trans. Amer. Math. Soc.* 342:827–854, 1994.

Index

Systems & Control: Foundations & Applications

Systems & Control: Foundations & Applications publishes research monographs and advanced graduate texts dealing with areas of current research in all areas of systems and control theory and its applications to a wide variety of scientific disciplines.

We encourage the preparation of manuscripts in TEX, preferably in Plain or AMS TEX—LaTeX is also acceptable—for delivery as camera-ready hard copy which leads to rapid publication, or on a diskette that can interface with laser printers or typesetters.

Proposals should be sent directly to the editor or to: Birkhäuser Boston, 675 Massachusetts Avenue, Cambridge, MA 02139, U.S.A.

Estimation Techniques for Distributed Parameter Systems
H.T. Banks and K. Kunisch

Set-Valued Analysis
Jean-Pierre Aubin and Hélène Frankowska

Weak Convergence Methods and Singularly Perturbed Stochastic Control and Filtering Problems
Harold J. Kushner

Methods of Algebraic Geometry in Control Theory: Part I
Scalar Linear Systems and Affine Algebraic Geometry
Peter Falb

H_∞-Optimal Control and Related Minimax Design Problems
Tamer Başar and Pierre Bernhard

Identification and Stochastic Adaptive Control
Han-Fu Chen and Lei Guo

Viability Theory
Jean-Pierre Aubin

Representation and Control of Infinite Dimensional Systems, Vol. I
A. Bensoussan, G. Da Prato, M. C. Delfour, and S. K. Mitter

Representation and Control of Infinite Dimensional Systems, Vol. II
A. Bensoussan, G. Da Prato, M. C. Delfour, and S. K. Mitter

Mathematical Control Theory: An Introduction
Jerzy Zabczyk

H_∞-Control for Distributed Parameter Systems: A State-Space Approach
Bert van Keulen

Disease Dynamics
Alexander Asachenkov, Guri Marchuk, Ronald Mohler, and Serge Zuev

Theory of Chattering Control with Applications to Astronautics, Robotics, Economics, and Engineering
Michail I. Zelikin and Vladimir F. Borisov

Modeling, Analysis and Control of Dynamic Elastic Multi-Link Structures
J. E. Lagnese, Günter Leugering, and E. J. P. G. Schmidt

First Order Representations of Linear Systems
Margreet Kuijper

Hierarchical Decision Making in Stochastic Manufacturing Systems
Suresh P. Sethi and Qing Zhang

Optimal Control Theory for Infinite Dimensional Systems
Xunjing Li and Jiongmin Yong

Generalized Solutions of First-Order PDEs: The Dynamical Optimization Process
Andreí I. Subbotin

Finite Horizon H_∞ and Related Control Problems
M. B. Subrahmanyam

Control Under Lack of Information
A. N. Krasovskii and N. N. Krasovskii

H^∞-Optimal Control and Related Minimax Design Problems: A Dynamic Game Approach
Tamer Başar and Pierre Bernhard

Control of Uncertain Sampled-Data Systems
Geir E. Dullerud

Robust Nonlinear Control Design: State-Space and Lyapunov Techniques
Randy A. Freeman and Petar V. Kokotović

Adaptive Systems: An Introduction
Iven Mareels and Jan Willem Polderman

Sampling in Digital Signal Processing and Control
Arie Feuer and Graham C. Goodwin

Ellipsoidal Calculus for Estimation and Control
Alexander Kurzhanski and István Vályi

Minimum Entropy Control for Time-Varying Systems
Marc A. Peters and Pablo A. Iglesias

Chain-Scattering Approach to H^{∞}-Control
Hidenori Kimura

Output Regulation of Uncertain Nonlinear Systems
Christopher I. Byrnes, Francesco Delli Priscoli, and Alberto Isidori

High Performance Control
Teng-Tiow Tay, Iven Mareels, and John B. Moore

Optimal Control and Viscosity Solutions of Hamilton-Jacobi-Bellman Equations
Martino Bardi and Italo Capuzzo-Dolcetta

Stochastic Analysis, Control, Optimization and Applications
William M. McEneaney, G. George Yin, and Qing Zhang, Editors

The Dynamics of Control
Fritz Colonius and Wolfgang Kliemann

Optimal Control
Richard Vinter

Advances in Mathematical Systems Theory
Fritz Colonius, Fabian Wirth, Uwe Helmke, and Dieter Prätzel-Wolters